Supercritical Water Processing Technologies for Environment, Energy and Nanomaterial Applications

Shuzhong Wang · Donghai Xu · Yang Guo et al.

Supercritical Water Processing Technologies for Environment, Energy and Nanomaterial Applications

西安交通大学出版社
XI'AN JIAOTONG UNIVERSITY PRESS

Springer

Shuzhong Wang
Key Laboratory of Thermo-Fluid Science and Engineering, Ministry of Education
Xi'an Jiaotong University
Xi'an, Shaanxi, China

Yang Guo
Key Laboratory of Thermo-Fluid Science and Engineering, Ministry of Education
Xi'an Jiaotong University
Xi'an, Shaanxi, China

Yuzhen Wang
Department of Printing, Packaging Engineering and Digital Media Technology
Xi'an University of Technology
Xi'an, Shaanxi, China

Honghe Ma
Department of Thermal Engineering
Taiyuan University of Technology
Taiyuan, Shanxi, China

Yanhui Li
Key Laboratory of Thermo-Fluid Science and Engineering, Ministry of Education
Xi'an Jiaotong University
Xi'an, Shaanxi, China

Donghai Xu
Key Laboratory of Thermo-Fluid Science and Engineering, Ministry of Education
Xi'an Jiaotong University
Xi'an, Shaanxi, China

Xingying Tang
School of Marine Sciences
Guangxi University
Nanning, Guangxi, China

Jie Zhang
Institute of Water Resources and Hydro-electric Engineering
Xi'an University of Technology
Xi'an, Shaanxi, China

Lili Qian
Department of Energy and Power Engineering
Jiangsu University
Zhenjiang, Jiangsu, China

ISBN 978-981-13-9328-0 ISBN 978-981-13-9326-6 (eBook)
https://doi.org/10.1007/978-981-13-9326-6

Jointly published with Xi'an Jiaotong University Press
The print edition is not for sale in China. Customers from China please order the print book from: Xi'an Jiaotong University Press.

This Springer imprint is published by the registered company Springer Nature Singapore Pte Ltd.
The registered company address is: 152 Beach Road, #21-01/04 Gateway East, Singapore 189721, Singapore

Preface

Supercritical water (SCW: T > 374.15 °C, P > 22.1 MPa) possesses specially physical and chemical properties such as high diffusivity, low viscosity, low dielectric constant, and small amounts of hydrogen bonds. These good properties make SCW not only a good solvent for organic matter and gases in supercritical water oxidation (SCWO) but also a good reactant for hydrogen production in supercritical water gasification (SCWG). SCWO is a promising technology to deal with organic wastes which can be oxidized into harmlessly small molecular compounds such as carbon dioxide, nitrogen, water, and inorganic salts by excess oxidant in SCW. SCWG means that organic matter takes place hydrolysis and pyrolysis reactions and so on in SCW to form a hydrogen-rich mixing gas. SCWG is regarded as a promising technology for hydrogen production from wet biomass. Moreover, the supercritical hydrothermal synthesis of inorganic nanomaterials in SCW has also attracted more and more attention.

In this book, we aim to provide readers with comprehensive and systematic introduction and summary of supercritical water treatment process applications above in energy, environment, and material fields. For these technologies, we further investigate innovative reactor designs, construction materials, corrosion and salt precipitation problems, and industrial applications. More in depth, we propose the reaction mechanisms and reaction kinetics of SCWO and SCWG for some feedstocks. The reactions occurring in SCW are very complicated, which makes the studies of reaction mechanisms significant for the developments of supercritical water technologies. Relatively harsh reaction conditions in supercritical water technologies easily cause reactor corrosion and salt deposition problems in the reactor. We attempt to analyze the causes, mechanisms, and improvement measures of these two problems. On the one hand, this book gives a detailed and comprehensive introduction to the processes and mechanisms of supercritical water reaction systems herein. On the other hand, it also provides defensive measures and innovative designs for the problems arising from the processes above.

More specifically, there are nine chapters in this book, which can be roughly divided into five parts:

Part I: A brief introduction of the supercritical water technologies and their problems and industrial applications (Chap. 1).
Part II: Reaction mechanism and reaction kinetics of SCWG and SCWO (Chaps. 2–3).
Part III: Hydrothermal combustion and hydrothermal synthesis in SCW (Chaps. 4–5).
Part IV: Existing problems and innovative improvements in SCW (Chaps. 6–8).
Part V: A proposal of near-zero emission systems of pollutants based on SCWO (Chap. 9).

Generally speaking, the content of this book is very suitable for students and researchers who are related to their majors. People from all walks of life are also very welcome to read. At last, there are many people who have helped a lot in the process of writing this book we wish to thank. Many students in the Department of Energy and Power Engineering, Xi'an Jiaotong University, have made great contributions to the preparation of this book, including data collection and update, text entry and proofreading, and case studies.

Xi'an, China

Shuzhong Wang
Donghai Xu
Yang Guo

Acknowledgements

In addition, Ph.D. candidates (including Panpan Sun, Jianqiao Yang, Jianna Li, Shuwei Guo, Mengmeng Ren, Yishu Zhang, Ning Wei et al.) and master candidates (Jiandong Zhang, Mei Yang et al.) in the Department of Energy and Power Engineering, Xi'an Jiaotong University, have made great contributions to the preparation for this book, including data collection and update, text entry and proofreading, case studies, and writing. Yanmeng Gong and Lu Zhou et al. also had contribution to this book.

This book was supported by the projects from National Natural Science Foundation of China (21576219, 51871179, 51876174), the National Key Research and Development Program of China (2016YFC0801904, 2017YFB060360), the National High Technology Research and Development Program of China (2006AA06Z313), the Fundamental Research Funds for the Central Universities (xjj2016116, xjj2018006, xjh012019005, xtr042019015), Natural Science Basic Research Plan in Shaanxi Province of China (2018JM5011, 2019JLM-21), and China Postdoctoral Science Foundation (2019TQ0248).

Contents

Chapter 1
Introduction

1.1 Properties of Supercritical Water

Supercritical water (SCW) refers to water whose temperature and pressure are above its critical point (374.15 °C, 22.1 MPa). The phase diagram of water is shown in Fig. 1.1, the left side of the blue line is the solid area, the area between the blue line and the red line is the liquid area, the area below the red line is the gas area, and the red area in the figure is the supercritical area. The normal water is a polar solvent, which can dissolve most polar substances such as salts but is slightly soluble or insoluble to nonpolar substances such as gases and most organic substances.

As the temperature and pressure of water rise to supercritical conditions, their physical properties such as density, viscosity, dielectric constant, hydrogen bonding, and ionic product will change greatly between normal water and steam. Table 1.1 shows the physical properties of normal water, subcritical water, supercritical water, and superheated steam. The density of supercritical water is about 1/3 of the normal state, the hydrogen bonding degree of water molecules is reduced to 1/3 of ice, the dielectric constant is reduced from 87.7 at 0 °C to 2–30, and the ion product is increased 10–100 times that of normal water. In addition, the supercritical water has a significantly lower viscosity than that of the normal water, the diffusion coefficient is improved, the mass transfer performance is enhanced, and it can be completely miscible with nonpolar gases and hydrocarbons. However, the polar substances such as salts have low solubility in SCW, making itself easily be separated.

Hydrogen bonds, which still exist under supercritical conditions, affect many special properties of water such as increasing melting point, boiling point, solubility, and viscosity. Hydrogen bonds are very sensitive to temperature, and as the temperature increases, the hydrogen bonds will always decrease rapidly. In the range of 298–773 K, the hydrogen bond degree of water decreases linearly with the increase of temperature, because of the break of the hydrogen bond in water.

S. Wang et al., *Supercritical Water Processing Technologies for Environment, Energy and Nanomaterial Applications*,
https://doi.org/10.1007/978-981-13-9326-6_1

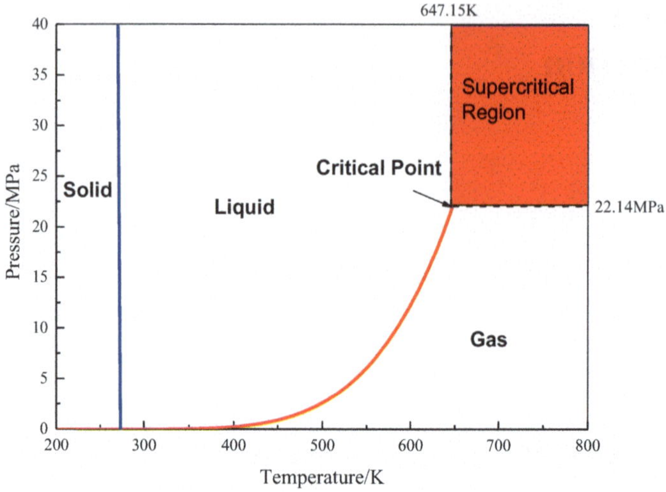

Fig. 1.1 Phase diagram of water

Table 1.1 Physical properties of water at different temperatures and pressures [1]

Parameter	Normal water	Subcritical water	Supercritical water	Superheated steam
Temperature/°C	25	250	400	400
Pressure/MPa	0.1	5	25	0.1
Density/g cm^{-3}	0.997	0.8	0.17	0.0003
Dielectric constant	78.5	27.1	5.9	1.0
Ionic product	10^{-14}	$10^{-11.2}$	$10^{-19.4}$	–
Viscosity/mPa s	0.89	0.11	0.03	0.02

If the temperature is 673 K, the hydrogen bond degree is about 0.3. Hydrogen bonds still exist under high-temperature conditions. Supercritical water can be miscible with nonpolar organic substances (such as benzene, alkanes, CO_2, N_2, O_2), but inorganic salts such as Na_2SO_4, NaCl, and Na_2CO_3 have extremely low solubility in supercritical water. This extremely low solubility property can be utilized to remove inorganic salts from supercritical water. The density of water does not change substantially with pressure under normal temperature conditions, but as the temperature increases, the density of water changes with pressure. Under supercritical conditions, density is sensitive to temperature changes, especially when near the critical point, small temperature changes can cause large changes in density. For example, when the pressure is 25 MPa and the temperature changes from 650 to 660 K, the water density will vary from 0.489 to 0.258 g/cm^3. The change of water density with temperature under different pressures is shown in Fig. 1.2. The water

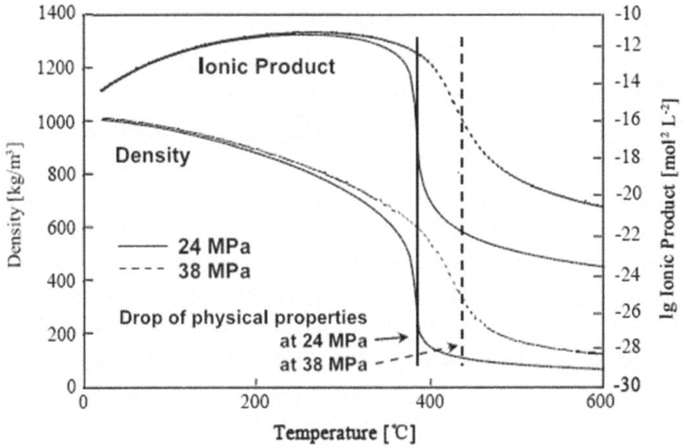

Fig. 1.2 Changes in ionic product and density of high-temperature water at different pressures [3]

density and ionic product decrease sharply near the critical point, while the critical temperature of water increases with the increase of pressure. The ionic product of water under standard conditions is 10^{-14}, and it determines the electrolysis and conductivity of water. Under supercritical conditions, the density of water decreases rapidly with the increase of temperature, resulting in a decrease in ionic product.

The value of the dielectric constant of supercritical water is close to the value of polar organic matter at normal temperature and pressure. For example, when the water is at 873 K and 24.6 MPa, the dielectric constant is only 1.2. Due to the small dielectric constant, it is difficult for supercritical water to shield the electrostatic potential energy among ions. Ionic compounds in water are presented in the form of ion pairs, and polar compounds precipitate in supercritical water due to extremely low solubility [2]. In contrast, supercritical water has a low dielectric constant and is extremely soluble to nonpolar compounds such as organic matter and oxygen. The viscosity of water reflects the comprehensive effect of energy transfer caused by frequent collisions between individual molecules and surrounding molecules during the free translation of molecules in water. The viscosity of water decreases with increasing temperature under high-density conditions, but increases with increasing temperature under low-density conditions. The viscosity increases with temperature under supercritical conditions [2]. The decrease in the density and viscosity of water under supercritical conditions also leads to an increase in the diffusion coefficient of supercritical water.

In view of the special properties of supercritical water, nonpolar molecules such as O_2 and organic matter have strong solubility in supercritical aqueous solution, which can be mutually dissolved in any ratio to form a homogeneous phase, and the mass transfer resistance at the interface is eliminated. Supercritical water can be used as an excellent reaction medium for organic matter and oxygen [4].

1.2 Supercritical Water Oxidation (SCWO) and Supercritical Water Gasification (SCWG)

SCWO is a promising technology to deal with organic wastes. With this technology, organic wastes can be thoroughly oxidized and decomposed into harmlessly small molecular compounds such as CO_2, N_2, water, and so on under excess oxidant condition in single-phase supercritical water. Heteroatoms in organic matters are mineralized into corresponding acids or inorganic salts, and the formation of nitrous oxides is inhibited owing to the low reaction temperature. SCWO is particularly suitable for disposing organic wastewaters with high toxicity, high-concentration, and biorefractory components. It can also recover energy and achieve heat self-sufficiency to ensure an economic advantage [5, 6]. When mass concentration of organic matters in feedstock is in the range of 3–4%, maintaining the whole reaction process does not require an extra energy input process commonly [7, 8]. Furthermore, compared with incineration, SCWO does not have the problems of high cost, public resentment, and secondary pollutants [6] like dioxins [9]. Hence, SCWO has attracted much attention in the past three decades.

Nowadays, SCWO plants have been commercialized by several famous companies [6, 10] such as General Atomics, EcoWaste Technologies, Chematur, and Supercritical Fluids International [11] that have summarized SCWO results of real waste materials including textile wastewater, wastewater from terephthalic acids, food wastes, municipal excess sludge, and alcohol distillery wastewater. Furthermore, B. Veriansyah [6] has also systematically introduced SCWO experiments of toxic organic wastes such as pesticide DDT, bacteria and dioxins, chlorophenol and chlorobenzene, pharmaceutical and biopharmaceutical wastes, chlorocarbon γ-hexachlorocyclohexane and hexachlorocyclohexane, highly contaminated activated sludge, pond sludge, chemical weapon stockpiles, stockpiled chemical warfare agents, smokes, dyes, pyrotechnics, VX [O-ethyl S-(2-diisopropylamino)ethyl methylphospho-nohioate] and GB [Sarin, isopropyl-methylphosphonofluoridate] [12], solid rocket propellants, military wastes, and so on. Some other feedstocks like acrylic acid production wastewater [13], industrial dyeing wastewater, polychlorinated biphenyls (PCBs) [14, 15], chlorinated wastes [16], and oily wastes have also been reported to dispose via SCWO. In 2001, two commercial-scale SCWO plants were installed in Harlingen (Texas, USA) to treat sewage sludge with the solid-containing rates of 7–8 wt% [7, 17]. Another SCWO plant with a treatment capacity of 7 m^3/h was also built by Chematur Engineering AB (Sweden) to deal with sewage sludge [18]. Besides, there were some other pilot- or commercial-scale SCWO plants for the harmless treatment of sewage sludge or organic pollutants. Nevertheless, the three key problems of concerning corrosion, plugging, and high running cost still exist [6] and even make some commercial-scale SCWO plants inactive [19, 20].

Supercritical water gasification (SCWG) refers to the process of dissolving all kinds of organic compounds in feedstock by using the high solvent power of supercritical water and finally producing a mixture of gases rich in hydrogen.

Gaseous products obtained from SCWG mainly include H_2, CO_2, CH_4, and CO. Besides, short-chain hydrocarbons, such as ethane, ethylene, propane, and propylene, are all present in extremely small amounts [21]. The generation of gases in SCWG is the result of the formation of water-soluble oligomers through hydrolysis followed by further decomposition of these oligomers [22].

The overall chemical conversion of biomass SCWG can be simplified as Eq. (1.1), where x and y are the mole ratios of H/C and O/C in biomass, respectively. Equations (1.2)–(1.4), a group of competing intermediate reactions essential for successful gasification, should also be taken into consideration [23].

$$CH_xO_y + (2 - y)H_2O \rightarrow CO_2 + (2 - y + x/2)H_2 \tag{1.1}$$

Steam reforming:

$$CH_xO_y + (1 - y)H_2O \rightarrow CO + (1 - y + x/2)H_2 \tag{1.2}$$

Water–gas shift:

$$CO + H_2O \leftrightarrow CO_2 + H_2 \tag{1.3}$$

Methanation:

$$CO + 3H_2 \leftrightarrow CH_4 + H_2O \tag{1.4}$$

Compared with traditional technologies, SCWG has many advantages, mainly including:

(1) Biomass SCWG is a homogeneous reaction because most organic compounds can be dissolved in SCW. Therefore, a decrease in mass transfer resistance between phases can be achieved in SCWG.
(2) The reaction temperature is much lower relative to conventional gasification and pyrolysis.
(3) The gas product with high pressure is beneficial to its further treatment such as transportation, usage, and carbon capture [24].
(4) There is no need to dry the feedstock.
(5) Because water acts both as reaction medium and as reactant of reforming, the yield of H_2 is high.
(6) The amount of tar and char formed is less [25].

1.3 Problems in SCWO and SCWG

To date, some pilot-scale plants and commercial plants have been employed to deal with different varieties of organic pollutants such as wastewaters and sewage sludge [19, 20, 26]. However, harsh reaction conditions (high temperature, high pressure, excessive oxygen, corrosive ions, etc.) easily induce severe reactor corrosion problems, meaning a shorter reactor life and a poorer treatment result due to the formation of corrosion products. On the other hand, inorganic salts will precipitate in supercritical water due to its extremely low dielectric constant, which will result in reactor plugging owing to their deposition and further cause expensive and frequent shutdowns of the whole SCWO plant. These two key problems are still not effectively solved and seriously hinder the extensive commercialization of SCWO. Thus, pilot-scale and industrial-scale SCWO plants for real wastewater treatments are still scarce now [26].

1.3.1 Corrosion

Harsh operation conditions (high concentration of oxidant, extreme pH value, high temperature and pressure) together with reaction intermediate/ultimate products (high concentrations of ionic species, free radicals, acids, and inorganic salts) result in severe corrosion problem in SCWO. Corrosion mainly occurs on reactor's inner wall, and it also appears in heat exchanger and cooler on inlet and outlet pipelines of reactor [27]. Corrosion in SCWO circumstance depends on parameters concerning both the aqueous solutions (chemical dissolution process, electrochemical dissolution process, and influence of anions) and materials (alloy composition, surface condition, material purity, and heat treatment) [28, 29]. For a specific reactor material, corrosion is influenced by the dissociations of acids, salts, and bases, the solubilities of corrosion products and gases, and the stabilities of protective oxide layers, and these characteristics are affected by density and ionic product of solution [11]. In order to decrease corrosion rate, it is better to ensure that the solution density everywhere of reactor is below 200 kg/m^3 [27]. Corrosion forms under supercritical water conditions are categorized into general corrosion, dealloying, pitting, stress corrosion cracking, intergranular corrosion, crevice corrosion, under-deposit corrosion, hydriding, galvanic corrosion, non-coupled corrosion [30], or parts of them [11, 20].

Corrosion is a key obstacle to limit the commercial application of SCWO, and it not only shortens reactor life but also induces a bad treatment effect of feedstock due to the formations of corrosion products. Materials serving for SCWO include stainless steels, nickel-based alloys, titanium, tantalum, and noble metal ceramics [11]. Many researchers have studied the corrosion resistance properties of these materials under supercritical and/or subcritical conditions for the aim of minimizing corrosion rate [31–38]. It can be concluded from Table 1.2 that each material shows

Table 1.2 Corrosion resistance performances of metals against different media at subcritical and supercritical temperatures [27]

Materials	$T < T_c^a$; high density		$T > T_c^a$; low density	
	Good resistance	Poor resistance	Good resistance	Poor resistance
Nickel-based alloys	H_3PO_4, HF, Alkaline solutions	HCl, HBr, H_2SO_4, HNO_3	All acids	$[H_3PO_4] > 0.1$ mol/kg, NaOH
Titanium	All acids	F^-	HCl	H_2SO_4, H_3PO_4

[a]The critical temperature of water

an individual corrosion resistance property under the same reaction conditions. Not one kind of material can withstand corrosion at all conditions, but some have perfect corrosion resistance properties under specific conditions [11]. Thus, appropriate reaction conditions such as heteroatom types in feedstock, reaction temperature, and pressure should be fixed in order to minimize corrosion rate for a chosen reactor construction material. Generally speaking, nickel-based alloys show a good corrosion resistance performance among all the acids listed in Table 1.2 under supercritical conditions. Titanium is fit to be employed under subcritical conditions. Preheater and cooler are installed before and after reactor.

Marrone et al. [30] have detailedly reported some corrosion control approaches including utilization of high corrosion resistance materials, liners, coatings, transpiring wall/film-cooled wall reactors, adsorption/reaction on fluidized solid phase, avoidance of corrosive feeds, pretreatment to remove corrosive species, vortex/circulating flow reactors, preneutralization, cold feed injection, feed dilution with non-corrosive wastes, effluent dilution/cooling, and optimization of operation conditions. The most significant option may be to adopted develop specific reactor configurations such as transpiring wall reactors [39, 40], cooled wall reactors [41–43], or double-wall reactors [44, 45].

1.3.2 Salt Deposition Problems

Organic waste feedstock like sewage sludge commonly contains partial solid particles such as sand, clay, and rust, which are much less prone to deposit on process surfaces than nascent inorganic salt particles in SCW [46]. Thereby, in this article, salt is defined as polar inorganic salt only, excluding the solid particles above.

1.3.2.1 The Influence of Salt Deposition on Reactor Plugging

Solubility of salt is reduced evidently in SCW, which is usually lower than 100 mg/L. An approach based on the description of phase equilibrium between solid salt phase and SCW phase was chosen by Leusbrock et al. [47] to quantify solubility of salt. Precipitated salt from SCW often forms agglomerates and deposits on internal surface of reactor. Reactor plugging will take place when salt deposition is left uncontrolled [11], especially in a small diameter reactor with large and/or sticky salt crystals precipitated at a low flow velocity [48]. Kawasaki et al. [49] mentioned that reactor plugging occurred within 30 and 15 min for NaCl solutions of 1.0 wt% at 450 and 500 °C in SCWO, respectively. Sewage sludge was also proved to result in reactor plugging when treated in SCWG. If plugged, reactor has to be shut down, rinsed, and then restarted, which will increase running cost and hinder commercial applications of SCWO and SCWG [11, 27, 48, 50, 51]. Sticky salts described above are mainly calcium sulfate, aluminum phosphate, and so on, which can be rinsed by acid after reaction [52]. Moreover, salt deposition also decreases heat transfer efficiency of reactor wall, increases pressure drop of reactor, and accelerates deactivation rate of catalyst. Experimental methods of investigating salt nucleation and growth in SCW have been reported by Armellini and Tester [53]. Hodes et al. [54] have also reviewed fundamental principles and researches concerning about salt precipitation in SCWO. Salt deposition is a key problem resulting in reactor plugging in SCWO or SCWG, and fundamental investigation needs to be conducted further in the future.

1.3.2.2 The Influence of Salt Deposition on Reactor Corrosion

Comparing with salt of ion form in subcritical water, salt in the form of molecule in SCW is relatively less corrosive [27, 55–57]. Salt precipitated in SCW mainly results in chemical corrosion through oxidation reaction. Salt in subcritical water mostly promotes power–chemical corrosion and may result in intergranular corrosion starting from the edge of metal grain. Some salts are generated in SCWO process when organic matter includes heteroatoms such as chlorine, sulfur, and phosphorus [46]. They will severely erode reactor when depositing on the internal surface of reactor, especially when reactor material is sensitive to these corrosive species. Reactor corrosion should be prevented as much as possible, because it will lead to a short reactor life and a bad fluid treatment result. Nowadays, salt deposition is considered as an important factor causing reactor corrosion.

Due to the very low solubility of salt, two contradictions of preventing salt deposition and minimizing reactor corrosion rate are displayed as follows: First, preventing salt deposition needs high SCW density because it will exhibit a relatively improved solvent property for precipitated salts [2, 58]. However, minimizing corrosion rate requires low SCW density [29, 59] to decrease the content of salt in the form of ion. Oelkers et al. [58] and Su et al. [60] controlled SCW density at less than 250 kg/m^3 in order to reduce reactor corrosion rate. Second, adding alkali compounds independently or in feedstock before reactor is helpful for inhibiting reactor corrosion, but the possibility of reactor plugging increases because of salt deposition [27]. That is why some alkali compounds are delivered into reaction system from reactor outlet.

1.3.2.3 The Influence of Salt Deposition on Catalyst Deactivation

Catalyst can be employed to decrease reaction temperature and pressure, compensate reaction dynamics [61], improve reactant conversion efficiency, and accelerate reaction rate. Catalysts served for SCWO include zeolite, CuO, MnO_2, V_2O_5, nickel, platinum, and so on. Heterogeneous catalysts such as nickel, ruthenium, ZrO_2, active carbon, and olivine can be chosen for catalytic SCWG. Catalyst deactivation caused by char deposition can be avoided efficiently because of the benign solvent and flow properties of SCW. But these catalysts may be poisoned and/or polluted quickly by precipitated salts [62–64] in SCW in real feedstock. It is also difficult to replace catalyst in traditional reactor configuration. That may be the reason why Savage [65] has reported that no catalyst is implemented commercially for catalytic SCWO for organic waste treatment or catalytic SCWG for renewable energy production. Thereby, it is important to separate these precipitated salts before they contact heterogeneous catalyst.

1.3.3 Plugging

Plugging is another main obstacle to hinder SCWO commercialization, which is induced by sticky salt agglomeration and deposition on internal surface of reactor. Salt deposition also accelerates reactor corrosion rate and catalyst inactivation rate, and reduces heat transfer coefficient of reactor wall. When plugging takes place, SCWO plant has to be shut down, washed, and restarted, which will directly decrease the reliability of the plant and increase its running cost. Inorganic salts, whether soluble or not, may come from feedstock or be reaction by-products [11], and their viscosities decide the tendency of depositing on reactor wall. In general, different salts own different deposition characteristics [49]. Armellini and Tester [53] have explored experimental methods to research salt nucleation and growth in supercritical water. Salt deposition principles and corresponding solutions have been systemically presented in the previous literatures [46, 54]. In a commercialized SCWO plant, reactor plugging can be avoided by particular reactor designs, special instruments, and operation means [46]. Possible solutions are to adopt reverse-flow tank reactor, reverse-flow tubular reactor, transpiring wall reactor, centrifuge reactor, mechanical brushing, rotating scraper, reactor flushing, cross-flow filtration, density separation, additive, high flow velocity, homogeneous precipitation, and extreme pressure [46]. In fact, precipitated salts are relatively difficult to be removed out of SCWO plant during operation process; thus, most SCWO plants do not have salt removal function. Salts in SCWO are classified into type 1 or type 2 according to their solubilities [66]. Type 1 salt has a high solubility in the range of water's critical temperature, while type 2 salt has a low solubility in this region. Type 2 salt can be separated by properties of supercritical water [67], and the suitable supercritical conditions should be controlled at approximately 400 °C and 25 MPa according to the solubilities of salts mentioned in the previous studies

[54, 68–70]. Hydrocyclone or centrifugal reactor helps to remove soluble salts under the above supercritical conditions, but severe wearing problem is inevitable. At present, not an approach or a reactor configuration can really solve reactor plugging problem effectively and economically. Of course, low salt concentration in feedstock is helpful for avoiding reactor plugging, whereas this may need to dilute feedstock. According to the investigations of [19], aluminum concentration in energetic hydrolysates should be limited to 300–400 ppm to avoid plugging.

In addition, when the preheating temperature of feedstock is in the range of 200–450 °C [71, 72], some undesired intermediate products like tar and char will be generated, which may plug preheating pipeline. To overcome this problem, the following options are taken into consideration, which include enlarging the inner diameter of the preheating pipeline, adding a small amount of oxidant into the preheating pipeline to inhibit the formations of tar and char, increasing the heating rate of feedstock, and preheating feedstock up to a lower temperature even with out preheating [73]. However, the improvement of the heating rate is limited by space setting of heating part, heating methods, and withstanding temperature of preheater wall. The fourth approach above means that a large amount of heat will not be recovered into SCWO system, which will undoubtedly reduce energy efficiency and increase the running cost. Meanwhile, supplemental fuel should be injected into reactor to maintain a high reaction temperature [20, 43, 74–76]. But this alternative is less favorable for the energetic balance of the process [20].

1.3.4 High Running Cost

High reaction temperature and pressure result in a relatively expensive running cost, which also further affects the development of SCWO. The effective means may be to load proper catalyst to moderate the harsh reaction conditions, to minimize oxidant consumption, to recover heat of reactor effluent as well as to increase by-product income. First, catalyst can improve reaction conditions, but it is also easy to be inactive due to the critical reaction conditions. The employed homogeneous and heterogeneous catalysts are mainly heteropoly acids, alkali carbonates, carbons, transition metal oxides, and bulk MnO_2 [65, 77]. Besides, they also especially pointed out that the bulk MnO_2 has high activity, hydrothermal stability, ability of maintenance and resistance to metal leaching under supercritical reaction conditions. Second, oxidant consumption has an important influence on running cost [27]. Excessive oxidant should be recovered from SCWO process to improve the technology's sustainable potentialities [17]. If oxidant excess coefficient is controlled in the range of 1.0–1.3 to neglect oxidant recovery unit, organic matters might not be destructed thoroughly. So, the oxidant excess coefficient should be minimized as much as possible and meanwhile make sure that organic matters are decomposed completely. Third, designs and combinations of heat exchangers need to be optimized to recover the heat in reactor effluent because energy recycle utilization is beneficial to further reduce the running cost. Fourth, by-product income is also of importance for the economical improvement. For instance, when one ton of dry sewage (DS) sludge is treated by SCWO in Harlingen Water Works System, 120 dollars are available via selling CO_2 and hot water to a nearby clothing factory [7].

Heat self-sufficiency is a very significant condition to guarantee a low running cost, which means that no supplemental energy is provided in the operation process. Using the Aspen Plus (C) software and considering a turbine as a supercritical water detent [78], the simulation of the SCWO process found that it can achieve energetic self-sufficiency at above 600 °C, 23 MPa, and slightly excessive oxygen. The minimum heat value required for feedstock is 930 kJ/kg (equal to 2wt% n-hexane) to achieve a heat self-sufficient operation. In this case, reactor effluent can preheat the feedstock from room temperature to 400 °C and generate electric power equivalent to that consumed by high-pressure pump and air compressor in the SCWO plant [79, 80]. Therefore, heat in reactor effluent should be recycled into the reaction system as much as possible to accomplish heat self-sufficiency. Of course, auxiliary energy such as electricity, gas fuel, or liquid fuel ought to be supplied to maintain reaction temperature for low concentration of organic matter feedstock or to start SCWO plant. Here, heat supplying methods can be divided into internal heat source form and external heat source form. The internal heat source form means liquid fuel including methanol [74], alcohol, or isopropyl alcohol [20, 43]. It is pumped into reactor through a separate pipeline to combust with oxygen to compensate for the heat requirement. However, a special reactor configuration design requires that the pressure bearing wall can withstand a very high flame temperature [39, 70, 76]. The external heat source form without a combustion phenomenon in reactor means heat is supplied by an outer electric or gas-fired heater [7]. Obviously, the supplemental heat should be minimized as much as possible in order to obtain a low running cost.

1.3.5 Problems in SCWG

As a new technology of biomass gasification, SCWG has attracted great attention and has broad prospects. However, at present, there are still many problems to be solved. The problems of SCWG can be listed as follows:

(1) Corrosion

Corrosion problem of reactor materials exposed to reaction conditions in biomass SCWG has been reported in some literatures [81, 82]. Corrosion degree depends on feedstocks, reactor materials, reaction conditions, and operation times of the reactors [48]. However, comparing with SCWO, corrosion in SCWG is much less a problem [83]. Therefore, reports on corrosion in SCWG are not as common as that in SCWO. Research on corrosion resistance of reactors is of great significance undoubtedly.

(2) Plugging problem

In SCW, the solubility of salts will decrease sharply. Therefore, salt precipitation is another challenge for SCWG. If the tube diameters are small, crystals are large or sticky, heterogeneous catalysts are used, the velocity of flow is low, and plugging problem caused by precipitated salts is more likely to occur [84]. In addition, the formation of tar and char not only reduces the gasification efficiency of feedstock, but also increases the possibility of plugging.

(3) Catalyst deactivation

The usage of catalysts undoubtedly increases the cost. However, the deactivation and regeneration of catalysts will further increase the cost. Catalysts that have been shown to increase H_2 yield effectively can be classified into the following categories: alkali-based (such as Na_2CO_3, K_2CO_3, and NaOH), transition metal-based (Ni and Ru), and activated carbon catalysts. However, alkali catalysts may result in corrosion and plugging problems while transition metal and activated carbon-based catalysts can be easily deactivated [85]. The main reasons responsible for catalyst deactivation include: (i) catalyst surface fouling by precipitated salts and other minerals; (ii) coke formation and deposition on the catalyst surface; and (iii) active surface poisoning [1].

Catalyst poisoning means deactivation caused by adsorption of poisons on active catalyst sites. It has been reported that sulfur present in biomass can lead to catalyst poisoning during SCWG [86]. In order to explore the effect of sulfur on SCWG, the influence of reaction conditions in the presence of model compound containing sulfur [87] and the poisoning mechanism [88] have been extensively studied. In addition, regeneration methods of sulfur-poisoned Ru catalysts have also been proposed by Osada et al. [89] and Dreher et al. [90]. Nitrogen is also a potential poison. However, under supercritical conditions, catalyst poisoning caused by nitrogen existing in all kinds of proteins has hardly ever been reported. In order to study the possible inhibitory effect of nitrogen and sulfur on RuO_2 catalyst, carbazole and dibenzothiophene were selected as gasification feedstock, respectively. The results indicated that nitrogen did not deactivate RuO_2 catalyst while sulfur did [91]. Besides, the active surface of catalyst can also be poisoned by chloride ions during the SCWG of lignin [92].

1.4 Industrial Applications of SCWO and SCWG

1.4.1 Pilot Units

At Duke University, a SCWO system for the treatment of fecal sludge was designed, built, and placed in operation in December 2014 with a target capacity of ~1000 kg/day flow rate with 10–20% solids [93]. This unit was housed in a standard 20 ft shipping container and could be conveniently transported. The waste slurry was mixed with supercritical water and air at approximately 600 °C to reach supercritical conditions at approximately 400 °C and 24 MPa. After the reaction, heat recovery was achieved in a 40-m-long heat exchanger. Then, the effluent passed through a high-pressure gas–liquid–solid separator and the liquid was depressurized in a series of capillary tubes. In this process, organic matter was converted to CO_2 and clean water. This system had been tested with liquid fuel and with secondary sludge at supercritical conditions for over 200 h. The research was ongoing.

A continuous reaction system was designed and fabricated at the University of Missouri [93]. A simulant fecal sludge was mixed with a preheated supercritical fluid of mixture of water and air, which then entered in a 305-cm length of 316 stainless steel tubing. Although the reactor was not heated externally, the temperature in the reactor was higher than that at the mixing point, with an average temperature gain of 87 °C due to the exothermic nature of the SCWO reaction. Finally, the effluent was depressurized by back pressure regulators. The results from this bench-scale system show that the concentration of the feed significantly influences the temperature gain in the reactor while the stoichiometric excess of oxygen, n, and pressure affect the efficiency of heat transfer. Although nearly full conversion of organics can be obtained when n is as low as 1.1, further increasing n gains to about 1.5 in maximum temperature.

Xi'an Jiaotong University has constructed the first SCWO pilot-scale plant in China with a transpiring wall combined with a MODAR reactor made of 316 stainless steel to treat MSS with the solid content lower than 10 wt%. This system has a treatment capacity of 3 t/d and is composed of six skidded modules. It can reach a COD removal rate higher than 99.5% and surplus heat utilization efficiency higher than 80%. A brine pool is formed in the bottom of the reactor. Fluid in the pool with a high concentration of salts then enters the salt-removing device and is heated to a supercritical state. The salts are precipitated again and finally stored in the salt-collecting tank. The equipment investment is 0.58 million USD, and the operating cost is 76.56 USD/dry ton for MSS with a moisture content of 92 wt%, which is much lower than the cost of incineration [94, 95].

1.4.2 Commercial Systems

For SCWO, Table 1.3 summarizes the operation and economic parameters of three commercial systems.

1.4.2.1 HydroProcessing

HydroProcessing constructed the first commercial SCWO sludge processing equipment for the Harlingen Wastewater Treatment Plant, in Texas, USA [7]. It was built in 2001 and stopped operation in 2002 due to corrosion in the heat exchanger [26]. Its flow sheet is shown in Fig. 1.3.

Table 1.3 Operation and economic parameters of three commercial SCWO systems of MSS

Company	HydroProcessing	Chematur Engineering AB	SuperWater Solutions
Country	USA	Sweden	USA
Capacity	150 t/d	168 m^3/d	35 dry ton/d
Solid content (wt%)	6–9	15	10
COD(mg/L)	80000–120000	110000	n/a
Temperature (°C)	592	510–580	600
Pressure (MPa)	23.47	25	26
Residence time (s)	20–90	30–90	30–60
Oxidation consumption (kg/kg dry sludge)	1.5	1.05	n/a
Oxidation coefficient	1.125	1.43	n/a
Oxygen recovery	No	No	Yes
CO_2 recovery	Yes	No	Yes
Reactor	Tubular	Tubular	Tubular
Capital cost	3 million USD	5 million GBP	33.7 million USD
Operation cost	100 USD/dry ton	105 USD/dry ton	268 USD/dry ton
Depreciation cost	80 USD/dry ton	119 USD/dry ton	n/a

n/a, not available

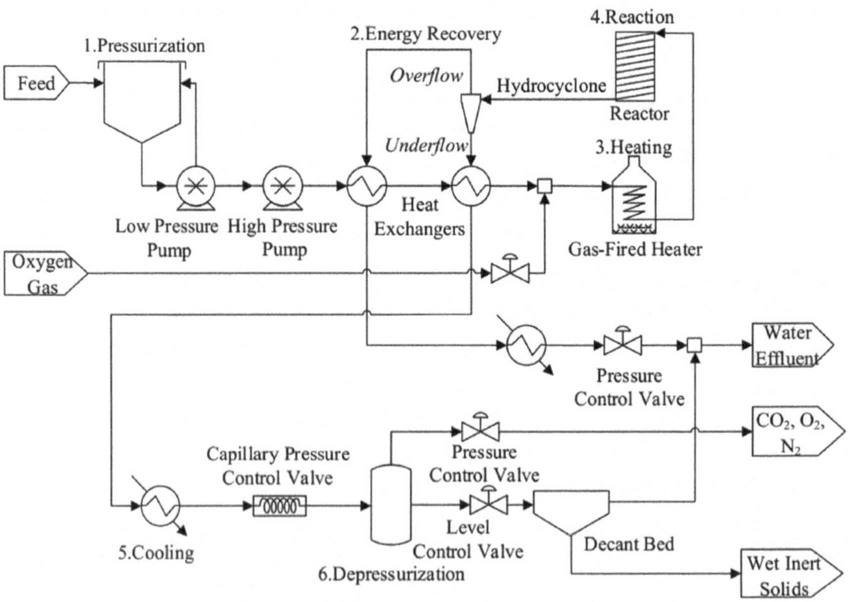

Fig. 1.3 Flow sheet of the HydroProcessing system [7]

As shown in Fig. 1.3, in the HydroProcessing system, a hydrocyclone is applied to separate the flow after the reactor and to maintain the solids in the underflow. The overflow and subsequent underflow are used to heat the feed. The heated feed mixes with oxygen, and they enter the gas-fired heater together for further heating; then, they enter a tubular reactor. It is interesting that the underflow with solids is depressurized by a capillary pressure reduction device with its frictional resistance.

The results from early operation show that the destruction efficiencies of sludge are 99.93–99.96%, 49.6–84.1%, and 92.89–98.90% for COD, ammonia, and total solid in the overflow, respectively. The destruction efficiencies of sludge are 99.92–99.93%, 49.6–86.4%, and 62.72–88.94% for COD, ammonia, and total solid in the underflow, respectively [7]. However, effluent ammonia ranges from 410 to 2075 mg/L, which is much higher than the related discharge standards, indicating the recalcitrant nature of ammonia [96]. Due to the high ammonia concentration of the effluent, it can be sent to municipal wastewater plants or to industrial plants as a nutrient. The heat from the gas-fired heater is approximately 4100 kWh/dry ton sludge, the oxygen consumption is 1500 kg/dry ton sludge, and the electricity consumed by the pumps is 550 kWh/dry ton sludge, producing a net operation and maintenance cost of approximately 100 USD/dry ton sludge [17].

1.4.2.2 Chematur Engineering AB and SCFI

For demonstration purposes, Chematur Engineering AB has built a 6 t/d unit in Karlskoga, Sweden, for the treatment of both undigested sludge and digested sludge, which has been in operation since 1998 [97]. Chematur Engineering AB licensed the SCWO process to the Shinko Pantec Co. of Japan, and they have built a pilot SCWO unit with a capacity of approximately 26.4 t/d for the treatment of MSS in Kobe, Japan [8]. They also evaluated the economic value of a unit treating 168 m^3 of sludge per day (see Table 1.3). Its flow sheet, called the Aqua Critox process, is presented in Fig. 1.4.

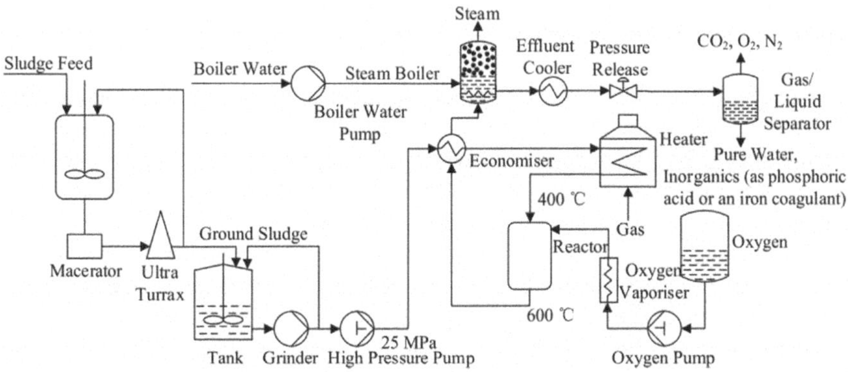

Fig. 1.4 Flow sheet of the Chematur Engineering AB system [97]

As depicted in Fig. 1.4, this system consists of a series of homogeneous sludge equipment including a feed tank with a paddle mixer, a macerator, an ULTRA-TURRAX, and a grinder. The first three facilities along with a mono pump constitute a re-circulation loop, which provides size reduction and homogenization during a batch operation. After homogenization, sludge enters a hose diaphragm piston pump pressurized to approximately 25 MPa and is pumped to a double-pipe economizer heated by the reactor effluent. Fouling problems in the economizer are avoided due to the high velocity and the reduced feed particle size. After heat exchange, sludge enters a heater for further temperature increase. Preheated sludge and oxygen subsequently enter the reactor. To accurately control the reaction temperature of high-concentration sludge, quench water is added, and oxygen is injected in two stages. Similar to the HydroProcessing system, the effluent is depressurized by capillaries in parallel. Many capillaries instead of a single one are applied to minimize the erosion and to control the velocity in the capillaries.

The tests show that all organics can be easily destroyed. The removal rate of COD is greater than 99.9% when $T > 520$ °C. A temperature higher than 540 °C is needed to completely destroy total nitrogen (TN). For a 168 m^3/d system, the natural gas consumed by the gas-fired heater is approximately 21.9 Nm^3/dry ton sludge, the oxygen consumption is 1048 kg/dry ton sludge, the electricity consumption is 229 kWh/dry ton sludge, the process water consumption is 1.7 m^3/dry ton sludge, the cooling water consumption is 100 m^3/dry ton sludge, and the steam generated by the heat of the reaction is 4200 kg/dry ton sludge, producing a net operation and maintenance cost of approximately 105 USD/dry ton sludge [8].

In 2007, the SCFI Group acquired patented supercritical water oxidation technology (Aqua Critox®) from Chematur Engineering AB of Sweden; therefore, Chematur Engineering AB is now SCFI [97]. Further work directed by the SCFI Group has emphasized the reduction of costs. The residual inorganic fraction of the sludge can be recovered as phosphoric acid and iron coagulant if required. The orthophosphates from the supernatant, representing 78% of the total influent phosphorus, can also be recovered by using fluidized bed crystallization technologies [98].

1.4.2.3 SuperWater Solutions

From 2009 to 2011, SuperWater Solutions installed and successfully tested a 5 dry t/d SCWO system for the Iron Bridge Regional Water Reclamation Facility in Orlando, FL, USA [99]. They also assessed the economic value of a unit that treats 35 dry tons of sludge per day (see Table 1.3). The flow sheet is illustrated in Fig. 1.5.

As presented in Fig. 1.5, the sludge and oxygen mix together enter a tubular system with a constant diameter pipe that includes a preheater, a reactor, and a cool-down heat exchanger. Pressurized pure water is applied to carry heat from the reactor effluent to the preheater. This structure can minimize solid and scale deposition. It is interesting that the SuperWater Solutions system has an oxygen recycling process based on the difference in liquefaction temperatures between O_2

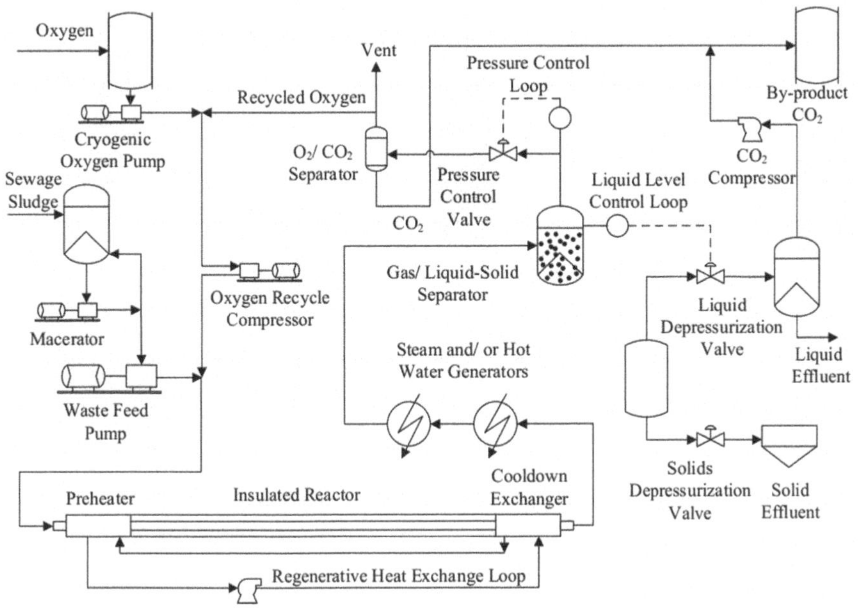

Fig. 1.5 Flow sheet of the SuperWater Solutions system [99]

and CO_2. Through the oxygen recycling process, sludge can be oxidized at a high *n* value with low oxygen consumption, thus increasing the effluent quality and decreasing operation costs.

1.4.3 Industrial Applications of SCWG

At present, batch reactors are wildly used for SCWG studies. However, only continuous process can be applied to industrial production. From intermittent process to continuous process, there are still many difficulties to overcome.

Pumpability of the feedstock is a challenge because the biomass must be processed into a pumpable slurry, dispersion, or aqueous solution. The type of biomass and its pretreatment determine the still pumpable dry matter content to a large extent. That the energy content of the feedstock is higher than the energy loss of the processing plant is the most basic condition to be satisfied. Consequently, the dry matter content should be high enough [48]. Meanwhile, considering the problem of possible clogging, the dry matter content should not exceed the pumpability limits [84].

Another challenge is to obtain sufficient process thermal efficiency. To apply SCWG into industry, thermal efficiency must be ensured to be high enough. The higher the reaction temperature is, the more important the problem of thermal

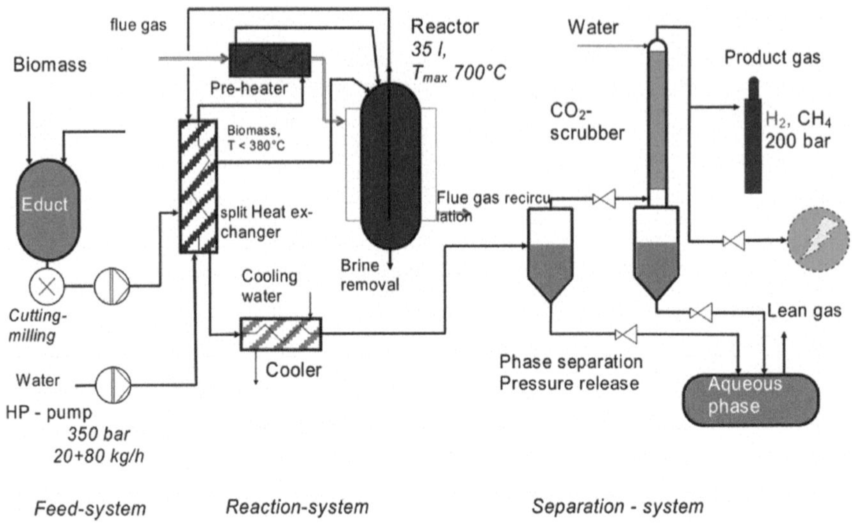

Fig. 1.6 Process scheme of VERENA

efficiency is [84]. In addition, effective energy recovery equipment is needed to reduce energy loss. Moreover, lower heating rate leads to the formation of tar and char. Therefore, heat exchanger which could ensure the heating rate must be developed.

The main reason why SCWG is difficult to achieve industrialization is that the cost is too high. Water maintained at supercritical conditions results in high energy consumption [85]. At the same time, the requirement of special corrosion, high-pressure, and high-temperature-resistant materials further increases its investment [100].

VERENA, a German acronym for "experimental facility for the energetic exploitation of agricultural matter," is regarded as the largest supercritical water gasification facility. Its process scheme is shown in Fig. 1.6. Maximum operating pressure and maximum temperature designed for VERENA pilot plant are 35 MPa and 700 °C, respectively. However, the plant is typically operated at a pressure of 28 MPa with a flow rate of 100 kg/h and a maximal reaction temperature of 660 °C. After the treatment with a cutting mill followed by a colloid mill, particle size of biomass can be less than 1 mm. One stream of biomass is heated up to critical temperature with a small part of the heat exchanger, while another stream of pure water is heated to high temperatures with the whole heat exchanger followed by a preheater, and these two streams are mixed at the top of the reactor. Mixed with the hot stream of water, the stream of biomass is heated up to reaction temperature in reactor. Due to gravitation, salts flow downward and will be removed by the brine removal system finally. The product mixture is transported by a small size tube from the lower part of the reactor to the heat exchanger. Subsequently, the product stream will be cooled by cooler, resulting in the separation of gases soluble in water under supercritical conditions from the water phase. In gas product, the decrease of CO_2 content is achieved by the addition of cold water in a scrubbing column filled with Pall rings. The gas product is stored in high-pressure bottles, while the excess amount of gas will be expanded to required pressure. The third separator used to remove lean gas from the effluent is run at ambient pressure, and the effluent is sent to a store tank finally [101].

References

1. K. C. Broell D, Kraemer A, et al., Chemistry in supercritical water. Angew Chem Int Ed, 1999, 38: 2998–3014.
2. Y. J. Z. R. T. R. J. Ruijie, A Pilot-plant-scale Study of Low Biodegradability Wastewater by Supercritical Water Oxidation. Journal of Green Science and Technology, 2014: 46–49.
3. W. G. Chen Zhong, Chen Hongzhen, Yang Shu, Yin Fengjun, Xu Yuanjian, Gas seal supercritical water oxidation system for treatment of concentrated organic wastes. Chinese Journal of Environmental Engineering, 2014, 8: 3825–3831.
4. L. Z. W. C. L. C. G. D. Z. K. C. H. Z. Yuezhao, Treatment and resource utilization of PTA dregs using supercritical water oxidation process. Industrial Water & Wastewater, 2014, 45: 1–4.
5. V. Vadillo, M. B. García-Jarana, J. Sánchez-Oneto, J. R. Portela, E. J. M. de la Ossa, Supercritical water oxidation of flammable industrial wastewaters: economic perspectives of an industrial plant. J Chem Technol Biotechnol, 2011, 86: 1049–1057.
6. B. Veriansyah, J.-D. Kim, RETRACTED: Supercritical water oxidation for the destruction of toxic organic wastewaters: A review. Journal of Environmental Sciences, 2007, 19: 513–522.
7. D. H. R. James W. Griffitha, The first commercial supercritical water oxidation sludge processing plant. Waste Manage (Oxford), 2002, 22: 453–459.
8. L. S. Anders Gidner, Supercritical Water Oxidation of Sewage Sludge State of the Art Chematur Engineering AB, 2001, 130.
9. Y. Kikuchi, K. Kurata, J. Nakatani, M. Hirao, Y. Oshima, Analysis of supercritical water oxidation for detoxification of waste organic solvent in university based on life cycle assessment. J Hazard Mater, 2011, 194: 283–9.
10. M. Dolores Bermejo, Daniel Rincon, Angel Martin, M. J. Cocero, Experimental Performance and Modeling of a New Cooled-Wall Reactor for the Supercritical Water Oxidation. Ind Eng Chem Res, 2009, 48: 6262–6272.
11. G. Brunner, Near and supercritical water. Part II: Oxidative processes. The Journal of Supercritical Fluids, 2009, 47: 382–390.
12. L. R. P. George W. Wagner, Richard J. O'Connor, Shekar Munavalli,, P. N. K. Corrie L. Carnes, Kenneth J. Klabunde, Reactions of VX, GB, GD, and HD with Nanosize Al2O3. Formation of Aluminophosphonates. J Am Chem Soc, 2001, 123: 1636–1644.
13. W.-J. Gong, F. Li, D.-L. Xi, Supercritical Water Oxidation of Acrylic Acid Production Wastewater in Transpiring Wall Reactor. Environ Eng Sci, 2009, 26: 131–136.
14. S. I. Kawasaki, T. Oe, N. Anjoh, T. Nakamori, A. Suzuki, K. Arai, Practical Supercritical Water Reactor for Destruction of High Concentration Polychlorinated Biphenyls (PCB) and Dioxin Waste Streams. Process Saf Environ Prot, 2006, 84: 317–324.
15. V. Marulanda, G. Bolaños, Supercritical water oxidation of a heavily PCB-contaminated mineral transformer oil: Laboratory-scale data and economic assessment. The Journal of Supercritical Fluids, 2010, 54: 258–265.

16. J. K. C. Schroer, J. Novotny, J. Hausselt, Aspects of the Corrosion Kinetics for Binary Nickel-Chromium Alloys Under Conditions Found in Supercritical Water Oxidation Plants Processing Chlorinated Substances. Corrosion, 2006.
17. M. Svanström, M. Fröling, M. Modell, W. A. Peters, J. Tester, Environmental assessment of supercritical water oxidation of sewage sludge. Resources, Conservation and Recycling, 2004, 41: 321–338.
18. L. S. Anders Gidner, Supercritical Water Oxidation of Sewage Sludge—State of the Art. Chematur Engineering AB, 2001.
19. S. D. C. Philip A. Marrone, Darren W. Dalton, SCWO System Designs for Waste Treatment: Application to Chemical Weapons Destruction. Ind Eng Chem Res, 2005, 44: 9030–9039.
20. M. D. Bermejo, M. J. Cocero, Supercritical water oxidation: A technical review. AIChE J, 2006, 52: 3933–3951.
21. C. Rodriguez Correa, A. Kruse, Supercritical water gasification of biomass for hydrogen production—Review. The Journal of Supercritical Fluids, 2018, 133: 573–590.
22. P. Azadi, R. Farnood, C. Vuillardot, Estimation of heating time in tubular supercritical water reactors. The Journal of Supercritical Fluids, 2011, 55: 1038–1045.
23. L. Guo, Y. Lu, X. Zhang, C. Ji, Y. Guan, A. Pei, Hydrogen production by biomass gasification in supercritical water: A systematic experimental and analytical study. Catal Today, 2007, 129: 275–286.
24. C. C. Liejin Guo, Youjun Lu, Supercritical Water Gasification of Biomass and Organic Wastes. Biomass Bioenergy, 2010: 165–182.
25. Q. Guan, C. Wei, P. Ning, S. Tian, J. Gu, Catalytic Gasification of Algae Nannochloropsis sp. in Sub/Supercritical Water. Procedia Environmental Sciences, 2013, 18: 844–848.
26. P. A. Marrone, Supercritical water oxidation—Current status of full-scale commercial activity for waste destruction. The Journal of Supercritical Fluids, 2013, 79: 283–288.
27. E. D. Peter Kritzer, An assessment of supercritical water oxidation (SCWO) Existing problems, possible solutions and new reactor concepts. Chem Eng J, 2001, 83: 207–214.
28. M. D. Bermejo, M. J. Cocero, Destruction of an industrial wastewater by supercritical water oxidation in a transpiring wall reactor. J Hazard Mater, 2006, 137: 965–71.
29. P. Kritzer, Corrosion in high-temperature and supercritical water and aqueous solutions: a review. The Journal of Supercritical Fluids, 2004, 29: 1–29.
30. P. A. Marrone, G. T. Hong, Corrosion control methods in supercritical water oxidation and gasification processes. The Journal of Supercritical Fluids, 2009, 51: 83–103.
31. E. Asselin, A. Alfantazi, S. Rogak, Corrosion of nickel–chromium alloys, stainless steel and niobium at supercritical water oxidation conditions. Corros Sci, 2010, 52: 118–124.
32. S.-H. Son, J.-H. Lee, C.-H. Lee, Corrosion phenomena of alloys by subcritical and supercritical water oxidation of 2-chlorophenol. The Journal of Supercritical Fluids, 2008, 44: 370–378.
33. R. Fujisawa, M. Sakaihara, Y. Kurata, Y. Watanabe, Corrosion behaviour of nickel base alloys and 316 stainless steel in supercritical water under alkaline conditions. Corrosion Engineering, Science and Technology, 2013, 40: 244–248.
34. G. S. Was, P. Ampornrat, G. Gupta, S. Teysseyre, E. A. West, T. R. Allen, K. Sridharan, L. Tan, Y. Chen, X. Ren, C. Pister, Corrosion and stress corrosion cracking in supercritical water. J Nucl Mater, 2007, 371: 176–201.
35. I. Betova, M. Bojinov, P. Kinnunen, S. Penttilä, T. Saario, Surface film electrochemistry of austenitic stainless steel and its main constituents in supercritical water. The Journal of Supercritical Fluids, 2007, 43: 333–340.
36. Q. Zhang, R. Tang, K. Yin, X. Luo, L. Zhang, Corrosion behavior of Hastelloy C-276 in supercritical water. Corros Sci, 2009, 51: 2092–2097.
37. M. Sun, X. Wu, Z. Zhang, E.-H. Han, Oxidation of 316 stainless steel in supercritical water. Corros Sci, 2009, 51: 1069–1072.
38. X. Gao, X. Wu, Z. Zhang, H. Guan, E.-h. Han, Characterization of oxide films grown on 316L stainless steel exposed to H2O2-containing supercritical water. The Journal of Supercritical Fluids, 2007, 42: 157–163.

39. B. Wellig, K. Lieball, P. Rudolf von Rohr, Operating characteristics of a transpiring-wall SCWO reactor with a hydrothermal flame as internal heat source. The Journal of Supercritical Fluids, 2005, 34: 35–50.
40. M. D. Bermejo, F. Fdez-Polanco, M. J. Cocero, Experimental study of the operational parameters of a transpiring wall reactor for supercritical water oxidation. The Journal of Supercritical Fluids, 2006, 39: 70–79.
41. D. R. M. Dolores Bermejo, Angel Martin, M. Jose Cocero, Experimental Performance and Modeling of a New Cooled-Wall Reactor for the Supercritical Water Oxidation. Ind Eng Chem Res, 2009, 48: 6262–6272.
42. M. J. Cocero, J. L. Martínez, Cool wall reactor for supercritical water oxidation. The Journal of Supercritical Fluids, 2004, 31: 41–55.
43. M. Bermejo, F. Cantero, M. Cocero, Supercritical water oxidation of feeds with high ammonia concentrations Pilot plant experimental results and modeling. Chem Eng J, 2008, 137: 542–549.
44. B. Veriansyah, J.-D. Kim, J.-C. Lee, A double wall reactor for supercritical water oxidation: Experimental results on corrosive sulfur mustard simulant oxidation. Journal of Industrial and Engineering Chemistry, 2009, 15: 153–156.
45. E. Fauvel, C. Joussot-Dubien, P. Guichardon, G. Charbit, F. Charbit, S. Sarrade, A double-wall reactor for hydrothermal oxidation with supercritical water flow across the inner porous tube. The Journal of Supercritical Fluids, 2004, 28: 47–56.
46. P. A. Marrone, M. Hodes, K. A. Smith, J. W. Tester, Salt precipitation and scale control in supercritical water oxidation—part B: commercial/full-scale applications. The Journal of Supercritical Fluids, 2004, 29: 289–312.
47. I. Leusbrock, S. J. Metz, G. Rexwinkel, G. F. Versteeg, Quantitative approaches for the description of solubilities of inorganic compounds in near-critical and supercritical water. The Journal of Supercritical Fluids, 2008, 47: 117–127.
48. A. Kruse, Hydrothermal biomass gasification. The Journal of Supercritical Fluids, 2009, 47: 391–399.
49. S.-I. Kawasaki, T. Oe, S. Itoh, A. Suzuki, K. Sue, K. Arai, Flow characteristics of aqueous salt solutions for applications in supercritical water oxidation. The Journal of Supercritical Fluids, 2007, 42: 241–254.
50. J. A. H. Schmieder, N. Boukis, E. Dinjus, A. Kruse, M. Kluth, G. Petrich, E. Sadri, M. Schacht, Hydrothermal gasification of biomass and organic wastes. J Supercrit Fluids, 2000, 17: 145–153.
51. A. A. Peterson, P. Vontobel, F. Vogel, J. W. Tester, In situ visualization of the performance of a supercritical-water salt separator using neutron radiography. The Journal of Supercritical Fluids, 2008, 43: 490–499.
52. P. Zhang, Wang J.C., Zhang, X.D., Liu, X.W., Xia, Y.J. and Li, Z.Y., Progress on Waste Water Treatment of Supercitical Water Oxidation. Environ. Prot. Sci., 2003, 29: 15–17.
53. J. W. T. Fred J. Armellini, Experimental Methods for Studying Salt Nucleation and Growth from Supercritical Water? The Journal of Supercritical Fluids, 1991, 4: 254–264.
54. M. Hodes, P. A. Marrone, G. T. Hong, K. A. Smith, J. W. Tester, Salt precipitation and scale control in supercritical water oxidation—Part A: fundamentals and research. The Journal of Supercritical Fluids, 2004, 29: 265–288.
55. L. B. Kriksunov, D. D. Macdonald, Corrosion in Supercritical Water Oxidation Systems: A Phenomenological Analysis. Chemphyschem A European Journal of Chemical Physics & Physical Chemistry, 1995, 142: 4069–4073.
56. P. Kritzer, Boukis, N. and Dinjus, E., Corrosion of Alloy 625 in Aqueous Solutions Containing Chloride and Oxygen. NACE International, 1998, 54: 824–834.
57. J. W. Tester, Holgate, H. R., Armellini, F. J., Webley, P. A., Killilea, W. R., Hong, G. T. and Barner, H. E., Emerging Technologies in Hazardous Waste Management III. ACS Symp Ser, 1993, 518: 35–76.
58. E. H. A. H. Oelkers, H. C., Multiple ion association in supercritical aqueous solutions of single electrolytes. Science, 1993, 261: 888–896.

59. N. B. P. Kritzer, E. Dinjus, Factors controlling corrosion in high-temperature aqueous solutions: a contribution to the dissociation and solubility data influencing corrosion processes. J Supercrit Fluids, 1999, 15: 205–227.
60. D. H. Su, Zheng, Z., Wang, Y., and Tang D. Y., Supercritical water oxidation technology. Ind. Water Treat., 2003, 23: 10–14.
61. D. C. A. S. Elliott, Jr. L. J., Low temperature gasification of biomass under pressure. Fundamentals of thermochemical biomass conversion, 1985, 29: 937–950.
62. M. J. Antal, Allen, S. G., Schulman, D., Xu, X. and Divilio, R. J., Biomass gasification in supercritical water. Ind Eng Chem Res, 2000, 39: 4040–4053.
63. K. S. Ro, Cantrell, K., Elliott, D. and Hunt P. G., Catalytic wet gasification of municipal and animal wastes. Ind Eng Chem Res, 2007, 46: 8839–8845.
64. M. H. Waldner, F. Krumeich, F. Vogel, Synthetic natural gas by hydrothermal gasification of biomass. The Journal of Supercritical Fluids, 2007, 43: 91–105.
65. P. E. Savage, A perspective on catalysis in sub- and supercritical water. The Journal of Supercritical Fluids, 2009, 47: 407–414.
66. M. Schubert, J. W. Regler, F. Vogel, Continuous salt precipitation and separation from supercritical water. Part 1: Type 1 salts. The Journal of Supercritical Fluids, 2010, 52: 99–112.
67. M. Schubert, J. W. Regler, F. Vogel, Continuous salt precipitation and separation from supercritical water. Part 2. Type 2 salts and mixtures of two salts. The Journal of Supercritical Fluids, 2010, 52: 113–124.
68. P. T. Steven N. Rogak, Deposition of Sodium Sulfate in a Heated Flow of Supercritical Water. AlChE J, 1999, 45: 240–247.
69. S. J. K. Stendahl, Phosphate recovery from sewage sludge in combination with supercritical water oxidation. Water Sci. Technol, 2003, 48: 185–191.
70. K. Príkopskı, Characterization of continuous diffusion flames in supercritical water. A dissertation submitted to ETH Zurich for the degree of Doctor of Technical Sciences, 2007.
71. Y. Matsumura, Minowa, T., Potic, B., Kersten, S. R. A., Prins, W., Swaaij W. P. M., Beld, B., Elliott, D. C., Neuenschwander, G. G., Kruse, A. and Antal, Jr M. J., Biomass gasification in near- and super-critical water: Status and prospects. Biomass Bioenergy, 2005, 29: 269–292.
72. S. Baur, H. Schmidt, A. Krämer, J. Gerber, The destruction of industrial aqueous waste containing biocides in supercritical water—development of the SUWOX process for the technical application. The Journal of Supercritical Fluids, 2005, 33: 149–157.
73. E. N. J. R. Portela, E. Mart′ınez de la Ossa∗, Kinetic comparison between subcritical and supercritical water oxidation of phenol. Chem Eng J, 2001, 81: 287–299.
74. A. Sobhy, R. I. L. Guthrie, I. S. Butler, J. A. Kozinski, Naphthalene combustion in supercritical water flames. Proceedings of the Combustion Institute, 2009, 32: 3231–3238.
75. C. Narayanan, C. Frouzakis, K. Boulouchos, K. Príkopský, B. Wellig, P. Rudolf von Rohr, Numerical modelling of a supercritical water oxidation reactor containing a hydrothermal flame. The Journal of Supercritical Fluids, 2008, 46: 149–155.
76. R. M. Serikawa, Usui, T., Nishimura, T., Sato, H., Hamada, S. and Sekino, H., Hydrothermal flames in supercritical water oxidation: investigation in pilot scale continuous reactor. Fuel, 2002, 81: 1147–1159.
77. P. E. Savage*, J. B. Dunn, J. Yu, Recent Advances in Catalytic Oxidation in Supercritical Water. Combust Sci Technol, 2006, 178: 443–465.
78. V. Lavric, V. Pleşu, J. De Ruyck, Chemical reactors energy integration through virtual heat exchangers—benefits and drawbacks. Appl Therm Eng, 2005, 25: 1033–1044.
79. E. A. M. J. Cocero, M. T. Sanz, F. Fdz-Polanco, Supercritical water oxidation process under energetically self-sufficient operation. J Supercrit Fluids, 2002, 24: 37–46.
80. A. Cocero M. J., E. and Fdez-Polanco, F., Supercritical water oxidation of wastewaters and sludges. Life Sci., 2002, 2: 195–200.

81. A. K. Andrea Kruse, Valentin Schwarzkopf, Celine Gamard, Thomas Henningsen, Influence of Proteins on the Hydrothermal Gasification and Liquefaction of Biomass. 1. Comparison of Different Feedstocks. Ind Eng Chem Res, 2013, 44: 3013–3020.
82. J. Michael Jerry Antal, Stephen Glen Allen, Deborah Schulman, Xiaodong Xu, Biomass Gasification in Supercritical Water. Ind Eng Chem Res, 2000, 39: 4040–4053.
83. A. Kruse, Supercritical water gasification. Biofuels, Bioproducts and Biorefining, 2010, 4: 241–241.
84. O. Yakaboylu, J. Harinck, K. Smit, W. de Jong, Supercritical Water Gasification of Biomass: A Literature and Technology Overview. Energies, 2015, 8: 859–894.
85. A. Arregi, M. Amutio, G. Lopez, J. Bilbao, M. Olazar, Evaluation of thermochemical routes for hydrogen production from biomass: A review. Energy Convers Manage, 2018, 165: 696–719.
86. A. G. Haiduc, M. Brandenberger, S. Suquet, F. Vogel, R. Bernier-Latmani, C. Ludwig, SunCHem: an integrated process for the hydrothermal production of methane from microalgae and CO2 mitigation. J Appl Phycol, 2009, 21: 529–541.
87. N. H. Mitsumasa Osada, Osamu Sato, Kunio Arai, Masayuki Shirai, Effect of Sulfur on Catalytic Gasification of Lignin in Supercritical Water. Energy Fuels, 2007, 21: 1400–1405.
88. M. Dreher, B. Johnson, A. A. Peterson, M. Nachtegaal, J. Wambach, F. Vogel, Catalysis in supercritical water: Pathway of the methanation reaction and sulfur poisoning over a Ru/C catalyst during the reforming of biomolecules. J Catal, 2013, 301: 38–45.
89. N. H. Mitsumasa Osada, Osamu Sato, Kunio Arai, Masayuki Shirai, Subcritical Water Regeneration of Supported Ruthenium Catalyst Poisoned by Sulfur. Energy Fuels, 2008, 22: 845–849.
90. M. Dreher, M. Steib, M. Nachtegaal, J. Wambach, F. Vogel, On-Stream Regeneration of a Sulfur-Poisoned Ruthenium-Carbon Catalyst Under Hydrothermal Gasification Conditions. ChemCatChem, 2014, 6: 626–633.
91. T. Yamamura, T. Mori, K. C. Park, Y. Fujii, H. Tomiyasu, Ruthenium(IV) dioxide-catalyzed reductive gasification of intractable biomass including cellulose, heterocyclic compounds, and sludge in supercritical water. The Journal of Supercritical Fluids, 2009, 51: 43–49.
92. N. H. Aritomo Yamaguchi, Osamu Sato, Mitsumasa Osada, Masayuki Shirai, Lignin Gasification over Supported Ruthenium Trivalent Salts in Supercritical Water. Energy Fuels, 2008, 22: 1485–1492.
93. A. E. Miller, R. Junker, A. Hendry, D. Wilkinson, N. Bollinger, D. Abelleira-Pereira, J. M. Deshusses, M. A. Inniss, E. Jacoby, W., Supercritical water oxidation of a model fecal sludge without the use of a co-fuel. Chemosphere, 2015, 141: 189–96.
94. D. H. Xu, S. Z. Wang, Y. M. Gong, Y. Guo, X. Y. Tang, H. H. Ma, A novel concept reactor design for preventing salt deposition in supercritical water. Chem Eng Res Des, 2010, 88: 1515–1522.
95. D. Xu, S. Wang, X. Tang, Y. Gong, Y. Guo, Y. Wang, J. Zhang, Design of the first pilot scale plant of China for supercritical water oxidation of sewage sludge. Chem Eng Res Des, 2012, 90: 288–297.
96. J. W. T. Paul A. Webley, H. Richard Holgate, Oxidation Kinetics of Ammonia and Ammonia-Methanol Mixtures in Supercritical Water in the Temperature Range 530–700 "C at 246 bar. Ind Eng Chem Res, 1991, 30: 1745-1754.
97. D. A. Patterson, Stenmark, L., Hogan, F., Pilot-scale supercritical water oxidation of sewage sludge. 6th European Biosolids and Organic Residuals Conference Wakefield, 2001.
98. P. O. Callaghan, Regan, J. O., Phosphorus Recovery from Sewage Sludge Using the Aquacritox Supercritical Water Oxidation Process. O2 Environmental and SCFI, 2010: 1–8.

99. D. S. Sloan, Pelletier, R. A., Modell, M., Sludge management in the city of Orlandod it's supercritical. Florida Water Resour. J., 2008, 7: 46–54.
100. S. Heidenreich, P. U. Foscolo, New concepts in biomass gasification. Prog Energy Combust Sci, 2015, 46: 72–95.
101. U. G. N. Boukis, H. Müller, E. Dinjus, Biomass gasification in supercritical water. experimental progress achieved with the verena pilot plant. 15th European Biomass Conference & Exhibition, 2007.

Chapter 2
Supercritical Water Gasification: Reaction Mechanism and Kinetics

2.1 Supercritical Water Gasification of Biomass

Hydrogen is defined as an attractive energy carrier due to its potentially higher energy efficiency and lower generation of pollutants, which can replace conventional fossil fuels in the future. The governments have invested huge funds and made great efforts on the research of hydrogen production. Among the various options, supercritical water gasification (SCWG) is the most promising method of hydrogen production from biomass. Supercritical water is the best reaction medium for biomass with high water content. Physicochemical properties of supercritical water are greatly different from gas to liquid at normal condition, which makes it possible not only to take part in the reaction as a reactant but also to affect the reaction pathway as a catalyst. Among the various options, supercritical water gasification (SCWG) is one of the most promising methods of hydrogen production from biomass. Supercritical water (SCW) has received a great deal of attention as one of the most suitable reaction medium for biomass gasification because it is safe, non-toxic, readily available, inexpensive, and environmentally benign [1].

Sewage sludge (SS), containing a large amount of organics such as carbohydrates, proteins, lipids, and so on, is a by-product of traditional domestic wastewater treatment and can be regarded as a wet biomass. It also contains heavy metals, pathogens, and other microbiological pollutants, etc., causing hazards to environment and human health, so a cost-effective harmless disposal has become a pressing environmental issue. To meet the challenge, hydrothermal gasification is alternative as a feasible technology for SS treatment in terms of H_2 production, energy recovery, and solid waste management. Typically, supercritical water gasification takes advantage of unique properties of supercritical water (T > 374 °C, P > 22.1 MPa) to convert organics present in SS into a hydrogen-rich gas via pyrolysis, hydrolysis, steam reforming, etc. Due to the prevention of energy-intensive dewatering, supercritical water gasification is a promising technology to produce H_2 cleanly and efficiently if taking the revenues associated with SS disposal into account.

S. Wang et al., *Supercritical Water Processing Technologies for Environment, Energy and Nanomaterial Applications*,
https://doi.org/10.1007/978-981-13-9326-6_2

Continuous research is necessary to get a deeper understanding of chemistry, which can be mastered with the development of efficient and stable catalysts. Given the opportunities that catalysis in SCWG provides green chemistry/engineering and sustainable energy in advance, we can expect even more research and development activity in this growing field. If the good cost performance can be provided, this technology holds great potential to be large-scale commercialized in the future. A co-precipitation method was adopted to make a –Mn–Cu– mixture catalyst, of which components were 50.9 wt% $CuMn_2O_4$, 46.1 wt% CuO and 3.0 wt% Mn_2O_3, and its average pore sizes of BJH adsorption and desorption were 50.6 and 41.2 nm, respectively. We combined it with AC and Na_2CO_3 to investigate their co-existence influences on SS SWPO at 450 °C, 25 MPa, OC = 0.3, and 5 h of residence time for the first time. The results showed that –Mn–Cu–AC–N had the best catalytic effect on SS SWPO among MnO_2–AC–N, CuO–AC–N and –Mn–Cu–AC–N, and made the yields of H_2, CO, CH_4 and CO_2, and XCOD and XTOC reach up to 202.4 ml/L, 68.5 ml/L, 140.0 ml/L, 373.2 ml/L, 95.6 wt%, and 88.0 wt %, respectively. –Mn–Cu–AC–N could obviously improve H_2 yield and COD removal of liquid products in SS SWPO, evidently higher than 48.7 ml/L and 86.2 wt% obtained without catalyst. However, in SS SWPO, porous catalyst (–Mn–Cu–) surface was easily plugged by pollutants and suffered slight dissolution issue. We also propose potential catalytic reaction pathways of SS SWPO with the multi-component –Mn–Cu–AC–N catalyst. Undoubtedly, reaction mechanism investigation, catalytic stability improvement, and long-term tests in a continuous flow reactor are further exploring directions in the future [2].

2.2 Supercritical Water Gasification of Model Compound

Industrial organic wastewater, such as the effluent from paper industry, pharmaceutical industry, and chemical industry usually contains high-concentration of hydrocarbons which may be difficult and uneconomic to be treated by conventional methods because many of them are toxic or inhibitory to anaerobes. However, it is appropriate for SCWG to produce hydrogen with high additional values. It can also be considered as a research highlight and application area with tremendous potential. Unlike the biomass feed, most wastewater is homogeneous liquid and can be easily pumped to desired critical pressure by conventional pressurization equipment. Typically, industrial wastewater has characteristic of high-concentration of pollutant and blending of many refractory organic compounds, such as HAc, propionic acid, 1,2-ethanediol, ketones, and phenols, which is very similar to the basic ingredients and intermediate products of biomass in SCWG reaction [3] (Fig. 2.1).

Acetic acid (HAc) and phenol are the most common and refractory compounds during supercritical water gasification (SCWG) of organic wastewater for hydrogen production. Here, we report SCWG of HAc and the two- and three-components mixture containing HAc and phenol with and without sodium hydroxide catalyst,

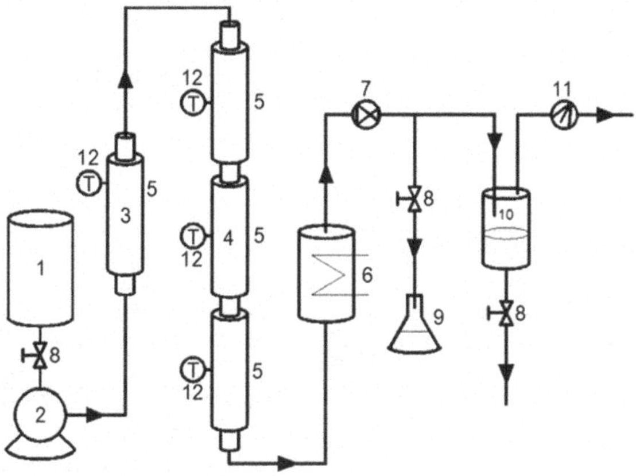

Fig. 2.1 Schematic diagram of experimental apparatus: (1) feed tank; (2) high-pressure pump; (3) pre-heater; (4) reactor; (5) electric heater; (6) quencher; (7) back pressure regulator; (8) valves; (9) liquid product collector; (10) gas–liquid separator; (11) wet type gas meter; (12) thermocouples

which was conducted in a tubular-flow reactor. Sodium hydroxide had a catalytic effect on the steam reforming reaction of methane during SCWG of HAc. For SCWG of HAc (0.5 wt%) and phenol (1.0 wt%) mixture, yield of hydrogen and carbon dioxide reached a maximum at 0.2 wt% of sodium hydroxide. The hydrogenation of phenol to form benzene and further to cyclohexane could be reinforced by the acceleration of NaOH on hydrogen production. The amount of undesired long-chain alkyl esters were also found in the liquid products. Large amount of sodium acetate and sodium phenolate formed at high NaOH concentrations, which could lead to lower reaction efficiency and more serious deposition and coking problem. For SCWG of the three-components (methanol, HAc, and phenol) mixture, total organic carbon (TOC) removal efficiency, H and C gasification efficiency could be strongly retarded by both HAc and phenol. Decarboxylation of HAc led to relatively higher C gasification efficiency. Besides, their empirical quadric polynomial regression models were provided via response surface method [3] (Figs. 2.2 and 2.3).

As the most common and refractory compounds during SCWG of organic wastewater, acetic acid, and phenol were primarily studied in this paper. The SCWG of HAc and multi-components mixture was conducted in a tubular-flow reactor under different conditions. For SCWG of HAc, the experimental values of methane and H_2 mol% in gas product showed a great deviation from their equilibrium values. It can be attributed to the incomplete reaction resulting from the non-ideal flow and thermal conduction and obstacle from intermediate products. NaOH had a catalytic effect on the steam reforming reaction between methane and H_2O; for SCWG of HAc (0.5 wt%) and phenol (1.0 wt%) mixture, yield of H_2 and CO_2 reached a maximum at 0.2 wt% of NaOH. The content % of phenol in liquid product was lower than that without adding NaOH. The hydrogenation of phenol to form benzene and further to cyclohexane could be reinforced because of the acceleration of NaOH on hydrogen production. The amounts of long-chain alkyl esters were also found in the liquid products. The esterification and polymerization reactions between carboxylic acid and ring-opening products such as long-chain

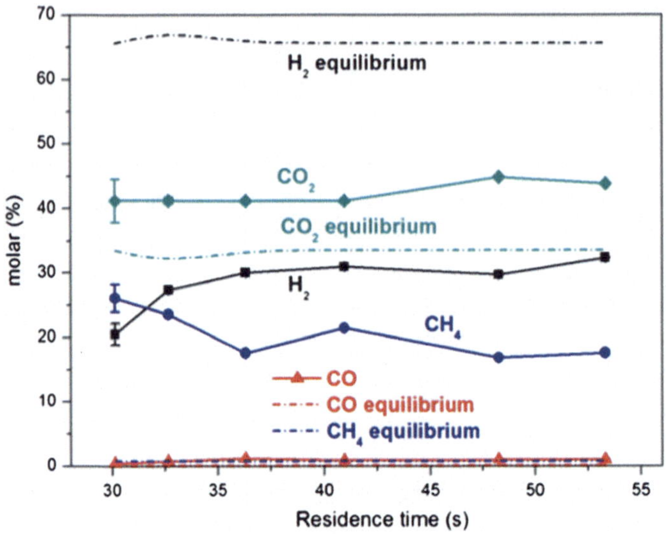

Fig. 2.2 Effect of residence time on gas composition for SCWG of acetic acid (1 wt%) without NaOH (550 °C, 25 MPa). Dashed lines represent the calculated equilibrium value for each gas

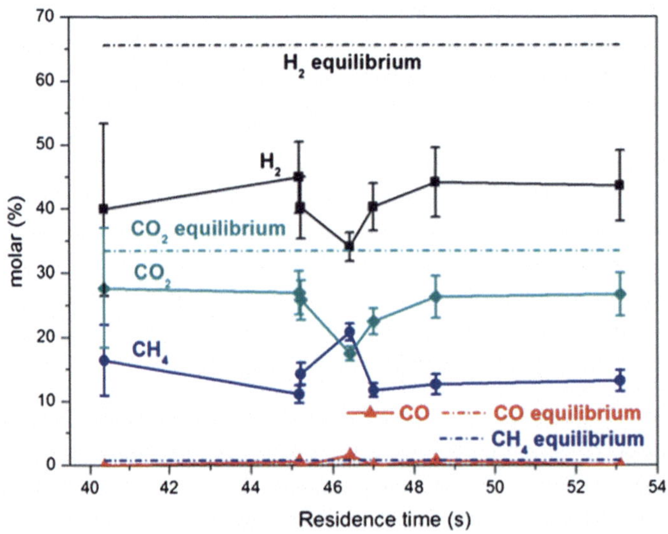

Fig. 2.3 Effect of residence time on gas composition for SCWG of acetic acid (1 wt%) with 0.1 wt% NaOH (550 °C, 25 MPa). Dashed lines represent the calculated equilibrium value for each gas

alcohols were responsible for this phenomenon. The TIC content showed a negative tendency as concentration of NaOH increased after reaching a maximum at 0.2 wt %. Na_2CO_3 and $NaHCO_3$ could be formed via CO_2-capture and neutralization reactions. In this process, large amount of undesired sodium acetate and sodium phenolate was formed at high NaOH concentrations, which could lead to lower reaction efficiency in SCWG and more serious deposition and coking. For SCWG of the three-components (methanol, HAc, and phenol) mixture, TRE could be strongly depressed by adding both HAc and phenol. But the former showed a slightly weak inhibiting effect on CGE. Decarboxylation of HAc to form CH_4 and CO_2 may have contribution to this result (Figs. 2.4 and 2.5).

A mixture of phenol, acetic acid, and naphthalene was partially oxidized in supercritical water in order to produce hydrogen. The effects of temperature, oxidation ratio (OR), reaction time, and reactant concentrations on gaseous distributions, gasification efficiencies, and reactants removal efficiencies were investigated. Furthermore, the effects of oxygen on the main intermediate products were analyzed, and possible degradation pathways were proposed. Results indicated that higher temperatures significantly promoted the H_2 yield. In addition, small amounts of oxygen (OR < 0.2) accelerated the production of H_2 and CH_4. The maximum H_2 gasification efficiency (240.25%) and H_2 yield (70.16 mmol/g) were obtained at 560 °C, 25 MPa, reaction time of 20 s and OR of 0.2. Longer reaction time enhanced the gasification efficiencies in 10 s, while the effect was little when reaction times longer than 10 s. The concentrations of the reactants did not significantly influence the gasification efficiencies [4]

Phenol, acetic acid, and naphthalene are the refractory intermediates in the SCWG of biomass process. The ternary mixture of these compounds for hydrogen

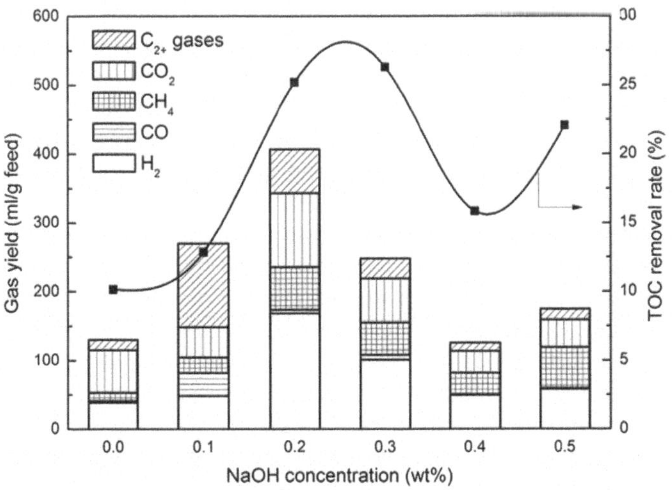

Fig. 2.4 Yields of different gaseous products and TOC removal rate for SCWG of mixture solution (0.5 wt% HAc and 1 wt% phenol)

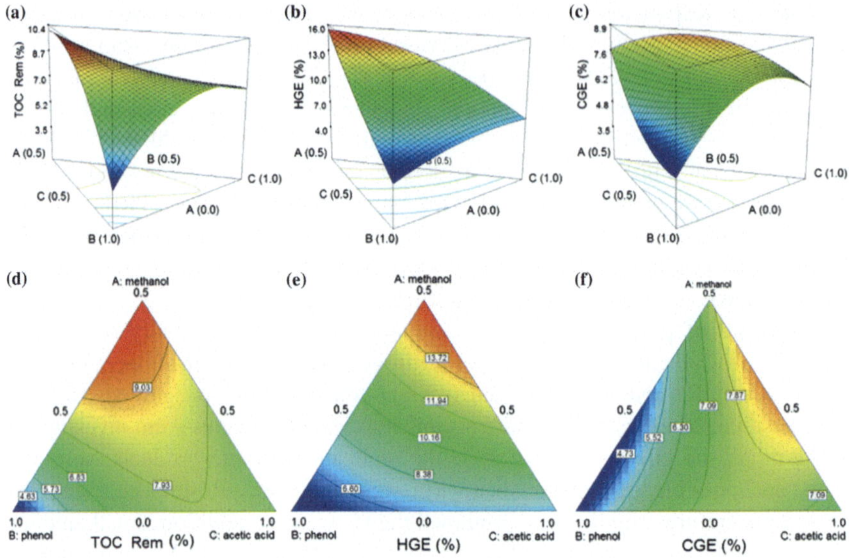

Fig. 2.5 Response surface (aec) and contour (def) plots for TRE, HE, and CGE in SCWG of mixture containing methanol, acetic acid, and phenol

in supercritical water with partial oxidation was tested in a continuous tubular reactor at 400–640 °C, reaction times of 5–40 s, mass concentrations of 0.5–3 wt %, and OR of 0–1. Both temperature and OR gave crucial effects on gas products distributions and gas yields. Increasing temperature enhanced the H_2 production and TOC removal efficiency greatly. Analysis of liquid products showed that the addition of small amounts of oxidant (OR < 0.2) promoted the ring-opening products generation and inhibited the polymer formation, which induced to the higher yield of H_2. At conditions of 560 °C, 25 MPa, and OR of 0.2, the reactions needed 10 s to reach equilibrium, while the effect of the reactants concentrations on H_2 yield was negligible, which followed the first-order kinetic reaction. The interactions between phenol, acetic acid, and naphthalene will be studied in our later research [4] (Figs. 2.6, 2.7, 2.8 and 2.9).

A quantitative kinetic model for the supercritical water partial oxidation of the mixture of acetic acid, naphthalene, and phenol at 560 °C, 25 MPa was established. The model consisted of a group of pathways that included two kind of compounds, which could be classified as stable cyclic compound (Int.1) and unstable ring-opening products (Int.2). The model was validated with the experimental data at oxygen ratio (OR) of 0 and 0.2, respectively. Results showed that the model could accurately predict the influences of reactants time on gas yields and intermediates concentrations. Reaction rate analysis indicated that the general trends of reaction rate at OR ¼ 0 and OR ¼ 0.2 were similar. The decomposition and steam reforming reactions of Int.2 were the main pathways for gas production. The oxygen during the gasification played a positive role in promoting the ring-opening

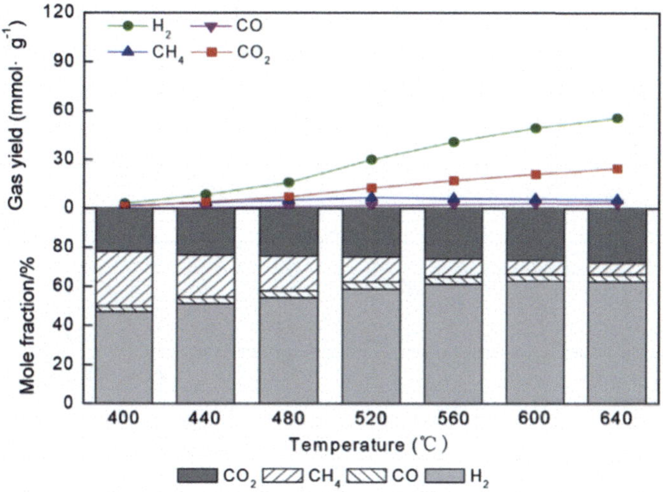

Fig. 2.6 Effects of temperature on gas distributions and yields (25 MPa, OR of 0, and reaction time of 20 s)

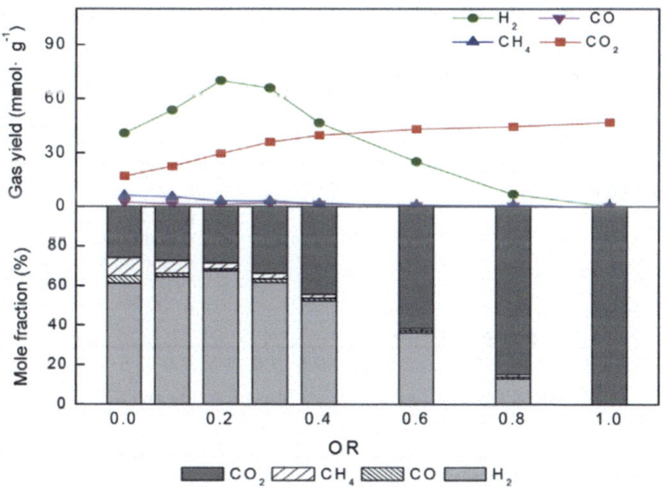

Fig. 2.7 Influence of OR on gas distributions and yields (560 °C, 25 MPa, and reaction time of 20 s)

reactions, which induced higher Int.2 production and increasing of steam reforming and decomposition rates, and finally increased the H_2 production [5].

Quantitative kinetic model for the influences of oxygen on SCWG of acetic acid, phenol, and naphthalene mixture at 560 °C, 25 MPa was established by classifying the intermediates as stable cyclic compound (Int.1) and unstable ring-opening

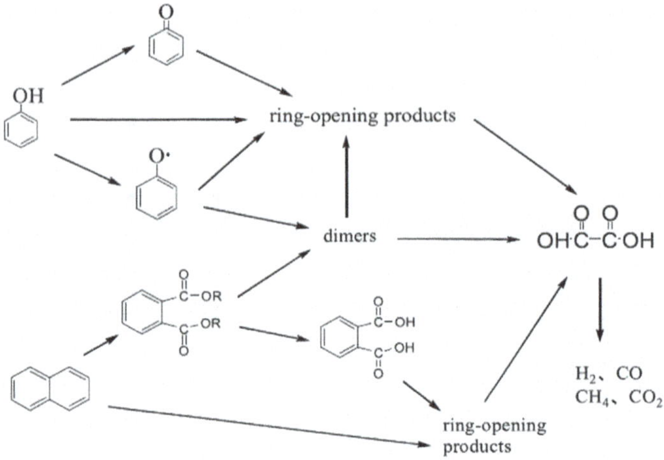

Fig. 2.8 Degradation pathways of the mixture of phenol, naphthalene, and acetic acid

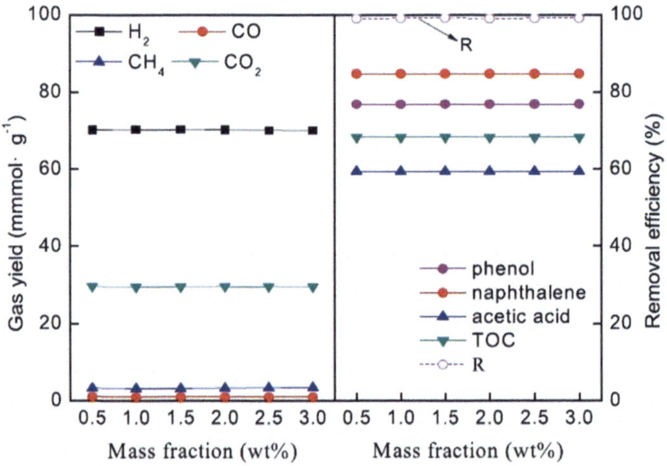

Fig. 2.9 Effects of reactants concentrations on gas yields and reactants removal efficiencies (560 °C, 25 MPa, OR of 0.2, and reaction time of 20 s)

products (Int.2). The kinetic model fitted the experimental data of gas yields well and predicted that the variation trends for Int.1 and Int. 2 generation at OR ¼ 0 and OR ¼ 0.2 were similar. Int.1 disappeared more slowly than Int.2, and the concentration of Int.1 at OR ¼ 0.2 was much lower than that at OR ¼ 0. Steam reforming and decomposition of Int.2 were the fast pathways to gas formation at short reaction times. Steam reformation reactions were the main pathways for H_2, while decomposition of intermediates was the main path for CO_2 and CH_4. The main effect of oxygen in the gasification lay in promoting the ring-opening

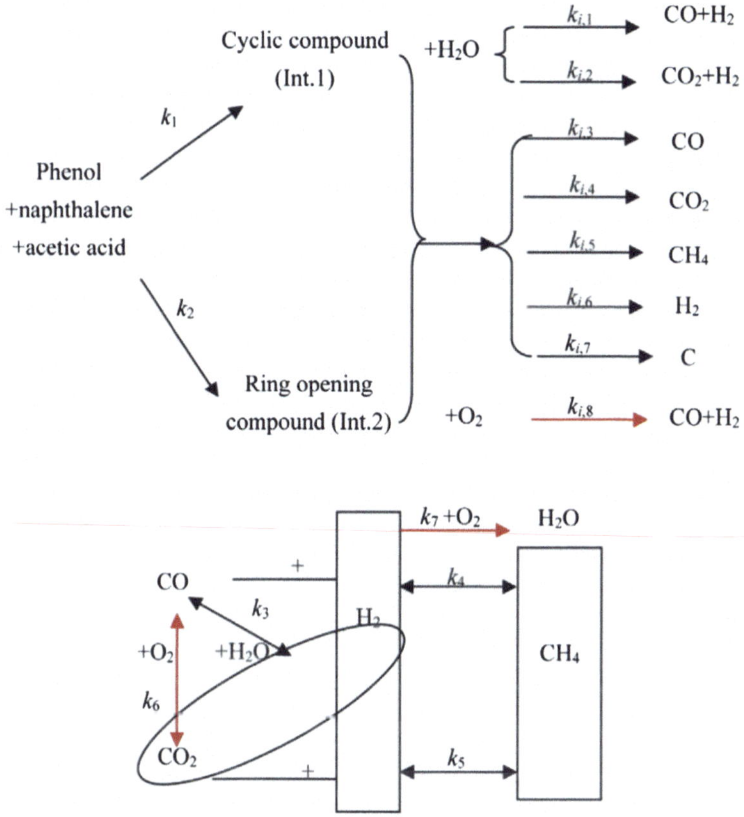

Fig. 2.10 Reaction pathways for supercritical water gasification of the mixture of phenol, acetic acid, and naphthalene at OR ¼ 0 and OR ¼ 0.2

reactions, which induced higher Int.2 production and increasing of steam reforming and decomposition rates, and finally increased the H_2 production [5] (Fig. 2.10).

Indole is commonly reported as a product from hydrothermal processing of algal biomass. The reactions of indole in supercritical water were investigated between 550 and 700 °C in quartz, mini-batch reactors. The indole disappearance rate followed first-order kinetics and the activation energy was 155 ± 10 kJ/mol. Methane and hydrogen were the most abundant gaseous products under most of the tested conditions, while benzene was the most abundant liquid-phase product. Hydrogen and carbon gasification efficiencies (HGE and CGE) exhibited values up to 79% and 20%, respectively. The influence of water density on the yields of H_2, CH_4, and C_2H_6 was negligible at densities above 0.081 g/ml, but the CO_2 yield increased with water density whereas the CO yield decreased. The yield of CH_4 increased significantly as the initial indole concentration increased. The collective results, which showed how the yields of numerous intermediate reaction products responded to changes in the process variables, permitted advancement of a potential reaction network [6] (Fig. 2.11).

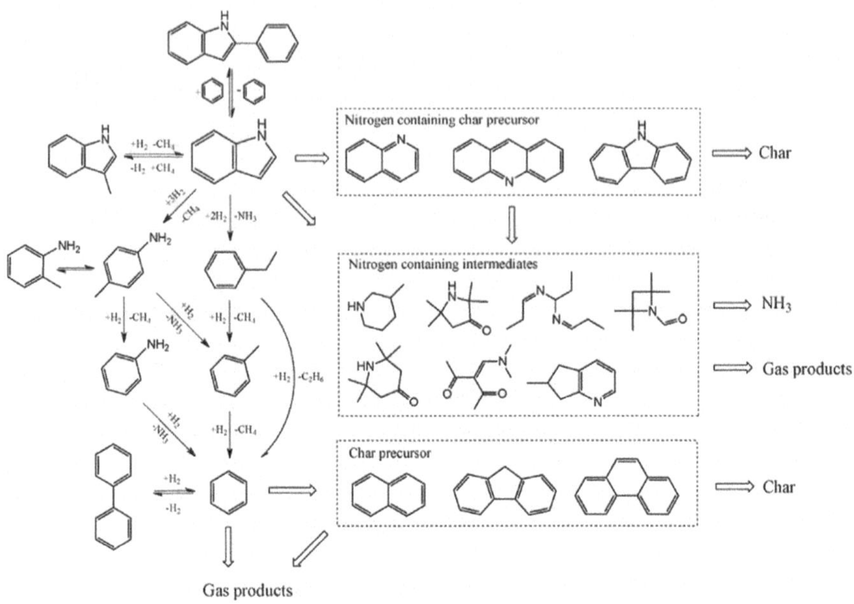

Fig. 2.11 Potential reaction network for indole in SCW

This article provides information about the reaction kinetics and pathways of indole in SCW at temperatures higher than any previously explored. Indole disappearance in SCW at 550–700 °C followed first-order kinetics and the activation energy was 155 ± 10 kJ/mol. Numerous liquid-phase intermediate products formed, both by addition to indole and by cleavage of the pyrrolic ring. Benzene was the most abundant liquid-phase product and it reached molar yields of nearly 15%. The other major liquid-phase products were toluene, aniline, and ammonia. Several different polycyclic aromatic compounds, some containing nitrogen, were also formed, but in molar yields of less than 1%. The variation of the molar yields of the various products with the indole conversion permitted identification of likely reaction pathways for each product. Methane and hydrogen were the most abundant gaseous products under nearly all of the investigated reaction conditions. The highest molar yields of H_2 and CH_4, 64% and 107%, respectively, appeared at 700 °C, which was the highest temperature examined. The hydrogen gasification efficiency always exceeded that for carbon and the highest values obtained were 79% and 20%, respectively. This value for the carbon gasification efficiency was much lower than that observed for other materials under similar conditions and it illustrated the difficulty of gasifying indole in supercritical water in the absence of a catalyst. There was little influence of water density on the yields of H_2, CH_4, and C_2H_6, but the yield of CO_2 increased whereas that for CO decreased as water density increased, which suggested that the water–gas shift reaction becomes more favorable at the higher densities. Char formation was observed on the inner wall of the quartz reactors, and char covered more of the interior surface at the higher initial indole concentrations. The yield of CH_4 increased significantly as the initial indole concentration increased whereas the yield of CO decreased, but other gases showed little change [6].

We developed a quantitative kinetics model for the homogeneous decomposition and gasification of indole in supercritical water at 550–650 °C based on a reaction network with 17 separate reaction pathways. The model accurately describes the effects of time, temperature, and initial concentrations on the concentration of intermediate products such as aniline, toluene, and benzene as well as the terminal gaseous products such as H_2, CH_4, and CO_2. The model revealed that three steps are most important for H_2 production. Ring-opening of indole to form aniline was the fastest such step during the first few minutes at 600 °C, depletion of the large variety of gasifiable products dominated at times between 5 and 55 min, and water–gas shift was the main H_2-producing reaction at longer times. The model also revealed that some of the potential pathways were kinetically insignificant at 600 °C. Such paths included methanation, gasification of benzene, steam reforming of indole, and formation of CO, CO_2, and C_2H_6 from intermediate products [7].

The hydrothermal reaction network and kinetics given herein for indole provide the foundation for a kinetics model that is quantitatively consistent with the experimental results. The model accurately correlated the effects of batch holding time and reaction temperature on the concentrations of indole, toluene, benzene, and aniline and the concentrations of six different gas-phase species. It also accurately predicted the effects of water density and indole's initial concentration on the yields of key gaseous products. The model showed that indole gasification to form H_2 or CH_4 occurs almost exclusively through paths that avoid prior formation of aniline, toluene, and benzene. That is, formation of these aromatic by-products makes gasification even less likely because they decompose even more slowly than indole does. Thus, these aromatic by-products are barriers rather than intermediates to gas formation. The model results suggest that one path to higher gas yields is to use higher gasification temperatures. The activation energies for the conversion of gasifiable products to H_2 or CH_4 are 116 and 145 kJ/mol, respectively, whereas the activation energy for making stable products (char precursors) from the gasifiable products is 51 kJ/mol. Accordingly, the rates of the desired gasification reactions would increase faster than the rate of char formation as temperature increasing [7] (Figs. 2.12, 2.13 and 2.14).

2.3 Supercritical Water Gasification in Catalytic Condition

High-temperature and pressure are required to meet the minimum reaction condition. Therefore, the high operating cost has become the biggest obstacle to the development of this technology. To overcome this bottleneck, many researchers

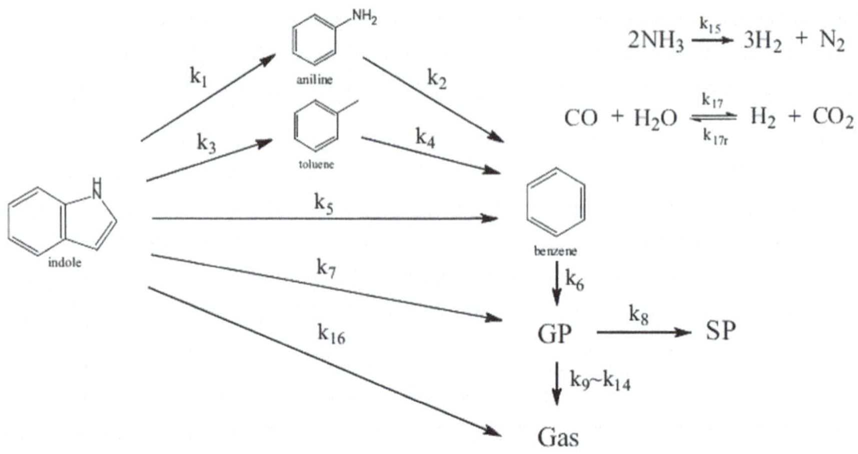

Fig. 2.12 Simplified reaction network for indole in supercritical water

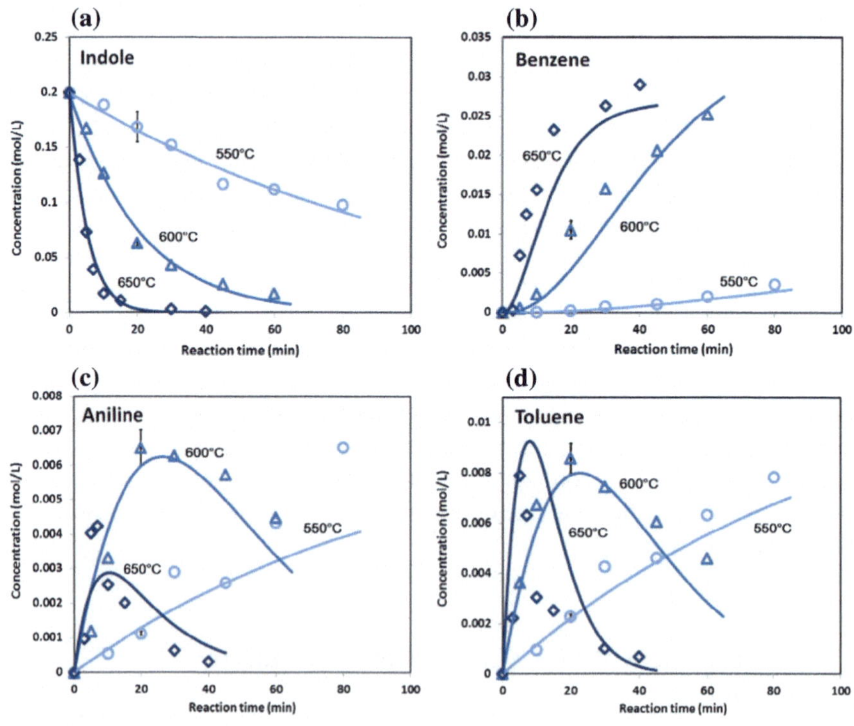

Fig. 2.13 Experimental and model results for temporal variations of indole (**a**), benzene (**b**), aniline (**c**), and toluene (**d**)

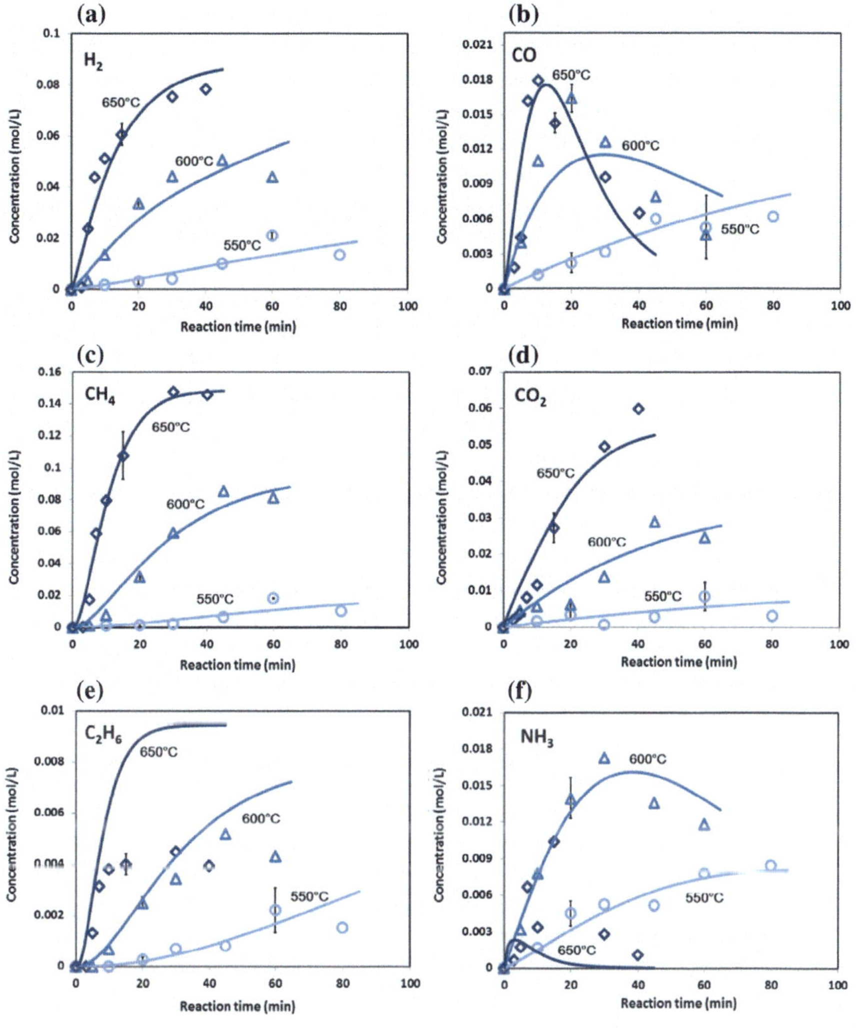

Fig. 2.14 Experimental and model results for temporal variations of H_2 (**a**), CO (**b**), CH_4 (**c**), CO_2 (**d**), C_2H_6 (**e**), and NH_3 (**f**)

have carried out intensive research work on the catalytic supercritical water gasification (CSCWG). Based on the previous studies stated in the literature, the authors try to give an overview (but not an exhaustive review) on the recent investigations of CSCWG. But how to reduce the activation energy and to moderate the violent reaction condition have become the pressing problems we have to deal with. It is available to overcome these shortcomings by adding catalysts which are suitable for hydrothermal reaction [1].

According to this summarization, the catalysts used in previous researches can be classified as alkali metals, transition metal (including "wall" effect), and activated carbon catalyst. During the processes, alkali metal catalyst can effectively improve the water–gas shift reaction but may cause corrosion, plugging or fouling of equipment. For transition metal catalyst, such as Ni, Pt, Ru, and Rh, most studies suppose that they improve the reaction by accelerating the steam reforming reaction, methanation reaction, and cleavage of C–O and C–C, and so on. In addition, activated carbon catalyst is also effective during water–gas shift reaction and methanation reaction. It is concluded that these catalysts can effectively increase conversion rate and hydrogen selectivity and decrease the activation energy in SCWG reaction. On the other hand, dozens of studies have confirmed the instability of most catalysts. Besides, destruction of their support is also observed [1].

Supercritical water gasification (SCWG) of nitriles was studied in a tubular-flow reactor at different temperatures. This article focuses on the product distributions and corresponding reaction pathways influenced by addition of Na_2CO_3 catalyst. Results showed that gas yield for both acetonitrile and acrylonitrile could be greatly enhanced by adding Na_2CO_3 catalyst. Especially, H_2 gasification efficiency could reach up to 55.4% and 123.3% at 550 °C, respectively. But the catalytic effect on the gas yield of benzonitrile was relatively insignificant. As a base catalyst, Na_2CO_3 could also accelerate the hydrolysis of cyanogen and amido. Benzene and acetic acid were the primary intermediate products during the SCWG of benzonitrile and acetonitrile, respectively. The conversion of acrylonitrile was more complicated because of the activity of double bond. It was possible that 3,30 iminodipropionitrile was formed by Na_2CO_3 catalyzed in the range of 490–520 °C, which dominated two-thirds of pathways for the subsequent formation of acetic acid. Ammonia-nitrogen content in the liquid effluent was limited by the hydrolysis degree of cyano-group and the possible polymerization reaction of intermediate products. There was no obvious trend to reveal that NH3 was converted into nitrogen under our experimental conditions [8].

The SCWG of benzonitrile, acetonitrile, and acrylonitrile were investigated in a tubular-flow reactor at different temperatures. The gas yield for both acetonitrile and acrylonitrile could be greatly enhanced by adding Na_2CO_3 catalyst, which might accelerate the decarboxylation reaction of intermediate products. However, the similar improvement was not observed for that of benzonitrile which might be attributed to the stable molecular structure of benzene ring. Char/coke might be formed during the preheating process. The major gaseous products for SCWG of acetonitrile and acrylonitrile were CO_2, CH_4, and H_2. Most of CO converted to CO_2 via water–gas shift reaction by Na_2CO_3 catalyst. The hydrogen gasification efficiency for both reactants increased with increasing temperature and reached 55.4% and 123.3% at 550 °C in the presence of Na_2CO_3, respectively. SCW could play a role as hydrogen source for SCWG of acrylonitrile via hydrolysis and water–gas shift reaction. Na_2CO_3 as a base catalyst could accelerate the hydrolysis of cyano-group. Benzene and acetic acid were likely to be the key intermediate products for SCWG of benzonitrile and acetonitrile, respectively. SCWG of acrylonitrile was more complicated because of the activity of double bonds. It is

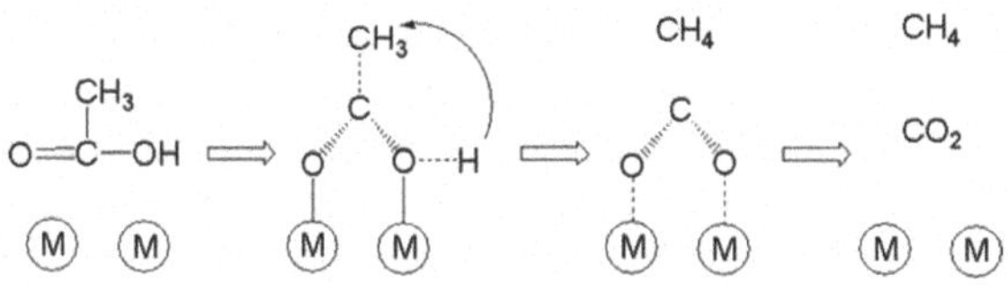

Fig. 2.15 Heterogeneous catalytic mechanism for decarboxylation reaction of acetic acid. The "M" represent the catalytic active sites

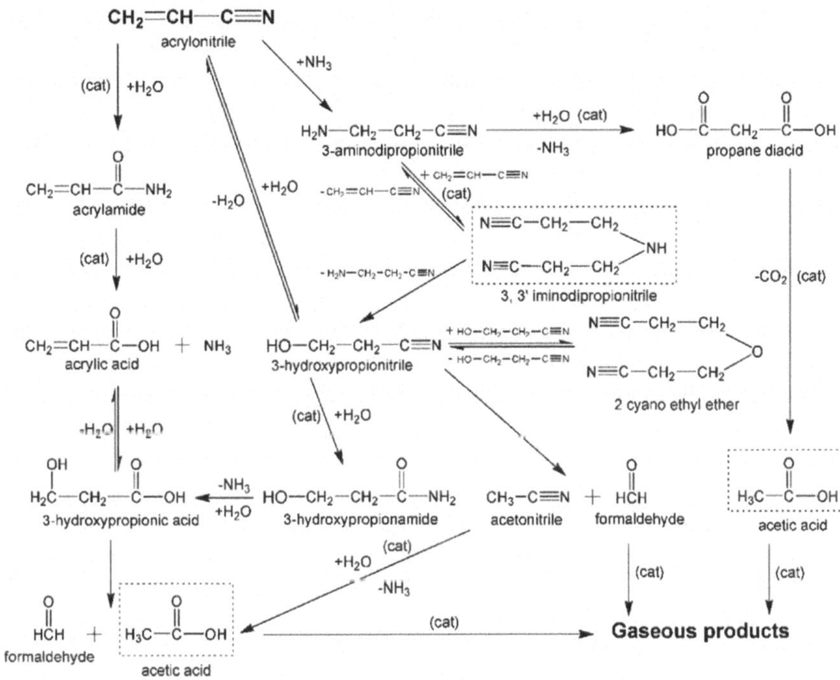

Fig. 2.16 Reaction pathway for SCWG of acrylonitrile. (The term "(cat)" designates reaction pathways influenced by the presence of catalyst. The dashed frame highlighted substances designate key intermediate products)

possible that 3,30 iminodipropionitrile was formed by Na_2CO_3 catalyzed, which dominated 2/3 pathways for the subsequent formation of acetic acid. The yield of C_{2p} gases may be caused by incomplete decomposition of 3,30 iminodipropionitrile. Ammonia-nitrogen contents in the liquid effluent were limited by the hydrolysis degree of cyanogen and possible polymerization reaction of intermediate products. There was no obvious trend to reveal that NH_3 was converted into nitrogen under our experimental conditions [8] (Figs. 2.15 and 2.16).

We tested Ru/C, Pt/C, Pd/C, Pt/Al_2O_3, Pd/Al_2O_3, activated carbon, and Ni–Ru/CeO_2 catalysts for the gasification of indole in supercritical water. All of the materials enhanced the gasification of this nitrogen-containing compound, and especially the bimetallic catalyst, Ni–Ru/CeO_2, showed the highest activity for producing CO and CH_4 and for converting indole. We focused on this catalyst in more detail. The catalyst loading in the reactor had no effect on the H_2 yield, but both CH_4 and CO_2 yields decreased with decreased loading. Water density had no statistically significant effect on the yield of gas products, but operating above the critical pressure of water was a prerequisite for the Ni–Ru composite catalyst to reach a high conversion of indole. The catalyst surface area and pore volume decreased from 5.91 m^2/g and 0.0169 cm^3/g to 2.86 m^2/g and 0.0137 cm^3/g, respectively, after an SCWG reaction time of 180 min. NiO species were detected in both fresh and used catalysts, which may be due to oxidation of the reduced Ni metal during sample preparation for XPS analysis, and/or to a strong interaction between metal and support possibly leading to the formation of cationic Ni species. These results point to potential benefits of using this bimetallic catalyst for SCWG of streams with recalcitrant N-containing compounds [9].

This first report on SCWG of a N-containing compound over supported metal catalysts showed that Pt, Pd, Ru, and Ni are all active for indole gasification. Bimetallic Ni–Ru/CeO_2 catalysts were even more active than any of the monometallic catalysts tested. The amount of this bimetallic catalyst loaded into the reactor had no effect on the H_2 yield, but both the CH_4 and CO_2 yields and the concentrations of benzene, toluene, and aniline increased with increased catalyst loading. The water density had no clear effect on the yields of the gas products, but a pressure above the critical pressure of water was required for the Ni–Ru composite catalyst to generate a high conversion of indole. The surface area and pore volume of the used catalyst were lower than in the fresh catalyst, but catalyst used at SCWG conditions induced no change in its surface composition. This work has identified a promising heterogeneous catalyst for SCWG of recalcitrant N-containing compounds that will be abundant in algal biorefinery process streams [9] (Figs. 2.17, 2.18 and 2.19).

Phenol, a substance difficult to be gasified, is often an intermediate product from biomass gasification in supercritical water. In this study, 1 wt% phenol was gasified

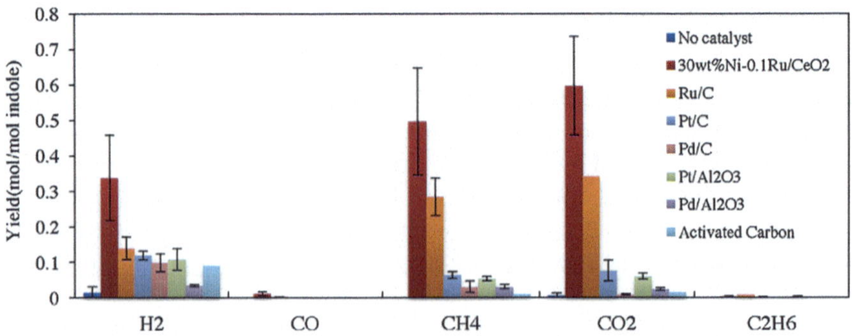

Fig. 2.17 Gas product yield over different catalysts

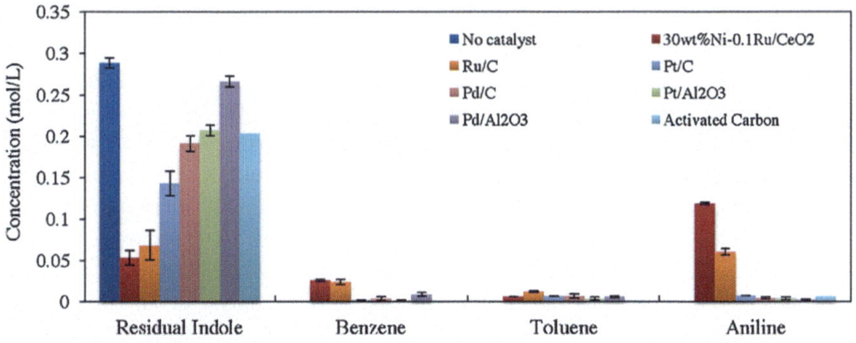

Fig. 2.18 Liquid product concentrations over different catalysts

Fig. 2.19 Gas yields over Ni–Ru/CeO_2 at different Ru/Ni ratios

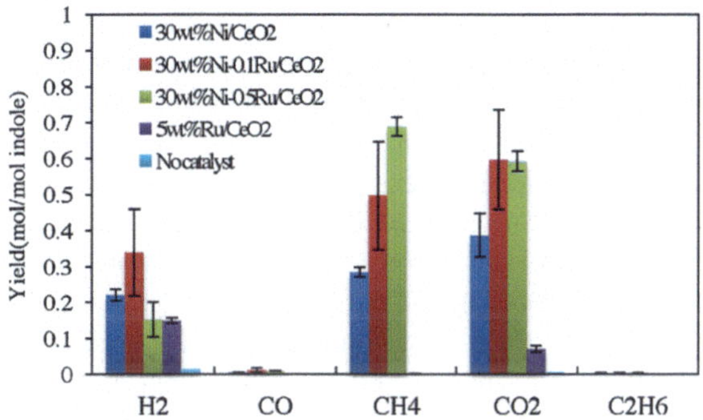

in supercritical water with different oxidation coefficients ($0 \leq n \leq 0.8$) at 430–550 °C, 25 MPa in a continuous tubular-flow reaction plant with or without 0.1 wt % Na_2CO_3. The results showed that total organic carbon (TOC) removal rate and chemical oxygen demand (COD) removal rate of reactor effluent rose with the increase of oxidation coefficient and reaction temperature. When n rose from 0.3 to 0.8 at 500 °C, hydrogen yield and hydrogen molar fraction first increased and then decreased. They reached their respective maximum values of 0.245 mol/mol and 46.2% at $n = 0.6$ with the residence time of 2.96 min. There is an appropriate oxidation coefficient for obtaining the highest hydrogen yield for phenol supercritical water partial oxidative gasification (SWPO). Na_2CO_3 can serve as a catalyst or an additive for phenol SWPO. When 0.1 wt% Na_2CO_3 is added, hydrogen yield is near 0.477 mol/mol at $n = 0.6$, 500 °C with the residence time of 2.96 min. Furthermore, the potential conversion pathways for phenol SWPO with Na_2CO_3 catalyst are also proposed [10].

Gasification experiments of 1 wt% phenol were conducted in a continuous tubular-flow reaction plant with different oxidation coefficients ($0 \leq n \leq 0.8$) in supercritical water at 430–550 °C, 25 MPa with or without 0.1 wt% Na_2CO_3. Some significant conclusions are obtained as follows: When n increases from 0.3 to

0.8, both TOC removal rate and COD removal rate of reactor effluent will rise. Hydrogen yield and the hydrogen molar fraction will first increase and then decrease. They reach their respective maximum values of 0.245 mol/mol and 46.2% with the residence time of 2.96 min at $n = 0.6$, 500 °C. There exists an appropriate oxidation coefficient to obtain the highest hydrogen yield for phenol SWPO. If reaction temperature increases, hydrogen yield, COD removal rate, phenol conversion efficiency, and TOC removal rate will increase. Oxidant ($n = 0.5$) or 0.1 wt% Na_2CO_3 can improve hydrogen yield, COD removal rate, and phenol conversion efficiency. The effect is more evident when Na_2CO_3 is added at $n = 0.5$ and 0.1 wt%. Na_2CO_3 can be employed as a catalyst or an additive for phenol SWPO. Hydrogen yield can reach up to 0.477 mol/mol with the residence time of 2.96 min at $n = 0.6$, 500 °C with 0.1 wt% Na_2CO_3, which is about two times that obtained without Na_2CO_3. Moreover, the potential conversion pathways for phenol SWPO with Na_2CO_3 catalyst are also proposed [10] (Fig. 2.20).

Glycine and glycerol were used as the model compounds of protein and fattiness, respectively. A continuous tubular-flow reactor was used for the gasification experiments operated at 380–500 °C and 25 MPa with or without Na_2CO_3 catalyst. Compared with a negative effect on glycerol gasification, Na_2CO_3 could increase hydrogen yield and chemical oxygen demand (COD) destruction efficiency, and the catalytic performance of 0.1 wt% Na_2CO_3 was better than that of 0.2 wt% for glycine gasification. When 1 wt% glycine solution with 0.1 wt% Na_2CO_3, or 1 wt % glycerol solution without Na_2CO_3 was gasified at 500 °C with the residence time of 0.98 min, their corresponding gasification efficiencies were up to 95.8 and 98%, and hydrogen yields could reach 4.14 and 5.08 mol/mol, respectively. Hydrogen molar fraction in gaseous product was about 60% and liquid effluents could be re-utilized. Correspondingly, the ideal overall reaction equations for glycine and glycerol gasification were proposed [11].

Na_2CO_3 could increase hydrogen yield and COD destruction efficiency of glycine SCWG products, and the catalytic performance of 0.1 wt% Na_2CO_3 was better than that of 0.2 wt%. However, Na_2CO_3 had a negative effect on glycerol gasification. Therefore, as a catalyst or additive, only Na_2CO_3 had a positive effect on some types of organic matters SCWG processes. A combined catalyst may be utilized as the catalyst for complex wet biomass such as sewage sludge SCWG in the future. When 1 wt% glycine solution with 0.1 wt% Na_2CO_3, or 1 wt% glycerol solution without Na_2CO_3 SCWG were conducted using a continuous tubular-flow reactor at 500 °C, 25 MPa, and residence time of 0.98 min, both of them could be gasified almost completely (95.8% and 98%, respectively) and hydrogen yields were 4.14 and 5.08 mol/mol, respectively. Additionally, hydrogen molar fraction was about 60% and the liquid effluents could be recycled into the reactor. Water participated in SCWG processes was an important resource of hydrogen.

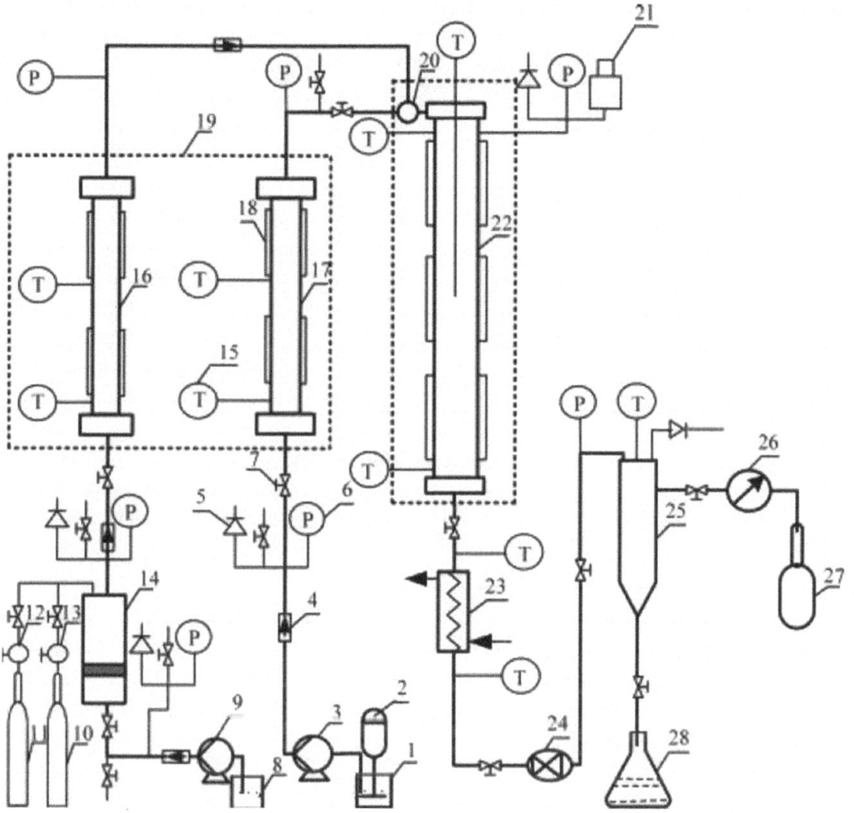

Fig. 2.20 Schematic diagram of tubular-flow reaction plant for phenol gasification in supercritical water: (1) feed tank, (2) homogenizer, (3) high-pressure metering pump, (4) check valve, (5) safety valve, (6) pressure gauge, (7) pin valve, (8) water tank, (9) high-pressure metering pump, (10) oxygen cylinder, (11) nitrogen cylinder, (12) nitrogen pressure reducing valve, (13) oxygen pressure reducing valve, (14) supercharger, (15) thermocouple, (16) oxidant preheater, (17) feedstock preheater, (18) heating wire, (19) safeguarding structure, (20) mixer, (21) bursting disk device, (22) tubular reactor, (23) cooler, (24) back pressure valve, (25) gas–liquid separator, (26) volumetric flow meter, (27) gas bag, (28) effluent tank

Correspondingly, the ideal overall reaction equations were proposed. Presentation above indicated that SCWG is feasible for converting biomass or organic pollution matter containing protein or fattiness to produce hydrogen [11] (Figs. 2.21, 2.22 and 2.23).

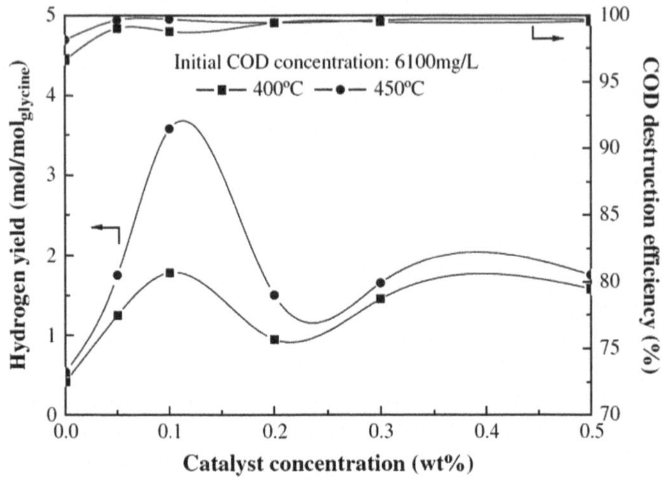

Fig. 2.21 Different concentrations of Na_2CO_3 dependences on hydrogen yield and COD destruction efficiency for glycine SCWG at different reaction temperatures

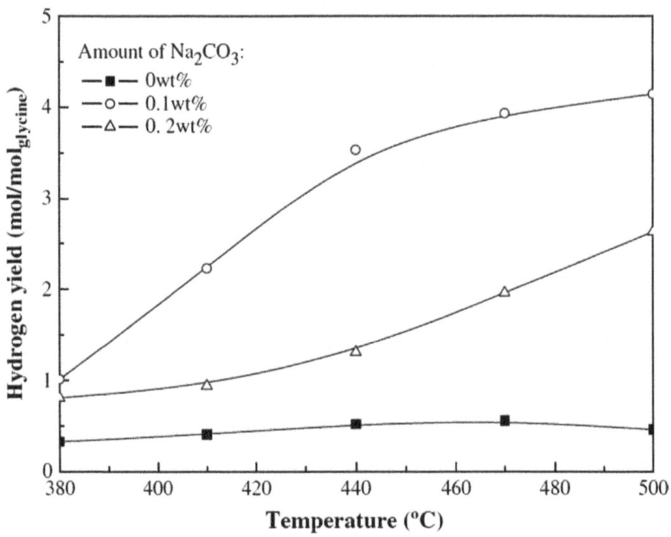

Fig. 2.22 Influences of different amounts of Na_2CO_3 on hydrogen yield of glycine SCWG

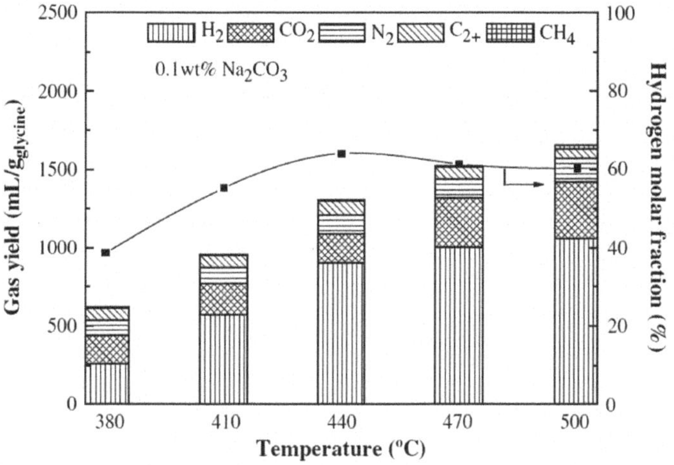

Fig. 2.23 Yields of different gas products and molar fraction of hydrogen measured at atmospheric pressure and 20 °C for glycine SCWG with 0.1 wt% Na_2CO_3

References

1. Y. Guo, S. Z. Wang, D. H. Xu, Y. M. Gong, H. H. Ma, X. Y. Tang, Review of catalytic supercritical water gasification for hydrogen production from biomass. Renewable and Sustainable Energy Reviews, 2010, 14: 334–343.
2. D. H. Xu, G. K. Lin, Z. J. Ma, Y. Guo, M. U. Farooq, S. Z. Wang, Partial oxidative gasification of sewage sludge in supercritical water with multi-component catalyst. Chemical Engineering Research & Design, 2017, 124: 145–151.
3. Y. Guo, S. Z. Wang, Y. Z. Wang, J. Zhang, D. H. Xu, Y. M. Gong, Gasification of two and three-components mixture in supercritical water: Influence of NaOH and initial reactants of acetic acid and phenol. International Journal of Hydrogen Energy, 2012, 37: 2278–2286.
4. Y. Z. Wang, S. Z. Wang, G. Y. Zhao, Y. F. Guo, Y. Guo, Hydrogen production by partial oxidation gasification of a phenol, naphthalene, and acetic acid mixture in supercritical water. International Journal of Hydrogen Energy, 2016, 41: 2238–2246.
5. Y. Z. Wang, F. Gao, J. Q. Yang, Y. Guo, Y. T. Zhu, G. Y. Zhao, C. Q. Fang, Y. F. Guo, S. Z. Wang, N. Guo, Partial oxidation kinetics of the mixture of acetic acid, phenol and naphthalene in supercritical water for hydrogen production. International Journal of Hydrogen Energy, 2017, 42: 19843–19850.
6. Y. Guo, S. Z. Wang, C. M. Huelsman, P. E. Savage, Products, pathways, and kinetics for reactions of indole under supercritical water gasification conditions. Journal of Supercritical Fluids, 2013, 73: 161–170.
7. Y. Guo, S. Z. Wang, C. M. Huelsman, P. E. Savage, Kinetic model for reactions of indole under supercritical water gasification conditions. Chemical Engineering Journal, 2014, 241: 327–335.
8. Y. Guo, S. Z. Wang, D. H. Xu, Y. M. Gong, X. Y. Tang, J. Zhang, Hydrogen production by catalytic supercritical water gasification of nitriles. International Journal of Hydrogen Energy, 2010, 35: 4474–4483.
9. Y. Guo, S. Z. Wang, T. Yeh, P. E. Savage, Catalytic gasification of indole in supercritical water. Applied Catalysis B-Environmental, 2015, 166: 202–210.

10. D. H. Xu, S. Z. Wang, Y. Guo, X. Y. Tang, Y. M. Gong, H. H. Ma, Catalyzed Partial Oxidative Gasification of Phenol in Supercritical Water. Industrial & Engineering Chemistry Research, 2011, 50: 4301–4307.
11. D. H. Xu, S. Z. Wang, X. Hu, C. M. Chen, Q. M. Zhang, Y. M. Gong, Catalytic gasification of glycine and glycerol in supercritical water. International Journal of Hydrogen Energy, 2009, 34: 5357–5364.

Chapter 3
Oxidative Mechanisms and Kinetics of Organics in Supercritical Water

3.1 Municipal Sludge

Sewage sludge (SS) with a moisture content of about 80 wt% is a byproduct of wastewater treatment. With the development of the urbanization process, this amount has been increasing with the construction and expansion of wastewater treatment plants. Containing complex organic matters as well as heavy metal and viruses, SS is not only a pollutant but also a wet resource. Conventional SS treatments such as landfill, composting, and incineration, all need a predrying procedure and can generate secondary pollution. Supercritical water oxidation (SCWO) is an efficient and environmentally benign method for degrading organic compounds completely with excessive oxidant. Goto et al. [1] studied the effects of reaction time, temperature, pressure and total solid content on the removal of TOC, organic acid and NH_3–N from MSS. Kinetic data of SCWO for SS with a moisture of 96.51 wt% were also analyzed [2]. However, little information is available on SCWO of SS with a moisture content lower than 90 wt%. We investigated the treatment process of SS with a moisture content of 87 wt% in supercritical water (SCW) with increasing oxidant as well as temperature.

Influences of temperature and oxidation coefficient (n) on liquid products are illustrated in Fig. 3.1. As shown in Fig. 3.1a, it can be observed that TOC concentration significantly decreases from 18,219 to 120 mg/L as the temperature increases from 673 to 873 K and n from 0 to 4 because increasing temperature can enhance the reaction rate and adding oxidant can produce more hydroxyl (OH·) and hydroperoxyl (HO_2·) radicals. However, the TOC value is as high as 4567 mg/L at 723 K even $n = 4$, which demonstrates that SS is difficult to be destructed completely at a moderate temperature condition owing to its high initial organic contents. Particularly, TOC variation is remarkable from 673 to 723 K but slows down at higher temperatures, which may be attributed to the formation of some refractory intermediates inhibiting the rapid degradation at higher temperatures. In addition, the TOC concentration at 573 K and $n = 0$ was examined, which is 40,173 mg/L

S. Wang et al., *Supercritical Water Processing Technologies for Environment, Energy and Nanomaterial Applications*,
https://doi.org/10.1007/978-981-13-9326-6_3

even higher than the supernatant, indicates that, at a low temperature, organics transferred from SS is more than those decomposed in liquid phase.

As presented in Fig. 3.1b, the NH_3–N concentration first increases and then decreases with a rising temperature. NH_3–N is an important intermediate product of nitrogenous organic compounds in SCW treatment through hydrolysis and pyrolysis [3, 4] with recalcitrant nature [5, 6]. Without oxidant, the NH_3–N concentration increases from 5251 to 7267 mg/L with temperature from 673 to 873 K, which is mainly produced by the dissolution and hydrolysis of protein in SS. Once NH_3–N is formed, it tends to be stable in SCW, especially at a low temperature and with starving oxidant. At 673 K, it is interesting to find that the NH_3–N concentration increases with an increase in n. This phenomenon is also found by Lee et al. [7] in the research of SCWO of ethylene diamine tetraacetic acid. A possible explanation is that, with increasing oxidant, oxidation of easy-degraded hydrolysis products enhances, which promotes the hydrolysis reaction, thus produces more NH_3–N. At temperatures lower than 723 K, the NH_3–N concentration increases with an increasing temperature regardless of n, indicating that the hydrolysis and dissolution of SS surpass the oxidation of NH_3–N in this condition. However, this effect is weakened with increasing n, and when n = 4, the NH_3–N concentration at 673 K is almost the same as that at 723 K. At 723 K, it is obvious that NH_3–N concentration is nearly unchanged with increasing n, showing that most nitrogenous organic

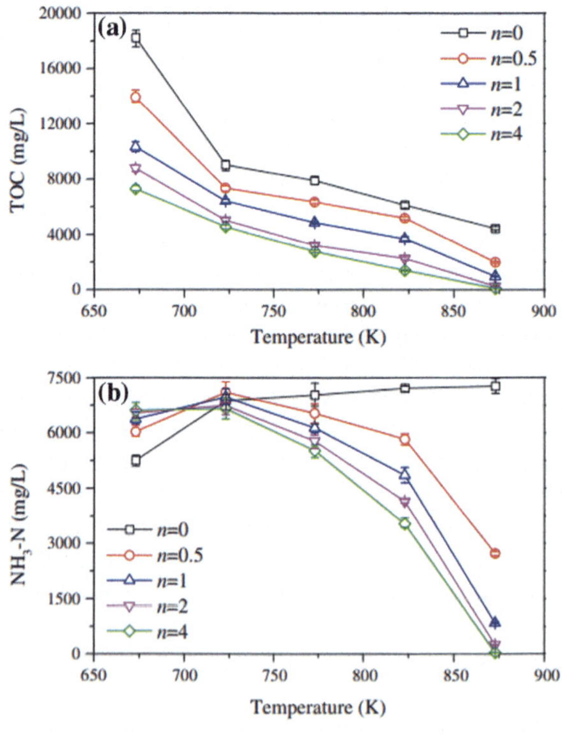

Fig. 3.1 Influences of temperature and n on liquid products after treating in SCW at 25 MPa for 20 min: **a** TOC concentration; **b** NH_3–N concentration

compounds have converted to ammonia, but ammonia cannot be oxidized further at this temperature. At temperatures higher than 773 K, the NH_3–N concentration decreases with increasing n, because NH_3–N can be oxidized to N_2 at a high temperature. However, when the reaction temperature is 823 K, NH_3–N is higher than 3000 mg/L even if n is up to 4, but it reduces to 48 mg/L at 873 K, which is consistent with the investigation of Du et al. [8] with respect to the coking wastewater. They reported that NH_3–N conversion increases dramatically from 14 to 76% as the temperature increases from 823 to 848 K. Therefore, temperature has a dominating effect on the degradation of NH_3–N and NH_3–N is an important intermediate product of SS after treatment in SCW.

In order to better understand the process from SCWG to SCWO with gradually increasing temperature and oxidant, SS and three typical liquid samples were analyzed by GC/MS. Compounds were divided into five types, including non-nitrogen open chain compounds, nitrogen open chain compounds, non-nitrogen aromatic compounds, nitrogen aromatic compounds and non-aromatic ring compounds, and their relative proportion by area and the total average RMW are exhibited in Table 3.1.

With the temperature and n increasing, total average RMW decreases, indicating that macro-molecular compounds changed to small molecules via the cleavage of the bond. In SS, non-nitrogen open chain compounds make up the largest proportion. Whereas after treatment in SCW, non-nitrogen aromatic compounds and nitrogen aromatic compounds dominate. Typical organics of these three kinds of main compounds and their conversion paths are shown in Fig. 3.2. It can be observed that non-nitrogen aromatic like phenols and nitrogen aromatic compounds like pyridines are main refractory intermediates in liquid products.

Non-nitrogen open chain compounds consist of aliphatic acids and a small proportion of aliphatic hydrocarbons and esters. In SS, non-nitrogen open chain compounds account for 55.035% by area. Among it, acids, hydrocarbons and esters are 50.234%, 4.517% and 0.284%, respectively. Among aliphatic acids, volatile fatty acids are only 5.439% including propanoic acid, butanoic acid, pentanoic acid,

Table 3.1 GC/MS qualitative analysis results for SS and three liquid products

	Relative proportion by area (%)					Total average RMW
	NNO[a]	NO[b]	NNA[c]	NA[d]	NAR[e]	
SS	55.035	6.378	24.979	5.144	8.465	299.158
723 K, $n = 0$	0.879	7.529	40.750	39.546	11.293	158.710
723 K, $n = 0.5$	0	8.335	41.076	43.293	7.294	104.599
823 K, $n = 4$	0	0	64.314	35.688	0	100.654

[a]Non-nitrogen open chain compounds (NNO)
[b]Nitrogen open chain compounds (NO)
[c]Non-nitrogen aromatic compounds (NNA)
[d]Nitrogen aromatic compounds (NA)
[e]Non-aromatic ring compounds (NAR)

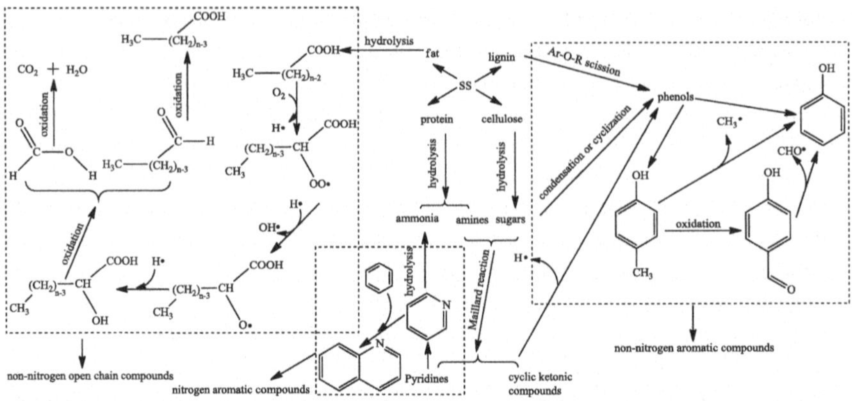

Fig. 3.2 Typical organics of non-nitrogen open chain compounds, non-nitrogen aromatic compounds, nitrogen aromatic compounds, and their conversion paths

and their methyl substituendums, and the rest are long-chain aliphatic acids accounting for 44.795%. Long-chain aliphatic acids mainly originate from fats and most of them have even number of carbon atoms such as hexadecanoic acid (26.861%), octadecanoic acid (7.456%), and so on because they are synthesized by two carbon units. These long-chain aliphatic acids contribute to the high viscosity of SS and bring about the corresponding transportation problems. After treatment in SCW, they are decomposed to small molecular substances, and the viscosity decreases from 7960 mPa s to less than 10 mPa s. At supercritical conditions, aliphatic acids disappear to form gases mainly by the abstraction of hydrogens near a –COOH group [9] (see Fig. 3.2). With oxidant, acids and oxidant can be completely mixed with SCW to form CO_2 and H_2O and release immense energy.

The relative proportion by area of nitrogen open chain compounds is low, and nearly all of them are amides. Most detected amides in SS are macro-molecules such as hexadecanamide (2.775%), 9-octadecenamide (1.633%) and tetradecanamide (0.889%), which originate from proteins and vary to small molecular amides such as butanamide and pentanamide at 723 K via the abstraction of hydrogens near a $-CONH_2$ group acting as high molecular weight carboxylic acids or the direct scission of C–C bond. They subsequently hydrolyze to form acids, gases, and ammonia, and finally disappear at 873 K and $n = 4$. However, ammonia produced by hydrolysis cannot fully react with oxidant to form N_2 and H_2O at this condition.

Non-nitrogen aromatic compounds consist of phenols (phenol and its methyl substituendums) and a small proportion of aromatic acid (benzeneacetic acid and benzenepropanoic acid), aromatic alcohol (phenylethyl alcohol), and aromatic ketone (acetophenone). Non-nitrogen aromatic compounds change dramatically with temperature and n from 24.979% in SS to 64.314% in the liquid product at 823 K and $n = 4$. Phenols are generated by two routes: one is the cleavage of the aryl ether linkages during the disassembly process of lignin in SCW [10, 11], the other is the condensation or cyclization of sugars produced by the hydrolysis of cellulose [12]. Phenols are mainly composed of phenol and *p*-cresol. The proportion of *o*-cresol is small because the elimination of the ortho-substituent to form phenol is easier than that of the para-substituent. The relative proportion of phenol increases from 13.394% at 723 K and $n = 0.5$ to 37.510% at 823 K and $n = 4$, but that of *p*-cresol decreases from 25.104 to 18.355%, suggesting that *p*-cresol may transform into phenol. As depicted in Fig. 3.2, phenol can be generated from *p*-cresol by the direct demethylation or by the oxidation of methyl to form *p*-hydroxybenzaldehyde. Subsequently, *p*-hydroxybenzaldehyde can produce phenol by the loss of –CHO group. As a result, phenol accounts for the largest share, indicating its recalcitrant nature.

In SS, nitrogen aromatic compounds are macro-molecular substances and account for only 5.144%, but after treated in SCW, they become the maximum categories and their relative proportion increases significantly by the dissolution and hydrolysis of protein. After disposed in SCW, they mainly change into pyridine, pyrazine, and their methyl, ethyl, or amino substituendums. However, pyridine and its substituendums seem more difficult to be oxidized than pyrazine and its substituendums because at 823 K and $n = 4$, most nitrogen aromatic compounds are pyridine and its derivatives. As Fig. 3.2 shows, amines as degradation products of proteins reacting with cellulose splitting products can form N-heterocyclic compounds like pyridines via the Maillard reaction, and N-heterocyclic compounds are stable for acting as free radical scavengers to reduce the number of free radicals [13] With temperature and *n* increasing, parts of pyridines further react with water to form ammonia. Other parts of pyridines combine with benzene ring to form fused ring compounds like quinolines.

Same as nitrogen open chain compounds, the relative proportion by area of non-aromatic ring compounds is low and they disappear at 823 K and $n = 4$. Most non-aromatic ring compounds are cyclic ketonic compounds including cyclopentanone, cyclohexanone, pyrrolidinone, and piperidinone. They are the intermediates of the Maillard reaction or of the hydrolysis reaction by sugars and further dehydrogenation of them can form phenols.

3.2 Coal

The rapid increase in energy consumption recently has led to the large consumption of low-grade resources like coal all over the world, especially in China, which is one of the countries with the biggest reverse of coal. However, the negative impact of coal power engineering on the environment and related expenses for environmental safety impedes the growth of coal consumption [14]. New technology of using coal–supercritical water oxidation (SCWO), is under development, and as an innovative technology to completely decompose hazardous materials, and to recover combustion energy from low-quality fuels in supercritical water (SCW). An experimental investigation was undertaken to characterize the kinetics of SCWO of coal water slurry. Another major goal of this study was to do a tentative exploration with regard to the reaction mechanisms of coal under SCWO condition.

The products of organics are complicated during oxidative degradation, so it is difficult to deduce kinetic formula from elementary reactions. Therefore, global power-law reaction rate is adopted for an overall oxidation reaction, which can be described as follows

$$r = -\frac{\mathrm{d}[\mathrm{C}]}{\mathrm{d}t} = k[\mathrm{C}]^m[\mathrm{O}]^n[\mathrm{H_2O}]^p \tag{3.1}$$

where [C] is concentration of organics (mg/L), [O] is the concentration of oxidant (mg/L), $[H_2O]$ is the concentration of water, t is the reaction time (s), m, n, and p are the reaction orders of organic concentrations, oxidant and water concentration, respectively. k is the rate constant, which can be expressed in Arrhenius equation as follows:

$$k = k_0 \exp\left(-\frac{E_\mathrm{a}}{\mathrm{RT}}\right) \tag{3.2}$$

where k_0 is the pre-exponential factor ($(\mathrm{mol/L})^{1-m-n-p}\ \mathrm{s}^{-1}$), E_a is the activation energy (J/mol), R is the universal gas constant (8.314 $\mathrm{J/mol^{-1}\ K^{-1}}$), and T is the temperature (K), respectively.

In this study, the concentration of organics is represented by COD (mg/L). Besides, due to the excessive water in reaction system, the water concentration can be regarded as a constant and is integrated into k_0. Thus, the global reaction rate can be expressed as follows:

$$r = -\frac{\mathrm{d}[\mathrm{COD}]}{\mathrm{d}t} = k_0 \exp\left(-\frac{E_\mathrm{a}}{\mathrm{RT}}\right)[\mathrm{COD}]^m[\mathrm{O}]^n \tag{3.3}$$

In our discussion, the loss ratio X is used to evaluate the extent of oxidative reaction. [COD] for different conversion is defined as follows:

$$[\mathrm{COD}] = [\mathrm{COD}]_0(1 - X) \tag{3.4}$$

where $[\mathrm{COD}]_0$ is the initial COD concentration of coal water slurry, it is regarded as the theoretical amount of oxygen needed for complete oxidation (mg/L). Oxygen content used in our experiments was largely excessive. Therefore, [O] is substituted by $[\mathrm{O}]_0$. Thus, Eq. (3.3) can be simplified as follows:

$$r = -\frac{\mathrm{d}(1 - X)}{\mathrm{d}t} = k[\mathrm{COD}]_0^{(m-1)}(1 - X)^m[\mathrm{O}]_0{}^n \tag{3.5}$$

Table 3.2 Kinetic parameters at various temperatures

T/°C	k	ln$k$	m	n
560	6.615×10^{-4}	−7.321	1.756	0.232
580	1.022×10^{-3}	−6.886	1.791	0.2767
600	1.857×10^{-3}	−6.289	1.834	0.318

If $m \neq 1$, by the integral to Eq. (3.5), Eq. (3.6) is obtained as follows:

$$X = 1 - \{1 + (m-1)k[\mathrm{COD}]_0^{m-1}[\mathrm{O}]_0^n t\}^{1/(1-m)} \tag{3.6}$$

Base on the experimental data, the kinetic parameters under the pressure of 28 MPa, oxygen excess of 120% and initial coal concentration of 3% at different temperatures are obtained by parametric fitting which is solved by Gauss–Newton method and listed in Table 3.2.

The rate constants summarized in Table 3.2 at various temperatures are used to determine Arrhenius activation energy and pre-exponential factor for the reaction by linear regression of ln (k) versus $1/T$. The apparent activation energy and pre-exponential factor of the reaction are listed in Table 3.3.

Therefore, the global power-law reaction rate of oxidation of coal in SCW can be expressed as follows:

$$r = -\frac{\mathrm{d}C}{\mathrm{d}t} = 4.12 \times 10^2 \exp(-11.23 \times 10^4/\mathrm{RT})[\mathrm{C}]^{1.79}[\mathrm{O}]_0^{0.28} \tag{3.7}$$

The parity plot is shown in Fig. 3.3, which shows the comparison between the experimental conversion rate and those predicted by the global power-law reaction rate (Eq. 3.5). A ±9% deviation of reaction rate from the perfect match provided a quite satisfactory with the experimental data.

The reaction mechanism of SCWO is complicated for the reason that water is not only a reaction medium but also a reactant involved in the reaction [15]. Even if SCW does not take part in the reactions, it cannot be seen as a normal solvent for the reason that its density varies with temperature and pressure. It means that the both parameters may have a great influence on the concentrations of organics and oxygen. This phenomenon not only affects the oxidative degradation of organics but also multiplies the difficulties for simulation or description of the reaction mechanisms.

Water density is the most important factor to determine the reaction mechanism for hydrothermal reaction: the reaction is controlled by ionic reaction mechanism at high water density and low temperature. On the contrary, free radical reaction mechanism is dominant at the condition of low water density and high temperature

Table 3.3 Hydrothermal kinetic parameters in SCW at 28 MPa

m	n	k_0	E_a/kJ mol^{-1}	R^2
1.79	0.28	$(4.12 \pm 0.35) \times 10^2$	112.3 ± 5.6	0.99592

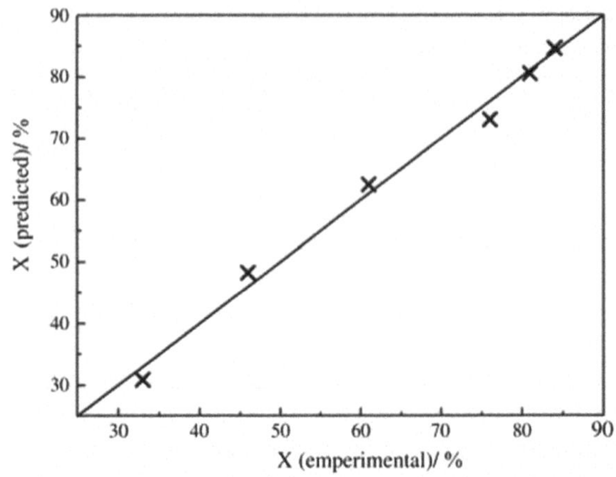

Fig. 3.3 Parity plot for power-law rate equation for the conversion of coal

[16]. SCW behaves as perfect gas in the vicinity of the critical point [17]. So, it is understandable that SCWO of hydrocarbon can be explained perfectly by free radical reaction mechanism. Because of some aliphatic structures and average structure unit [18] in coal, free radicals can be easily produced by the attack of oxygen at weak bonds, such as C–H bond, in these structures:

$$RH + O_2 \rightarrow R\cdot + HOO\cdot \quad \text{(RH represents aliphatic structure)} \tag{3.8}$$

$$RH + HOO\cdot \rightarrow H_2O_2 + R\cdot \tag{3.9}$$

Then, the hydrogen peroxide can be easily decomposed to hydroxyl radical:

$$H_2O_2 + M \rightarrow 2HO\cdot \tag{3.10}$$

where M represents homogeneous or heterogeneous collision partner. The hydroxyl radical (OH·) was considered to be the most important and effective oxidant in SCWO reaction, for the reason that its strong electrophilic reactivity can react with almost all of the hydrogen-containing compounds. In many cases, the formation of hydroxyl radical is a control step of overall reaction kinetics. Dagaut et al. [19] suggested that the SCWO reaction kinetic of methanol is controlled by the formation and decomposition of H_2O_2. Because the H_2O_2 can be decomposed to hydroxyl radical which controls the consumption of other intermediate products [20]. R· can be produced and then combine with oxygen to form peroxy radical, which can further abstract hydrogen for the production of peroxide:

$$RH + HO\cdot \rightarrow R\cdot + H_2O \tag{3.11}$$

$$R\cdot + O_2 \rightarrow ROO\cdot \tag{3.12}$$

$$ROO\cdot + RH \rightarrow ROOH + R\cdot \tag{3.13}$$

The peroxide can be decomposed to organic compounds with lower molecular weight, such as acetic acid and formic acid, which can be finally oxidized to CO_2 and H_2O. Oxygen and hydrogen peroxide employed in our tests were interchangeable in SCW. Therefore, whether the initial added oxidant was oxygen or hydrogen peroxide, the equilibrium between H_2O_2 and O_2 could be achieved easily in the mixed system. It provided an opportunity for oxidation process to undergo at their equilibrium state. The H_2O_2 and O_2 shown in the six reaction equations mentioned above should be considered as the state of oxidant when the equilibrium was reached.

Although oxidation is the main reaction during the SCWO process of coal, other reactions like hydrolysis, pyrolysis, dehydration, polymerization, and isomerization will proceed as well. Especially, hydrolysis reaction can play an important role in the overall reaction under different conditions. The aliphatic carbon chains connected with electrophilic group in coal may be broken because of hydrolysis, as shown in the following:

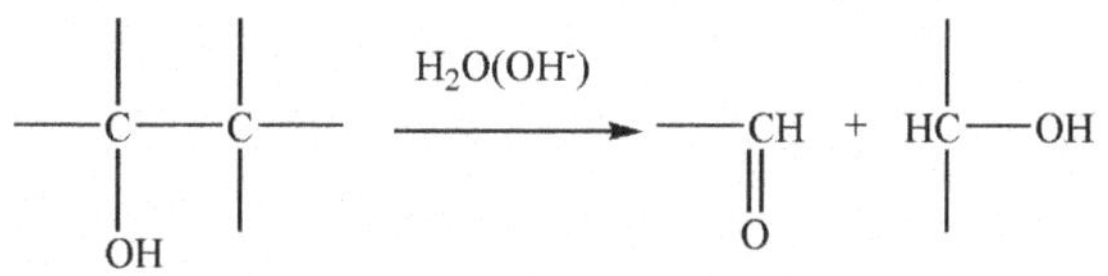

(3.14)

If phenolic hydroxyl in coal is blocked by methylation reaction, the oxidation rate would decrease dramatically. At the early stage of the reaction in our tests, content of phenolic hydroxyl increased with the increase of reaction time but further decreased sharply after reaching a maximum value. It indicates that phenolic hydroxyl was the intermediate product during SCWO of coal. The possible reaction mechanism is shown in Fig. 3.4. According to this mechanism, polyaromatic hydrocarbons can be first oxidized to phenolic hydroxyl followed by aromatic ring opening with simultaneous formation of carboxyl.

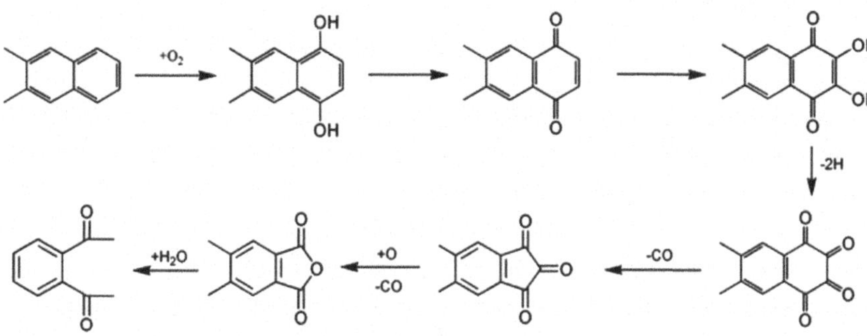

Fig. 3.4 Reaction pathway of aromatic compounds in SCWO of coal

It was found that mechanisms mentioned above occurred simultaneously during SCWO of coal. The aliphatic chains were easily oxidized, which can be explained by free radical and oxidation-hydrolysis mechanism. The intermediate products were formic and acetic acid, which were finally decomposed to CO_2 and H_2O. As the main structures in coal, aromatic compounds were hard to be oxidized relatively. Their reactions can be explained by phenol hydroxyl oxidation mechanism. Phenol hydroxyl was the intermediate products, and the oxidation of which was considered to be the reaction controlling step. With the deepening of oxidation degree, the reaction was driven further to completion. The final products of them were CO_2 and H_2O.

3.3 Landfill Leachate

Landfill leachate is generated from decomposing municipal solid waste (MSW) seeping into surface, ground, and rainwater. A wide range of pollutants are presented in landfill leachate, which includes dissolved organic matter, inorganic macro-components (e.g., ammonium, chloride, sulfate), heavy metals, and xenobiotic organic compounds [21]. For landfill leachate, biological treatment is most widely used for its reliability, simplicity, and high cost-effectiveness. However, biological treatment is hampered by specific toxic substances (such as PAHs—polyaromatic hydrocarbons, AOXs—adsorbable organic halogens, PCBs—polychlorinated biphenyls) and/or by the presence of biorefractory organics (such as humic substance or surfactants) [22]. We formulated the kinetics on the removal efficiencies of organic matters and ammonia nitrogen. We also prepared a composite catalyst ($CeMnO_x/TiO_2$) which consisted of cerium, manganese, and titanium oxides and examined the activity and stability of the synthesized catalyst in leachate SCWO. Finally, a possible catalytic mechanism was proposed.

Figure 3.5 shows the temporal variation of TOC removal efficiency (TRE), NH_3–N removal efficiency (NRE) at 500, 550, and 600 °C, when reaction times increased from 60 to 600 s at 1.7 OC. TRE and NRE increased significantly with rising temperature and reaction time. Nevertheless, the dependency of NRE on reaction time followed a nearly linear relationship, while TRE showed a trend that it increased rapidly at the beginning and subsequently leveled off after 240 s, especially at 600 °C.

According to the existing literatures [23, 24], reaction rate equation on TOC can be described as follows:

$$-\frac{d[\mathrm{TOC}]}{dt} = A\exp\left(\frac{-E_a}{\mathrm{RT}}\right)[\mathrm{TOC}]^a[\mathrm{O_2}]^b[\mathrm{H_2O}]^c \tag{3.15}$$

where [TOC] is the concentration of TOC (mol L^{-1}), $[O_2]$ is the concentration of O_2 generated from H_2O_2 (mol L^{-1}), $[H_2O]$ is the concentration of H_2O (mol L^{-1}), a, b, c are reaction orders of [TOC], $[O_2]$, and $[H_2O]$, respectively, A is the

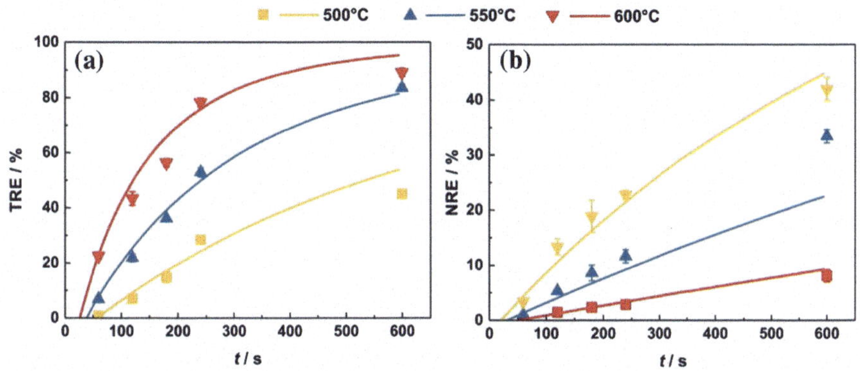

Fig. 3.5 Effects of time and reaction kinetic models on **a** TRE and **b** NRE at 1.7 OC. The individual points represent experimental data while the continuous curves are model calculated values

pre-exponential factor ($(\text{mol L}^{-1})^{1-a-b-c}\ \text{s}^{-1}$), E_a is the activation energy (J mol^{-1}), R is the gas constant (8.314 J mol^{-1} K^{-1}), and T is temperature (K).

H_2O is in excess during SCWO and its impact on organic disappearance rate can be omitted. Thus, the reaction order of [H_2O] can be regarded as zero [25]. Then, we can obtain

$$-\frac{\text{d}(1-\text{TRE})}{\text{d}t} = A\exp\left(\frac{-E_a}{\text{RT}}\right)[\text{TOC}]_0^{(a-1)}(1-\text{TRE})^a[\text{O}_2]^b \tag{3.16}$$

Detailed reaction mechanisms of organic matter in SCWO can be considered to be free radical chain reaction, including initiation, propagation, branching, and termination [26]. In general, the conversion of reactant can be neglected and a critical concentration of radicals (mainly HOO· and OH·) is built up during the induction period [26]. Vogel et al. [26] fitted an Arrhenius-type model for induction time (t_{ind}) for tubular reactors, in which t_{ind} represents a time constant for the induction period, and thus the inverse of t_{ind} can be regarded as a rate constant exhibiting an Arrhenius-type dependence on temperature:

$$t_{\text{ind}} = \frac{1}{B}\exp\left(\frac{E_b}{\text{RT}}\right) \tag{3.17}$$

where t_{ind} is induction time (s), B is pre-exponential factor of induction time (s^{-1}), and E_b is the activation energy of induction time (J mol^{-1}).

We attempted to modify the kinetic equation by introducing the concept of t_{ind}. By using the initial condition of TRE = 0 at $t = t_{\text{ind}}$, submitting Eq. (3.17) into Eq. (3.16) and then integrating, we can obtain Eq. (3.18) at the condition of $a \neq 1$.

Table 3.4 Kinetic equation parameters on TRE and NRE

	a	b	$A/(\text{mol L}^{-1})^{1-a-b}$ s^{-1}	E_a/ kJ mol^{-1}	B/s^{-1}	E_b/ kJ mol^{-1}	R^2
TRE	1.24	0.10	1561	90.50	15.38	43.65	0.99
NRE	1.35	0.26	2601	101.0	41.35	48.94	0.97

$$\text{TRE} = 1 - \left\{1 + (a-1)A\exp\left(\frac{-E_\text{a}}{8.314T}\right)[\text{TOC}]_0^{(a-1)}[\text{O}_2]^b\left[t - \frac{1}{B}\exp\left(\frac{E_\text{b}}{8.314T}\right)\right]\right\}^{\frac{1}{1-a}} \tag{3.18}$$

With the aid of genetic algorithm, coefficients of Eq. (3.18) can be solved directly from the experimental data by Visio studio 2010 software, aiming at minimum residual sum of squares between the calculated and the experimental TRE at 500, 550, and 600 °C, as shown in Table 3.4.

Thus, the kinetic equation of TOC can be expressed as Eq. (3.19),

$$-\frac{\text{d[TOC]}}{\text{d}t} = 1561\exp\left(-\frac{90,500}{\text{RT}}\right)[\text{TOC}]^{1.24}[\text{O}_2]^{0.10} \tag{3.19}$$

In a similar approach, we obtained the kinetic equation of NH_3–N, as shown in Eq. (3.20),

$$-\frac{\text{d}[\text{NH}_3-\text{N}]}{\text{d}t} = 2601\exp\left(-\frac{101,050}{RT}\right)[\text{NH}_3-\text{N}]^{1.35}[\text{O}_2]^{0.26} \tag{3.20}$$

Table 3.4 shows that the activation energy of NRE is 101.0 kJ mol^{-1}, which is higher than that of TRE, 90.50 kJ mol^{-1}. This explains why the abatement of NH_3–N is more difficult than TOC. In addition, reaction order of $[O_2]$ is 0.26 in the kinetic equation of NRE, while that of TRE is 0.1, indicating that the increase of oxidation coefficient has a greater impact on NRE than TRE.

As shown in Fig. 3.5, the derived kinetic equations can correlate TRE and NRE well in the experiment range. Moreover, the values at which the solid lines from the kinetic equations intersect with the horizontal axis are the t_{ind}. Table 3.5 shows t_{ind} of TRE and NRE in SCWO of landfill leachate. It can be seen that t_{ind} decreases

Table 3.5 Induction times of TRE and NRE

T/°C	t_{ind}/s	
	TRE	NRE
500	57.9	49.1
550	38.4	30.9
600	26.6	20.5

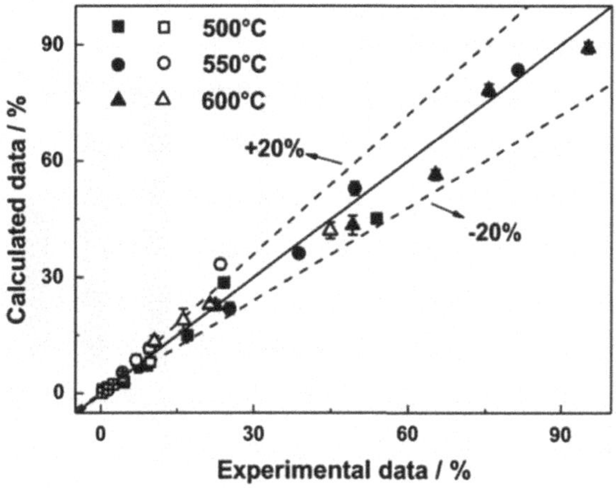

Fig. 3.6 Comparison between experimental values and those predicted from the kinetic models for TRE (black) and NRE (white)

with increasing reaction temperature, which has been confirmed by Eq. (3.17). Considering the existence of t_{ind}, the net reaction time (t_{net}) is defined as that the experimental reaction time (t) minus t_{ind} [23]. Generally, t_{ind} for SCWO of model compounds and wastewaters is less than 10 s in previous literatures [23, 26], which is below the t_{ind} of TRE and NRE. In fact, t_{ind} is not only influenced by the initial feed concentration, mixing of reactant and oxidant and reaction temperature but also strongly affected by reactor types, traces of metal ions, oxidation inhibitors, stabilizers, metal surface, and pH value [23, 26–28].

Figure 3.6 is a parity plot comparing the experimental values and those calculated from the kinetic models for TRE and NRE, respectively. It can be observed in Fig. 3.6 that the model calculated data was in good agreement with experimental data and the deviation was approximately ±20% compared with the experimental results.

Catalysts used were labeled by Ce/Mn molar ratio as follows C0M0, C3M0, C2M1, C1.5M1.5, C1M2, and C0M3, respectively. Here, C0M0 stands for the TiO_2 catalyst with no Ce and Mn. Six varieties of composite catalysts (C0M0, C3M0, C2M1, C1.5M1.5, C1M2, and C0M3) were employed to evaluate the catalytic activity on TRE and NRE at 600 °C and 600 s and OC of 1.7 and when the mass ration of composite catalysts to landfill leachate was set to be 2.0 wt%. These results were compared to a control test without catalyst and the effects of catalysts on TRE and NRE are shown in Fig. 3.7.

The catalysts highly improved TRE and NRE in SCWO of landfill leachate, compared with the non-catalytic reaction. Especially, the C1M2 composite catalyst displayed the highest catalytic activity compared with other catalysts. At conditions of 600 °C, 600 s and 1.7 OC, the concentrations of TOC and NH_3–N in liquid product were 95.73 mg L^{-1} and 18.71 mg L^{-1}, respectively, in the presence of 2 wt% C1M2 and the corresponding TRE and NRE were both 99.0%.

Generally, the neutral oxygen molecule adhering to the surface of metallic oxide can be activated into negatively charged O_2^-, O^-, and O^{2-} under aerobic

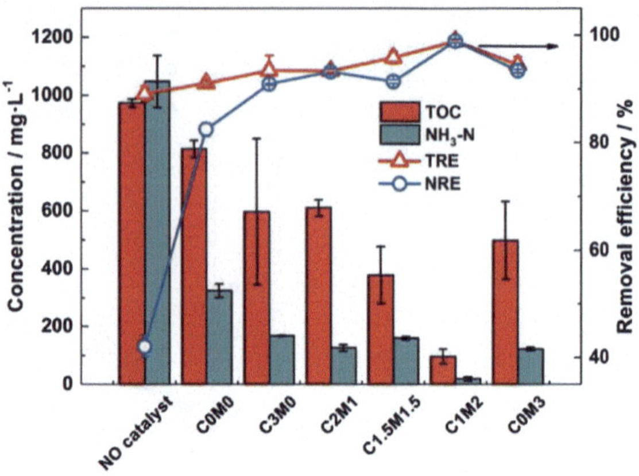

Fig. 3.7 Effects of catalysts on the degradation of TOC and NH_3–N (T = 600 °C, t = 600 s, OC = 1.7, 2 wt% catalysts)

conditions. Equilibrium processes of various kinds of oxygen species are shown in Eq. (3.21). Since CeO_2 has two stable oxidation states, Ce^{4+} and Ce^{3+}, oxygen could be stored and released by CeO_2 via the redox shift between Ce^{4+} and Ce^{3+} [29]. Thus, the addition of Ce could enhance the redox activity of composite catalyst, improve the catalyst oxygen storage capacity, and facilitate the oxygen mobility over catalyst [29]. The ionic radiuses of Ce^{4+} and Ti^{4+} are 0.87 and 0.68 Å, respectively. Ce^{4+} may replace Ti^{4+} leading to the TiO_2 lattice deformed, which broadens the characteristic peak of rutile TiO_2. In addition, CeO_2 is fluorite structure, while Mn_2O_3 is fluorite structure with anionic defects. The ionic radius of Mn^{3+} is 0.645 Å, so Mn^{3+} can replace Ce^{4+} to form a new solid solution with the fluorite-type structure [30], whose oxygen and electronic migration abilities are improved.

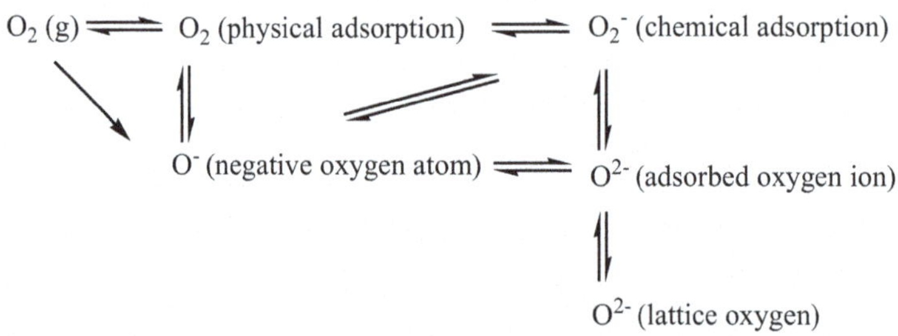

(3.21)

According to Mars–van Krevelen redox mechanism, the organics first react with lattice oxygen at the surface of the catalyst, second the surface is re-oxidized by gas-phase O_2 [31]. Under the assumption that rutile TiO_2 is the carrier and inert in catalytic reaction, we can speculate the catalytic mechanism of $CeMnO_x/TiO_2$ in the process of landfill leachate SCWO, as shown in Eq. (3.25).

The first step, oxidation and reduction reactio proceeds between CeO_2 and Mn_2O_3. The oxygen transfers from CeO_2 to Mn_2O_3 with the generation of MnO_2.

$$2CeO_2 + Mn_2O_3 \rightarrow Ce_2O_3 + 2MnO_2 \tag{3.22}$$

The second step, the produced MnO_2 releases active oxygen (O^*), forming Mn_2O_3 [32]. Here, O^* includes O_2^-, O^- and O^{2-}. As the electron relay, the roles of MnO_2 and Mn_2O_3 are to promote the generation of strong oxidizing agents [33]. Moreover, the existence of Mn_2O_3 is largely dependent on the oxygen partial pressure and reaction temperature of system [32]. Although Mn_2O_3 can be oxidized to MnO_2 under aerobic condition, the lower oxygen concentration and higher the temperature favor the existence of Mn_2O_3 [33]. Because OC was relatively low (1.7) and the reaction temperature was relatively high (600 °C) in the current study, Mn_2O_3 is not easily to be oxidized to MnO_2. That is the reason why we did not detect characteristic peak of MnO_2 in the XRD pattern of used C1M2 catalyst.

$$2MnO_2 \rightarrow Mn_2O_3 + 2O^* \tag{3.23}$$

The third step, O^* attacks organics and reacts with them to generate CO_2, H_2O and N_2 on the active sites of catalyst.

$$C_mH_nO_pN_q + O^* \rightarrow CO_2 + H_2O + N_2 \tag{3.24}$$

The fourth step, Ce_2O_3 captures O from oxygen on the surface of catalyst and is re-oxidized to CeO_2.

$$Ce_2O_3 + 1/2O_2 \rightarrow 2CeO_2 \tag{3.25}$$

Through above analysis, we can speculate that MnO_x is the active component; CeO_2 is additive and oxygen supplier, respectively. MnO_2 releasing active O^* in the second step is the key step in the whole catalytic process and the catalytic reaction takes place on the active sites of catalyst.

3.4 Phenolic Wastewater

Coal-gasification industry in China has been developing rapidly in recent years. And the Lurgi gasification technology is extensively used because of its technological maturity and capability to produce large amounts of methane [34]. However, the wastewater discharged from Lurgi process contains high concentration of toxic and refractory compounds, such as phenolics, ammonia, and pyridine posing a huge challenge to environmental safety [35]. The SCWO technology offers two advantages over similar treatment processes. First, it occurs at much faster reaction rate and more complete reaction than wet air oxidation. Second, it produces less harmful byproducts than incineration for its lower operating temperature [36].

Central composite design (CCD) in RSM was used to evaluate the interactive effects of temperature, pressure, OR and reaction time on COD and NH_3–N

Table 3.6 Independence factors and corresponding levels in response surface design

	Codes	Levels				
		−2	−1	0	+1	+2
Temperature (°C)	x_1	440	480	520	560	600
Pressure (MPa)	x_2	22.2	24.2	26.2	28.2	30.2
OR	x_3	1	2	3	4	5
Reaction time (s)	x_4	20	90	160	230	300

removal efficiencies. We also focused on the catalytic SCWO process of LCGW treatment and chose the more stable manganese oxide—Mn_2O_3, CuO and Co_2O_3 as catalysts. The investigation involved probing the catalytic effects on TOC and NH_3–N removal efficiencies along with temperature and oxygen ratio.

The LCGW sample used for non-catalytic experiments was collected from a coal-gasification plant located in North China. The initial COD and NH_3–N in the LCGW were about 14,000 and 1500 mg/L, respectively. More than 70% of the organic compounds in LCGW were phenolics. In this study, response surface methodology (RSM) was used to analyze and model the interactive responses of independent variables, which in turn aided in the determination of optimized conditions [37]. Independent factors and corresponding levels in response surface design are listed in Table 3.6.

The experimental design and results are shown in Table 3.7. The number of experiments was $2f + 2^f + n$, where f and n are the number of factors and center runs, respectively ($f = 4$, $n = 6$). The repetition of the central runs was carried out to provide information on the variation of the responses about the average, residual variance, and eventually to estimate the pure experimental uncertainty.

The ultimate models of COD and NH_3–N removal efficiencies resulting from ANOVA are given in Eqs. (3.26) and (3.27), respectively, in terms of coded factors. The predicted values for COD and NH_3–N removal efficiencies (Y_1 and Y_2) were calculated using these models, and the resulting values are also shown in Table 3.7.

$$\begin{aligned} Y_1 = {} & 99.16 + 1.45x_1 + 0.011x_2 + 0.65x_3 + 0.017x_4 \\ & - 4.375 \times 10^{-3}x_1x_2 - 0.61x_1x_3 - 9.375 \times 10^{-3}x_1x_4 \\ & - 0.014x_2x_3 - 1.875 \times 10^{-3}x_2x_4 - 6.875 \times 10^{-3}x_3x_4 \\ & - 0.68x_1^2 - 0.0027x_2^2 - 0.22x_3^2 - 0.025x_4^2 \end{aligned} \tag{3.26}$$

$$\begin{aligned} Y_2 = {} & 92.28 + 16.28x_1 + 0.34x_2 + 4.79x_3 + 2.15x_4 \\ & - 0.23x_1x_2 - 3.64x_1x_3 - 1.38x_1x_4 + 0.012x_2x_3 \\ & - 0.077x_2x_4 - 0.62x_3x_4 - 6.28x_1^2 - 0.49x_2^2 - 2.41x_3^2 - 1.46x_4^2 \end{aligned} \tag{3.27}$$

The LCGW sample used for catalytic study was collected from a coal-gasification plant located in North China. The initial COD, TOC, and NH_3–N

Table 3.7 Experimental design and corresponding results for response surface analysis

Runs	x_1	x_2	x_3	x_4	TOC concentration in the effluent (mg/L)		NH_3–N concentration in the effluent (mg/L)	
					Experimental	Predicted	Experimental	Predicted
1	−1	−1	−1	−1	228	228	715	718
2	1	−1	−1	−1	21	21	57	72
3	−1	1	−1	−1	227	225	703	699
4	1	1	−1	−1	18	19	55	67
5	−1	−1	1	−1	101	100	447	446
6	1	−1	1	−1	15	16	22	19
7	−1	1	1	−1	99	100	418	427
8	1	1	1	−1	16	16	16	13
9	−1	−1	−1	1	227	226	592	591
10	1	−1	−1	1	20	19	35	28
11	−1	1	−1	1	224	223	572	576
12	1	1	−1	1	18	17	31	27
13	−1	−1	1	1	98	98	367	357
14	1	−1	1	1	14	14	14	12
15	−1	1	1	1	99	98	362	342
16	1	1	1	1	16	15	12	11
17	−2	0	0	0	324	324	973	981
18	2	0	0	0	35	34	10	5
19	0	−2	0	0	48	49	155	155
20	0	2	0	0	48	46	132	135
21	0	0	−2	0	151	150	415	404
22	0	0	2	0	20	20	103	116
23	0	0	0	−2	50	49	283	268
24	0	0	0	2	43	46	120	138
25	0	0	0	0	42	42	115	116
26	0	0	0	0	43	42	107	116
27	0	0	0	0	41	42	128	116
28	0	0	0	0	42	42	110	116
29	0	0	0	0	41	42	122	116
30	0	0	0	0	43	42	113	116

were about 25,500, 8520, and 7600 mg/L, respectively. Figure 3.8 shows the influences of temperature on TOC and NH_3–N removals. Results clearly show that the rising temperature had a positive effect on TOC and NH_3–N removal efficiencies. And the addition of catalysts dramatically improved the degradation of NH_3–N. For example, at temperature of 460 °C, the TOC and NH_3–N removal efficiencies were 98.03 and 81.52% without catalyst, while the value reached to 99.45 and 98.13% with Mn_2O_3, 99.94 and 98.89% with Co_2O_3, and 98.8 and

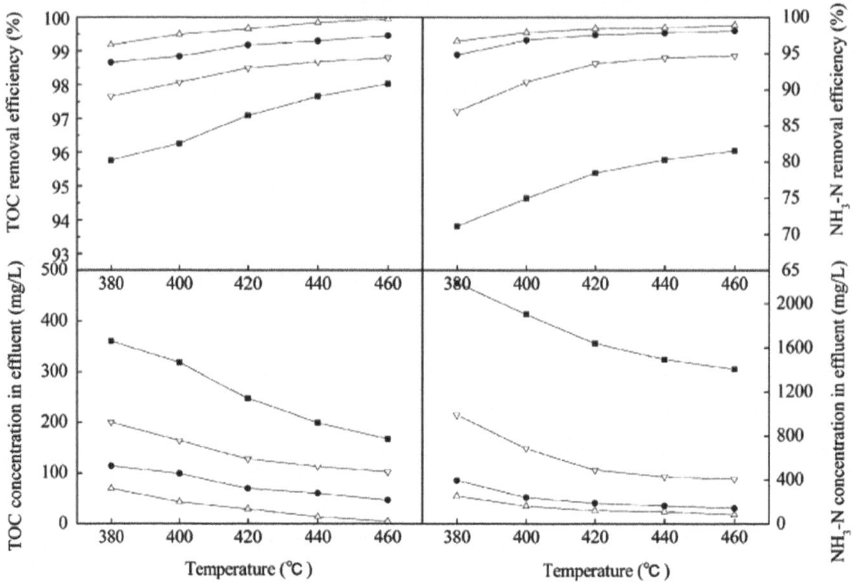

Fig. 3.8 Effects of temperature on TOC and NH_3–N removals at OR of 2.5: (filled square) without catalyst, (filled circle) with Mn_2O_3 catalyst, (open triangle) with Co_2O_3 catalyst, (open inverted triangle) with CuO catalyst

94.67% with CuO, respectively. At the temperature range of 380–460 °C, the catalytic effects were ordered as follows: Co_2O_3 > Mn_2O_3 > CuO.

NH_3–N removal efficiency reached up to 81% at 460 °C without catalyst, which was relatively higher compared with studies at similar conditions. This may account for the easier oxidized organic compound in LCWG. The oxidation of these organic compounds may supply as fuel [38]. Mizuno also reported that the activation energy in the destruction of NH_3 in a waste was slightly lower than that studied by Goto using both the same oxidant and similar experimental facility [39, 40]. They attributed this discrepancy to the differences in compositions. In addition, the catalytic effects of the reactor wall may also a favorable factor. Our reactor is made of SUS316 stainless steel, the main components Fe, Ni, and Cr, and many form transition metal oxide compounds during SCWO and play a role of catalyst during SCWO [41]. Similar study showed that when inconel beads were packed in the same reactor, ammonia conversion increased in fourfold [42].

Figure 3.9 clearly shows that the order of catalytic effects at 460 °C was Co_2O_3 > Mn_2O_3 > CuO at ORs from 1.5 to 3.5. The catalysis of Co_2O_3 was always kept at the highest value and the response to OR was small. However, the catalytic effect of Mn_2O_3 was almost as high as that of the CO_2O_3 when the OR reached 3.0. Except the promoting effect of oxygen, catalyst structure changes may contribute. In the test conditions, maximum TOC and NH_3–N removal efficiencies

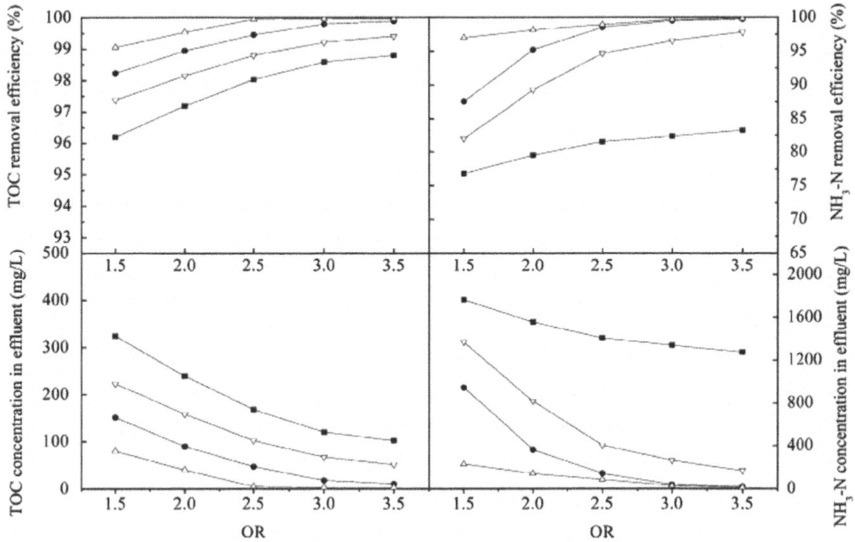

Fig. 3.9 Effects of OR on TOC and NH_3–N removals at 460 °C: (filled square) without catalyst, (filled circle) with Mn_2O_3 catalyst, (open triangle) with Co_2O_3 catalyst, (open inverted triangle) with CuO catalyst

of 99.97 and 99.83% were found with Co_2O_3 catalysis at 460 °C and OR of 3.5, where the residual concentrations in effluent were 3 and 13 mg/L, respectively.

Figure 3.9 also reveals that the increasing OR dramatically enhanced TOC and NH_3–N removal efficiencies, especially for NH_3–N. The reasons could be concluded as following: for the non-catalytic experiments, the increasing oxidant would increase the oxidation agent of OH· radicals, which are extremely powerful oxidizing species [43]. For the catalytic reactions, organic compounds reacted with the oxygen molecules activated on the catalyst surface, which are much more active than OH· or HO_2· radicals [44]. The increasing oxygen would increase the regeneration rate of active sites, thereby enhancing the reaction rate in SCWO [45]. In addition, increasing oxidant would inhibit the dimerization reactions of aromatic compounds, which consequently led to greater formation of light organic compounds and further to CO_2 [46].

Identification of reaction intermediates may provide some insights into degradation pathways for catalytic oxidation of LCGW. The incomplete oxidation products of LCGW treated at 380 °C and OR of 2.5 were detected by GC/MS. Results showed that the main compounds were cyclohexanone, derivatives of cyclohexanol, heterocyclic compounds and phenoxy in non-catalytic case. While in the presence of Mn_2O_3 and Co_2O_3, the 3-methyl-cyclopentanone was found. It could be explained by the isomerization of cyclohexanone which further underwent an oxidation to CO_2. The cyclohexanone found in CuO catalytic case reflected the relatively poor catalysis compared with Mn_2O_3 and Co_2O_3, which has been proved in the above study.

For the heterocyclic intermediates, the main products without catalyst were aniline, benzothiazole, and benzothiophene. While in the catalytic cases, the ring-opening products, such as tetramethyl-thiourea, 2,5-bis (2-methylpropyl)-thiophene, and tetramethyl-urea were found. This reflected that catalysts promoted the ring-opening reaction. And the similar intermediates found in catalytic reaction process indicated the similar degradation pathways.

For the phenoxy compounds, the kinds of compounds detected in all cases are similar. In SCWO process, derivatives containing phenoxy group have been demonstrated as the main intermediates in phenol oxidation. As more than 70% of the organic compounds in LCGW are phenolics, the degradation pathway of phenol oxidation in SCWO is representative. Studies of the phenol in catalytic SCWO showed that the reactions between adsorbed oxygen species and adsorbed phenol were the initial step to generate phenoxy radicals [47, 48]. Then radical–radical reactions and the combination of two phenoxy radicals followed by tautomerization formed the single-ring products and dimers. Krajnc concluded that the mechanism of phenol oxidation is similar in both non-catalytic and catalytic SCWO processes after the phenoxy radical has been formed [43]. Based on our experimental results, we speculate that the role of the catalysts in phenolics oxidation is to generate phenoxy radicals and to generate them more quickly [49]. And the diverse activities of catalysts may be explained by the differences in the production rate of phenoxy radicals.

3.5 Textile Wastewater and Sludge

The treatment of textile wastewaters has been identified as one serious environmental problem due to their refractory dyes, auxiliaries, heavy metals, etc. [50–53]. The present traditional treatment for this wastewater contains biological, physico-chemical methods or combined processes, in which the biological approaches dominate for the decontamination of the textile wastewater [54, 55]. However, whether the biological or chemical process (such as coagulation), sludge with a moisture content of about 80 wt% is always generated as byproduct. And, the amount of the textile sludge is larger than that of sewage sludge because of its high color and COD contents. Moreover, the dyes, sizing agents, additives, and heavy metals are the primary pollutants during the textile sludge, resulting in the significant biotoxicity and pollution. The conventional sludge treatment technologies such as incineration, landfill, and composting all have inevitable problems (predrying procedure, secondary pollution, high land area requirement). Therefore, the effective, economical, and clean treatment technologies for textile sludge destruction still need to be explored for potential commercialization. At present, the SCWO technology has been introduced into the harmless treatment for sludge, such as sewage sludge, oily sludge, and textile sludge [56–60]. In our research, the degradation reactions for textile sludge in subcritical water and supercritical water oxidation were investigated via experimental approach. Meanwhile, the research on the effects of sodium hydroxide was carried out.

The experiments for the textile sludge were conducted at variable reaction temperatures ranging from 250 to 600 °C with and without oxidant, experimental results and the corresponding treatment picture are demonstrated in Figs. 3.10 and 3.11, respectively. At first, the textile sludge was decomposed in subcritical water at 25 MPa and 250–400 °C with $n = 0$ in order to reproduce the reactions during the preheating process, as shown in Fig. 3.10a. It is obvious that COD and TOC removal efficiencies of 51–67% and 39–55% can be obtained, when the X_{solid} reached 31–47%, it reveals that sludge may occur pyrolysis and hydrolysis reaction in subcritical water. Additionally, samples b (250 °C) and c (300 °C) in Fig. 3.11 show that the black sludge varied into yellow supernatant with a part of residual black solid deposited at the bottom of the tube, and the color declined as temperature rose.

Figure 3.10b displays removal efficiencies for SCWO of the textile sludge as a function of the reaction temperature ranging from 450 to 600 °C at a constant oxidation coefficient of 2.0. Both the COD and TOC decomposition efficiencies for liquid products were enhanced by the increasing temperature, which reached up to 99.6% and 93.9% at 600 °C, respectively. On the other hand, the solid removal presented little variation with temperature during SCWO reactions, which was around 60%. It can be explained that organics in the solid of sludge were degraded completely. The residual solid was mainly inorganic substance, which cannot further decompose in SCW. As shown in Fig. 3.11, the products for textile sludge oxidation in SCW at 450–550 °C can be separated into the colorless clear supernatant and yellow sediment implying the sludge reaction without char and tar formations during these processes. In addition, solid products included not only the original inorganics of the sludge but also probably the corrosion products of the reactor inner-wall material in the SCWO environment.

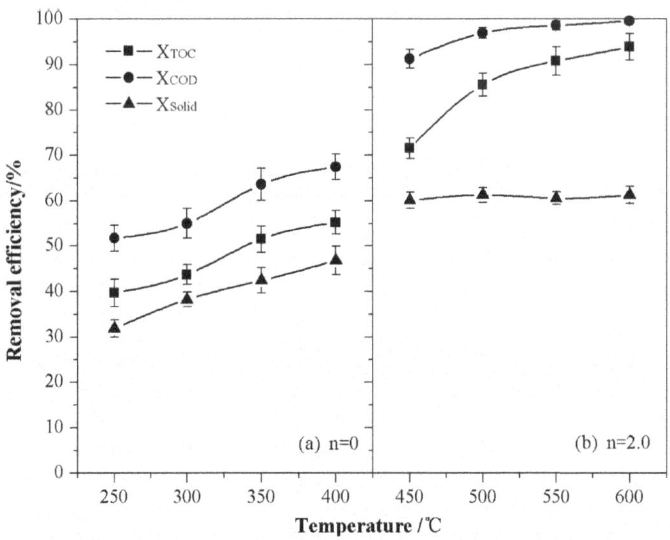

Fig. 3.10 Influence of reaction temperature on degradation of textile sludge: **a** $n = 0$, **b** $n = 2.0$

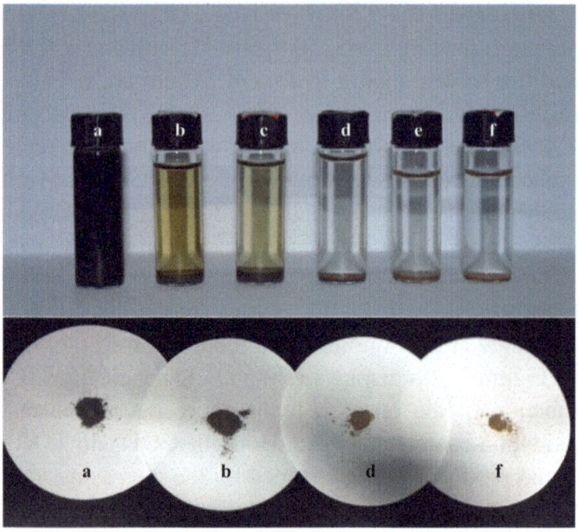

Fig. 3.11 Treatment of the textile sludge: **a** original sludge; **b** products at 250 °C (n = 0); **c** products at 300 °C (n = 0); **d** products at 450 °C (n = 2.0); **e** products at 500 ° C (n = 2.0); **f** products at 550 °C (n = 2.0)

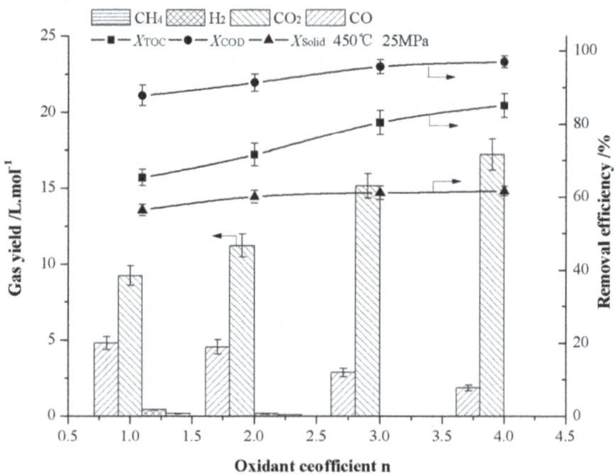

Fig. 3.12 Influence of oxidation coefficient n on the SCWO reaction at 450 °C, 25 MPa

The experimental results for the decomposition of textile sludge in supercritical water under 450 °C, 25 MPa are presented in Fig. 3.12, which demonstrates the pollutant removal efficiencies and the gaseous products' distribution as a function of the oxidation coefficient n. The TOC removal efficiency rose visibly as the oxidation coefficient increased from 1.1 to 4.0, while the amplification for COD removal efficiency was relatively weak. And, there were no obvious increases in the X_{solid} with the oxidation coefficient. Moreover, the addition of oxidants can lead to improvements in gas yield for textile sludge oxidation, and the gas yield was 19 L mol^{-1} TOC at n = 4.0. Small amounts of CH_4 and H_2 were observed during gaseous products at both oxidation coefficients of 1.1 and 2.0. The formation of methane can be attributed to the methanation reaction of organics during the preheat process or the pyrolytic reaction of tar in the oxidation reactor [61–63]. The CH_4

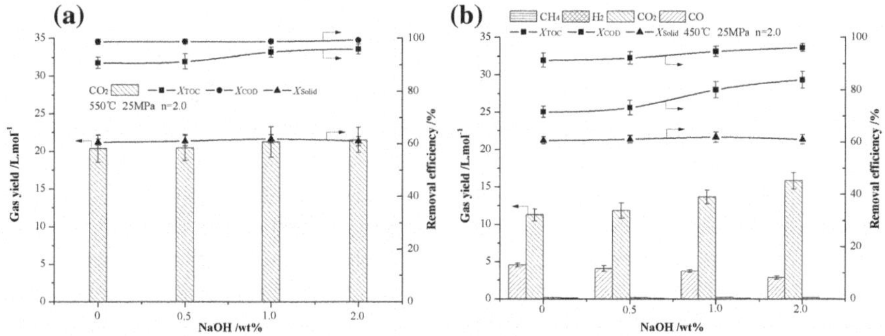

Fig. 3.13 Influence of NaOH addition on the SCWO reaction at 25 MPa, n = 2.0: **a** 450 °C, **b** 550 °C

and H_2 were not generated with the further increment of oxidants. And the yield of CO_2 increased while the yield of CO decreased. It indicates that the addition of oxidants would accelerate the complete disappearance of organic carbon and gaseous intermediates.

Figure 3.13a, b displays the dependence of NaOH concentration on the SCWO destruction of textile sludge at 450 °C and 550 °C, respectively. Figure 3.13a shows that as the addition of NaOH increased from 0 to 2.0 wt% not only the organics removals can be boosted but also the gaseous products can achieve complete decomposition at the reaction temperature of 450 °C. It appears that NaOH facilitated the decomposition for organic species of sludge in supercritical water, which can be explained by the fact that OH^- participated in the water-gas shift reaction (see Eq. 3.28) and the CO_2 capture reaction could be seen (Eqs. 3.29 and 3.30) [64, 65]. In addition, little CH_4 and H_2 were detected during the gaseous products under 450 °C, and the residual solids were basically the same, generally 40% of the solids in feedstock. As shown in Fig. 3.13b that CO_2 was the only component of the gaseous products at 550 °C. The organics decomposition and solid removal efficiencies showed a slight increasing trend with increasing addition of NaOH. It seems that the reaction temperature played a dominant role in the sludge degradation under 550 °C, and the TOC and COD removal efficiencies reached up to 90.8–95.9% and 98.6–99.3%, respectively.

$$CO + H_2O \rightarrow CO_2 + H_2 \tag{3.28}$$

$$CO_2 + 2NaOH \Leftrightarrow Na_2CO_3 + H_2O \tag{3.29}$$

$$Na_2CO_3 + CO_2 + H_2O \Leftrightarrow 2NaHCO_3 \tag{3.30}$$

3.6 Acrylic Acid Production Wastewater

Acrylic acid (CH_2=CHCOOH) is a colorless, corrosive, and irritant liquid, which is widely used in painting chemical fiber, adhesive, and cleaning agents industry. The production and application processes of acrylic acid can generate a large number of high concentration biorefractory organic wastewater. There are complex compositions in this type of wastewater, such as acetic acid, acrylic acid, methacrylic acid, formaldehyde, and acetaldehyde. Furthermore, the content of organic matter in the wastewater is very high, for example, the chemical oxygen demand (COD) content can be more than 100 g/L [66]. In addition, its pH value is mainly in the range of 1–2 [67]. Thus, there is no doubt that the wastewater will cause serious pollution to the environment if not be treated properly.

We have investigated the influences of key operation parameters on the destruction of COD and NH_3–N in SCWO reaction [23]. Furthermore, the corresponding kinetic analysis is carried out for the COD and NH_3–N removal rate via an additional induction time-modified global rate model. The characteristics of raw acrylic acid production wastewater are shown in Table 3.8.

COD removal efficiency (η) and NH_3–N removal efficiency (ψ) are defined as follows:

$$\eta\,(\%)=\left[1-\frac{[\mathrm{COD}]_{\mathrm{pro}}}{[\mathrm{COD}]_{\mathrm{raw}}}\times d\times\left(\frac{Q_1+Q_2}{Q_1}\right)\right]]\times 100\%, \tag{3.31}$$

$$\psi\,(\%)=\left[1-\frac{[\mathrm{NH_3{-}N}]_{\mathrm{pro}}}{[\mathrm{NH_3{-}N}]_{\mathrm{raw}}}\times d\times\left(\frac{Q_1+Q_2}{Q_1}\right)\right]\times 100\%, \tag{3.32}$$

Removal efficiencies of COD and NH_3–N at different temperatures and times when pressure (P) = 25 MPa, the oxidation coefficient (α) = 2.0 are shown in Table 3.9.

The global rate equation for COD removal can be written as follows:

$$-\frac{\mathrm{d[COD]}}{\mathrm{d}t}=A\exp\left(\frac{-E_{\mathrm{a}}}{\mathrm{RT}}\right)[\mathrm{COD}]^a[\mathrm{O_2}]^b[\mathrm{H_2O}]^c. \tag{3.33}$$

Herein, [COD], [O_2], and [H_2O] represent the concentrations of COD, oxygen, and H_2O, respectively. a, b, and c are reaction orders of COD, oxygen, and H_2O, respectively. A is the pre-exponential factor and E_a is the activation energy. R and

Table 3.8 Characteristics of raw acrylic acid production wastewater

COD (mg/L)	NH_3–N (mg/L)	pH	Color	Smell
291,500	2,8600	2.68	Black	Pungent

Table 3.9 Removal efficiencies of COD and NH_3–N at different temperatures and times when $P = 25$ MPa, $\alpha = 2.0$

Number	T (K)	t (s)	$[COD]_{pro}$ (mg/L)	$[NH_3–N]_{pro}$ (mg/L)	η (%)	ψ(%)
1	673	136.8	82.2	25.1	97.28	12.24
2	673	178.3	70.5	24.3	97.58	15.03
3	673	233.5	69.8	23.1	97.61	19.23
4	673	339.0	59.1	21.6	97.97	24.48
5	698	104.2	77.7	24.4	97.33	14.69
6	698	135.8	66.8	23.3	97.71	18.53
7	698	177.9	60.2	22.6	97.93	20.98
8	698	258.3	54.9	21	98.12	26.57
9	723	89.8	74.9	23.4	97.43	18.18
10	723	117.1	65.6	22.2	97.75	22.38
11	723	153.0	58.6	21.5	97.99	24.83
12	723	223.4	51.0	19.6	98.25	31.47
13	748	80.5	75.6	22.6	97.41	20.98
14	748	105.0	62.6	20.8	97.85	27.27
15	748	137.5	53.4	19.9	98.17	30.42
16	748	199.7	45.4	18.1	98.44	36.71
17	773	72.7	70.6	21	97.58	26.57
18	773	94.7	56.5	19.8	98.06	30.77
19	773	124.1	48.6	18.4	98.33	35.66
20	773	180.1	37.0	16.1	98.73	43.71

T are the universal gas constant and reaction temperature. It can be assumed that the order of H_2O is 0 because it is much excess [25, 68].

Substituting Eq. (3.31) into Eq. (3.33), we can obtain

$$-\frac{d[1-\eta]}{dt} = A\exp\left(\frac{-E_a}{RT}\right)[COD]_{raw}^{(a-1)} \times (1-\eta)^a[O_2]^b[H_2O]^c. \tag{3.34}$$

Detailed reaction mechanisms of organic matter in SCWO can be considered as free radical reaction with initiation, propagation, branching, and termination processes involved [68–71], as shown in Table 3.10. Initiation reactions among stable species, such as organic compounds (R–H), O_2, and H_2O, contribute to form free radicals like HOO· and OH· as in Eqs. (3.36)–(3.39). The OH· is considered to be the most important and effective oxidant in SCWO reaction, which can almost react with all hydrogen-containing compounds as in Eq. (3.35). Generated from Eqs. (3.36), (3.37), and (3.41), R· can combine with O_2 to form peroxy radical (ROO·), which can further abstract hydrogen to yield peroxide. The peroxide is very active and easy to decompose into small molecule organic compounds, such as formic acid and acetic acid, which can be finally oxidized to CO_2 and H_2O. When

Table 3.10 Possible free radical mechanism in SCWO

Initiation
$R{-}H + O_2 \rightarrow R\cdot + HOO\cdot$ (3.36) $R{-}H + HOO\cdot \rightarrow R\cdot + H_2O_2$ (3.37) $O_2 + H_2O \rightarrow HOO\cdot + OH$ (3.38) $H_2O \rightarrow H\cdot + \cdot OH$ (3.39)
Branching
$H_2O_2 \rightarrow 2OH\cdot$ (3.40)
Propagation
$R{-}H + OH\cdot \rightarrow R\cdot + H_2O$ (4.41) $R\cdot + O_2 \rightarrow ROO\cdot$ (4.42) $ROO\cdot + R{-}H \rightarrow ROOH + R\cdot$ (4.43)
Termination
$R\cdot + R\cdot \rightarrow R{-}R$ (4.44) $ROOH \rightarrow$ small molecule organic compounds $\rightarrow CO_2 + H_2O$ (4.45)

most of the organic compounds and oxygen have been converted, termination reactions among those free radicals, which yield stable products, become dominate, as shown in Eq. (3.44).

In general, the conversion of reactant can be negligible during the inductive period. Moreover, an induction time was observed in this stage, which was shortened with high reaction temperature and high reagent concentration [72]. An Arrhenius-type model for the induction time ($t_{\rm ind}$) was expressed as Eq. (3.35) for a tubular reactor, which was utilized to assess the order of magnitude effect of induction times on the apparent first-order rate constant.

$$t_{\rm ind} = \frac{1}{B}\exp\left(\frac{-E_{\rm b}}{\rm RT}\right). \tag{3.35}$$

Here, B and $E_{\rm b}$ represent pre-exponential factor and activation energy for the $t_{\rm ind}$, respectively.

Using the initial condition of $\eta = 0$ at $t = t_{\rm ind}$, and substituting Eq. (3.36) into Eq. (3.34), we can obtain the following equation:

$$\eta = 1 - \left\{1 + (a-1)A\ \exp\left(\frac{-E_{\rm a}}{8.314T}\right)[\mathrm{COD}]_{\rm raw}^{(a-1)}[O_2]^b \times \left[t - B\ \exp\left(\frac{E_{\rm b}}{8.314T}\right)\right]\right\}^{1/(1-a)} \quad \text{for } a \neq 1. \tag{3.46}$$

A nonlinear regression analysis was performed to fit the data for a, b, A, B, $E_{\rm a}$, and $E_{\rm b}$ by minimizing the sum of squared differences of the experimental and predicted removal efficiencies, with the help of genetic algorithm. The values of A and $E_{\rm a}$ for COD removal were 356.9 $\mathrm{mol^{-1.15}\ L^{1.15}\ s^{-1}}$ and 52,216 J $\mathrm{mol^{-1}}$,

Table 3.11 t_{ind} for COD and NH_3–N removal

T (K)	t_{ind} (s)	
	COD	NH_3–N
673	5.38	1.38
698	4.21	0.97
723	3.35	0.69
748	2.71	0.51
773	2.22	0.38

respectively. The experimental data sets led to reaction orders of a = 1.98 for COD and b = 0.17 for oxygen. Meanwhile, B and E_b for the t_{ind} were 173 s^{-1} and 38,250 (J mol^{-1}). t_{ind} in this work for COD removal is illustrated in Table 3.11. Considering the existence of t_{ind}, the net reaction time (t_{net}) is defined as the experimental residence time (t) minus t_{ind}. Thus, the kinetic equation for COD removal can be expressed below:

$$-\frac{d[COD]}{dt} = 356.9 \exp\left(-\frac{52216}{RT}\right)[COD]^{1.98}[O_2]^{0.17}. \tag{3.47}$$

The same procedure was conducted to obtain the global rate equation for NH_3–N removal, which was described as Eq. (3.48). A and E_a for NH_3–N removal were 500 $mol^{-0.91}$ $L^{0.91}$ s^{-1}) and 64,492 J mol^{-1}, and reaction orders were a = 1.88 for NH_3–N and b = 0.03 for oxygen. Additionally, B and E_b for t_{ind} were 15,231 s^{-1} and 55,690 J mol^{-1}, respectively. t_{ind} for NH_3–N removal are also illustrated in Table 3.4. All kinetic parameters of the global oxidation rate for COD and NH_3–N are summarized in Table 3.12.

$$-\frac{d[NH_3-N]}{dt} = 500 \exp\left(-\frac{64{,}492}{RT}\right)[NH_3-N]^{1.88}[O_2]^{0.03}. \tag{3.48}$$

Figure 3.14 compares the t_{ind} under different reaction temperatures in this work and previously reported literatures. It can be seen that most of t_{ind} decreased with increasing reaction temperature. The inverse of t_{ind} can be regarded as a rate constant exhibiting an Arrhenius-type dependence only on temperature. However, Gong and Duan [69] found that t_{ind} changed from 2.04 to 5.71 s at the temperature range of 658–713 K for SCWO of landfill leachate and reached the maximum of

Table 3.12 All kinetic parameters of the global oxidation rate for COD and NH_3–N

	a	b	A [(mol $L^{-1})^{1-a-b}$ s^{-1}]	E_a (kJ/mol)	B (s^{-1})	E_b (kJ/mol)
COD	1.98	0.17	356.9	52.216	173	38.25
NH_3–N	1.88	0.03	500	64.492	15,231	55.69

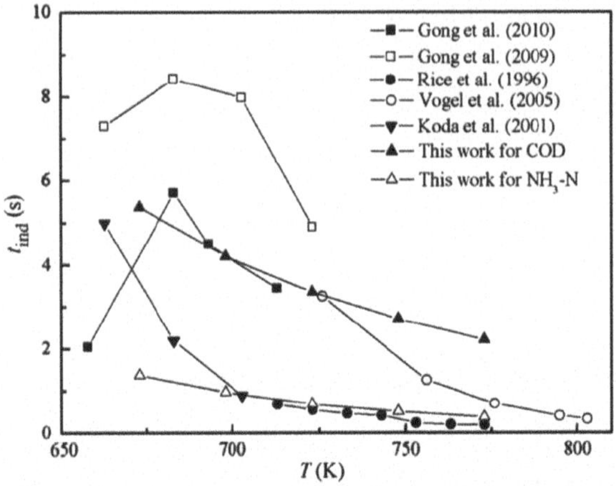

Fig. 3.14 t_{ind} in this work and previously reported literatures. Filled square: Gong and Duan [69] t_{ind} for SCWO of landfill leachate; open square: Gong et al. [66] t_{ind} for SCWO of acrylic acid production wastewater; filled circle: Rice et al. [73] t_{ind} for SCWO of methanol; open circle: Vogel et al. [26] t_{ind} for SCWO of methanol; filled inverted triangle: Koda et al. [72] t_{ind} for SCWO of methanol; filled triangle: t_{ind} for COD removal in this work; open triangle: t_{ind} for NH_3–N removal in this work

5.71 s at 683 K. In the SCWO of acrylic acid production wastewater, t_{ind} varied from 4.91 to 8.43 s as the reaction temperature raised from 663 to 13 K, with the maximum of 8.43 s at 683 K. These data reported by Gong et al. no longer fitted Eq. (3.36) obviously [66]. In fact, t_{ind} is not only influenced by the initial feed concentration, mixing of reactant and oxidant and reaction temperature but also strongly affected by traces of metal ions, oxidation inhibitors, stabilizers, metal surface, and pH value [66, 69]. Gong et al. estimated the values of t_{ind} via the same graphing method based on previous reports, but their experiments were carried out on a TWR, which had a porous ceramic tube and a stainless steel pressure vessel involved. In a TWR, pure water called transpiration water mixed with wastewater and oxidant finally. The quality value of transpiration water was still not mentioned in their research [26, 69].

3.7 Pesticide Wastewater

Pesticide wastewater, which usually comes from pesticide production processes, commonly contains a large quantity of refractory organic matters, biological inhibition, or toxicity matters. Traditional biological or physicochemical methods are insufficient or ineffective to deal with this wastewater, particularly for wastewater with high concentration of organic matters and complicated components [74]. We have used SCWO technology to treat this wastewater [75].

Table 3.13 Properties of the tested pesticide wastewater

Item	COD_0 (mg/L)	$(NH_3–N)_0$ (mg/L)	TN_0 (mg/L)	Colority (multiple)	Salt content (mg/L)	Chloride content (mg/L)	pH (−)
Value	66,300	210	7300	400	114,600	5700	10.55

Table 3.13 illustrates the properties of the wastewater. It can be observed that the wastewater contains high concentrations of COD, inorganic salts, TN, and chloride, which reach up to 66,300 mg/L, 114,600 mg/L, 7300 mg/L, and 5700 mg/L, respectively. Further identification indicates that these inorganic salts mainly consist of Na_2SO_4 and K_2SO_4. It can also be found that the initial ammonia–nitrogen (NH_3–N) content is much lower than the TN content, and the subsequent analysis displays that nitrates hardly exist in the wastewater. This suggests that N existence is mainly in the form of organic N. Moreover, the brownish-yellow wastewater has a colority of 400 multiple and a pH value of 10.55.

Table 3.14 elucidates the experimental results of pesticide wastewater SCWO in the batch and continuous-flow plants under different reaction conditions. We first performed experiments (OC=0) to determine the effect of the temperature on thermal decomposition of the pesticide wastewater. This is very necessary because

Table 3.14 Experimental results for pesticide wastewater SCWO in the batch and continuous-flow plants

No.	T (°C)	OC	t (min)	X_{COD} (%) in the batch plant	X_{COD} (%) in continuous-flow plant
1	400	0.0	1.5	74.1 ± 0.8	72.3 ± 0.9
2	400	1.1	1.5	82.5 ± 0.9	80.3 ± 0.8
3[a]	410	1.1	1.5	82.6 ± 0.8	80.5 ± 0.4
4	410	1.1	1.5	83.4 ± 0.7	81.6 ± 0.5
5[b]	410	1.1	1.5	82.9 ± 0.5	80.5 ± 0.7
6	430	0.0	1.5	75.1 ± 0.7	73.3 ± 0.6
7	450	0.0	1.5	76.5 ± 0.6	74.5 ± 0.5
8[c]	450	1.1	1.5	88.1 ± 0.7	86.7 ± 0.6
9	450	1.1	1.5	89.8 ± 0.5	88.2 ± 0.4
10	450	1.1	2.0	91.8 ± 0.6	89.2 ± 0.5
11	450	1.1	4.0	92.6 ± 0.7	91.3 ± 0.4
12	450	1.5	1.5	93.6 ± 0.4	92.2 ± 0.6
13	450	2.0	1.5	94.2 ± 0.5	93.1 ± 0.5
14	450	2.5	1.5	95.6 ± 0.7	94.0 ± 0.6
15	470	2.5	1.5	95.8 ± 0.3	94.8 ± 0.4
16	470	2.5	2.0	96.6 ± 0.4	95.5 ± 0.5

[a]Regulating the pH value of the initial wastewater to 7.5
[b]Adding 0.5 g Na_2CO_3
[c]Adding 0.3 g Na_2CO_3

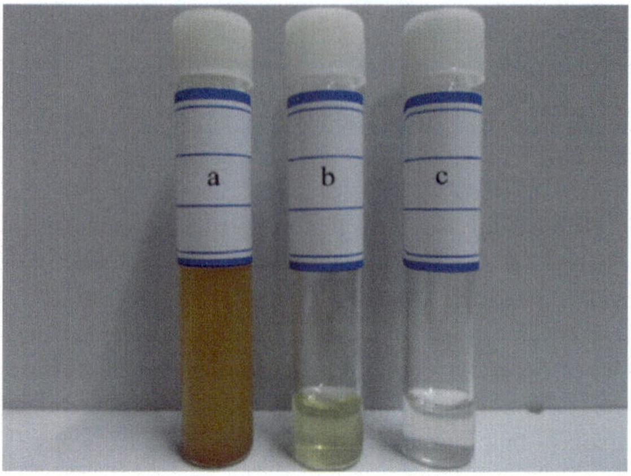

Fig. 3.15 Colority variations of the pesticide wastewater before and after SCWO treatment at OC = 3.0 and t = 2.0 min. **a** Initial pesticide wastewater, **b** reactor effluent of 550 °C, **c** reactor effluent of 600 °C

the thermal decomposition will occur at the preheating stage in SCWO with a preheating system. It can be observed that X_{COD} increased from 72.3 to 74.7% with the reaction temperature changing from 410 to 450 °C at t = 1.5 min in the continuous-flow plant. According to previous investigations concerning organic matter conversion in supercritical water [76, 77], the above relatively high X_{COD} can be explained by that organic matters in the pesticide wastewater were decomposed into CO_2, H_2O, H_2, CO, CH_4, and other small molecular compounds under above supercritical conditions without oxidant. Correspondingly, the several flammable gases in total gaseous products would further react with oxygen to release heat energy in the SCWO reactor. It can be supposed that consumed oxygen first converted these flammable gases and then further removed organic matters. This may be one of the reasons why X_{COD} at 400 °C and OC = 1.1 did not have a remarkable difference from that at OC = 0.

Figure 3.15 indicates colority variations of the pesticide wastewater before and after SCWO treatment at OC = 3.0, t = 2.0 min 550 °C and 600 °C. It can be observed that the initial wastewater with a brownish-yellow color turns into slightly yellow at 550 °C and becomes transparent like distilled water at 600 °C, without any suspended matters and solids.

Figure 3.16 displays C balance and X_{TOC} at 25 MPa, OC = 3.0, t = 2.0 min within 480–600 °C, and experiment errors are within 1%. It can be calculated that more than 92 wt% TOC_0 is converted into CO_2 in terms of CO_2 yields in the above temperature range. The ratio of the sum of C in gaseous products and TOC in liquid products to TOC_0 ranges from 94.21 to 95.65%. Hence, the relative errors of C balance calculation are less than 6%, which can be attributed to related measurement errors and a certain amount of CO_2 dissolving into liquid products. Additionally, it can be noticed that X_{TOC} increases from 5.22 to 99.67% within 480–600 °C. The CO_2 yield and X_{TOC} reach up to 32,713 mg/L and 99.67% at 600 °C, 25 MPa, OC = 3.0 and t = 2.0 min, respectively. This exhibits high CO_2 conversion and TOC removal, which is consistent with previous reports [74].

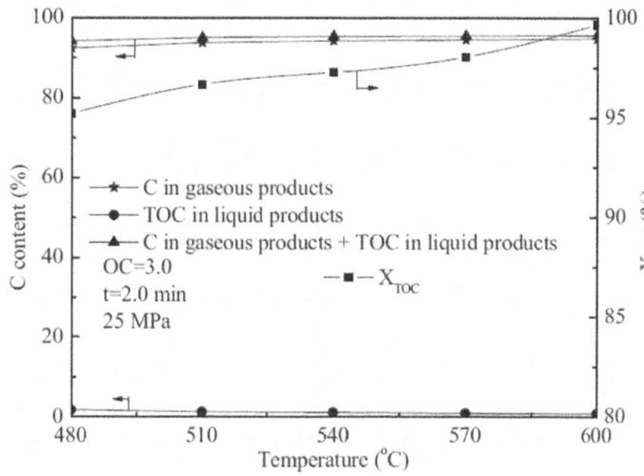

Fig. 3.16 C balance and X_{TOC} at different temperatures

Table 3.15 N balance analysis at 600 °C, OC = 3.0 and t = 2.0 min conditions

Item	Pesticide wastewater	Sample 1	Sample 2
TN (mg/L)[a]	7300 ± 120	742 ± 45	971 ± 55
NH_3–N (mg/L)[a]	210 ± 14	285 ± 17	351 ± 21
Nitrate (mg/L)[a]	12 ± 3	3 ± 2	4 ± 3
X_{NH_3-N} (%)[b]	–	−35.7	−67.1
N_2 yield (mg/L)[b]	–	6558	6329
G_N or X_{TN} (%)[b]	–	89.8	86.7

[a]Measured values
[b]Calculated values

It is well known that no NO_x is formed in organic matter SCWO, so the N_2 yield can be determined by N balance calculation. Samples 1 and 2 were liquid effluents when the pesticide wastewater was disposed at 600 °C, 25 MPa, OC = 3.0 and t = 2.0 min in the batch plant and the continuous-flow plant, respectively. Some measured values and calculated values regarding N balance calculation are listed in Table 3.15. Thus, we can neglect very low nitrate concentrations in the feedstock and in reactor effluents. It can be noticed that TN in the pesticide wastewater is much more than that in sample 1 or 2, meaning that most of N in N-containing organic matters are converted into N_2. There are no NO_x and SO_x gases formations because the oxidation reaction temperature is too low to form these pollutants. In terms of N balance calculation, the N_2 yield is 6558 mg/L in the batch plant and 6329 mg/L in the continuous-flow plant and corresponding GN are 89.8% and 86.7%, respectively. These data are valuable for the subsequent design of a large-scale SCWO plant.

Figure 3.17 depicts residual organic matter components in the reactor effluent for pesticide wastewater SCWO at 600 °C, 25 MPa, OC = 3.0 and t = 2.0 min. It contains formamide, N,N-dimethyl-, phenol, benzamide, 1(3H)-isobenzofuranone, benzoic acid, benzene, 1,3-bis(3-phenoxyphenoxy), etc., in trace, which can also be

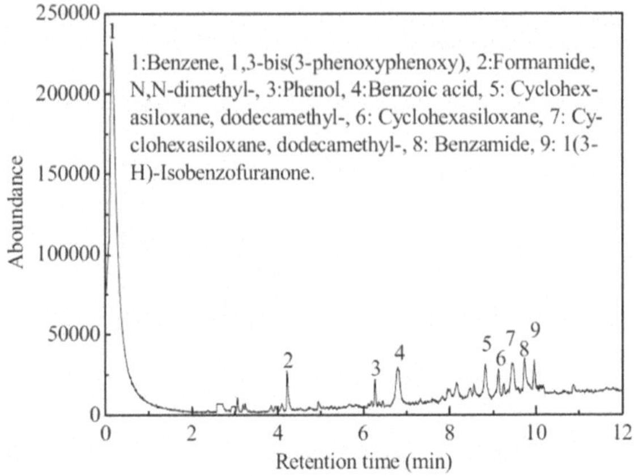

Fig. 3.17 Components of the reactor effluent for pesticide wastewater SCWO at 600 °C, OC = 3.0 and t = 2.0 min

regarded as the intermediate products of the longer reaction time (4 min). These produced intermediate matters lead to residual COD_1, $(NH_3–N)_1$, TN_1 in the reactor effluent at t = 2.0 min.

According to previous investigations, many researchers explain the SCWO process by the free radical reaction mechanism. For instance, this mechanism is used to reveal ethylenediamine SCWO [78]. It has been reported that organic matter degradation in SCWO occurs in two simultaneous steps, i.e., thermal decomposition and oxidation [79]. The former transforms organic matter into smaller gaseous hydrocarbons [80], and the latter leads to the formations of CO_2 and N_2, etc. These two processes take place in the whole SCWO process, and there are also lots of parallel reactions.

Pesticide wastewater SCWO involves very complicated reaction processes, which should also be explained by the radical reaction mechanism. In this research, we only find CO_2 and N_2 as main gaseous products, which is in accordance with previous findings for different organic wastes SCWO [78, 81]. Thus, it can be judged that most of the organic matters are converted into CO_2, H_2O, and N_2 in pesticide wastewater SCWO. There are some other formed matters in trace (see Fig. 3.17) leading to residual COD, TN, etc., in the reactor effluent at a shorter residence time conditions. According to product properties, the overall reaction equation at 600 °C, 25 MPa, OC = 3.0 and t = 2.0 min conditions can be expressed by the following equation:

$$\begin{aligned}&\text{Oragnic matters in pesticide waste water} + O_2 \rightarrow CO_2\\&+ N_2 + H_2O + \text{othersintrace}\end{aligned} \quad (3.49)$$

Detailed reaction pathways for SCWO of organic matters in the pesticide wastewater are too complicated for clear cognition. In fact, due to the polarity of the single-molecule [82], microscopic influences are very significant to relevant reactions in supercritical water. Therefore, a deep insight into all properties of the physicochemical interactions occurring at hydrothermal conditions on the molecular level is essential to reveal the reaction mechanisms in pesticide wastewater SCWO.

3.8 Reactive Orange 7

Effluent from the textile industry has become a serious environmental problem worldwide. Large quantities of textile wastewaters are produced every day and textile waste streams have the characteristics of high chemical oxygen demand, poor biodegradability and being variable in composition and pH value, making them difficult and expensive to treat by conventional methods [83, 84]. Most physical treatments, such as adsorption, coagulation, and precipitation, always perform at low levels of efficiency. In addition, such processes have the major drawback of yielding sludge as a secondary waste product, which needs further treatment and makes the process uneconomical. Biological methods are also unsuitable because of the biological resistance of some dyes. Consequently, the release of these wastewaters into the natural environment is very harmful to both aquatic life and human health.

SCWO is an efficient technology to treat the CI Reactive Orange 7 wastewater. The molecular structure of CI Reactive Orange 7 and some of the physicochemical properties of the 5 g/L dye solution are shown in Table 3.16 [85].

All experiments for supercritical water oxidation of textile wastewater were conducted at 25 MPa in a batch reactor. Table 3.17 showed the changing variations of COD and TOC removal efficiencies at different operation conditions. For the experimental results, removal of chemical oxygen demand varied between 55.48

Table 3.16 Molecular structure of CI Reactive Orange 7 and some physicochemical properties for the 5 g/L dye solution

Name	Properties		Chemical structure
CI Reactive Orange 7	COD (mg/L)	2038	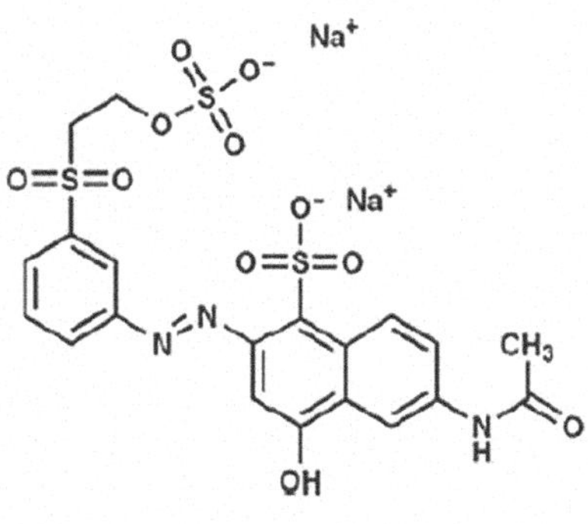
	TC (mg/L)	745.095	
Molecular formula	TOC (mg/L)	739.632	
$C_{20}H_{19}N_3O_{11}S_3Na_2$	TN (mg/L)	115	
	NH_3–N (mg/L)	<2	

Table 3.17 Changing variations of COD and TOC removal efficiencies with the at different operation conditions

						Experimental value				Predictive value		
Run	*A* (°C)	*B* (%)	*C* (mg/L)	*D* (min)	*E*	COD_{rem} (%)	TOC_{rem} (%)	TN_{rem} (%)	NH_3–N (mg/L)	COD_{rem} (%)	TOC_{rem} (%)	TN_{rem} (%)
1	400	175	4500	10.50	8 alk[a]	96.89	86.77	62.46	2.5	97.98	88.96	64.23
2	434	106	2578	5.28	5 aci[b]	85.92	72.10	51.99	2.4	84.70	69.71	50.03
3	434	106	6422	15.72	11 alk	88.09	77.76	64.70	2.0	87.48	76.58	63.73
4	434	244	2578	15.72	11 alk	93.63	83.79	53.17	2.6	93.02	82.61	52.20
5	434	244	6422	5.28	11 alk	87.91	76.35	61.55	2.3	87.30	75.17	60.57
6	434	244	6422	15.72	5 aci	92.63	77.80	57.08	3.0	92.02	76.61	56.10
7	475	50	4500	10.50	8 alk	55.48	43.02	34.16	1.5	56.58	45.22	35.94
8	475	175	1000	10.50	8 alk	80.41	45.36	20.10	1.4	81.51	47.56	21.88
9	475	175	4500	1.00	8 alk	96.67	86.44	56.82	4.0	97.77	88.63	58.60
10	475	175	4500	10.50	3 aci	96.24	85.48	56.74	4.0	97.36	87.72	58.57
11	475	175	4500	10.50	8 alk	93.37	82.49	58.78	2.0	92.60	81.00	57.55
12	475	175	4500	10.50	13 alk	93.08	82.03	58.38	2.6	94.26	84.41	60.29
13	475	175	4500	20.00	8 alk	97.66	86.91	58.34	3.5	98.75	89.11	60.11
14	475	175	8000	10.50	8 alk	94.73	84.93	64.16	1.0	95.82	87.12	65.94
15	475	300	4500	10.50	8 alk	97.98	88.25	59.45	2.5	99.07	90.44	61.23
16	516	106	2578	15.72	11 alk	94.23	84.10	50.05	1.0	93.62	82.93	49.07

(continued)

Table 3.17 (continued)

						Experimental value				Predictive value		
Run	*A* (°C)	*B* (%)	*C* (mg/L)	*D* (min)	*E*	COD_{rem} (%)	TOC_{rem} (%)	TN_{rem} (%)	NH_3–N (mg/L)	COD_{rem} (%)	TOC_{rem} (%)	TN_{rem} (%)
17	516	106	6422	5.28	11 alk	94.39	84.61	53.69	2.2	93.78	83.43	52.71
18	516	106	6422	15.72	5 aci	93.73	84.08	57.08	1.4	93.11	82.89	56.10
19	516	244	2578	5.28	11 alk	95.76	85.15	51.76	<1[c]	95.15	83.97	50.78
20	516	244	2578	15.72	5 aci	96.63	86.34	50.34	1.8	96.02	85.15	49.36
21	516	244	6422	5.28	5 aci	97.49	86.77	57.68	<1	96.88	85.58	56.70
22	550	175	4500	10.50	8 alk	98.11	88.42	62.80	<1	98.87	90.62	64.58

[a]Adjusting the pH value by means of adding sodium hydroxide alkali solution

[b]Adjusting the pH value by means of adding sulfuric acid solution

[c]'1' is the detection limit for NH_3–N content of the Spectroquant NOVA 60 minpectrophotometer

and 98.11%. Removal of total organic carbon was lower than that of chemical oxygen demand, which changed from 43.02 to 88.42%, while total nitrogen removal was the lowest and ranged from 20.10 to 64.70%. As can be seen, the total organic carbon removal was always lower than that of chemical oxygen demand, suggesting that complete oxidation of CI Reactive Orange 7 was more difficult to achieve than partial oxidation.

It is interesting to note that during SCWO, the disappearance of total nitrogen was evident, but the ammonia nitrogen production was very low, mainly ca. 1–4 mg/L. Thus, it is suspected that nitrogen was mainly converted to gases and discharged, while the ammonia was dissolved in liquid. However, many investigations [86–88] on the supercritical water oxidation of nitrogen-containing compounds have reported that ammonia was the major intermediate between nitrogen-containing reactants and their final product like nitrogen gas, which was always refractory during the supercritical water oxidation process. Thus, the results in the present work differed from previous research. It is speculated that the discrepancy was mainly attributable to the nitrogen-containing reactants, in which nitrogen existed in the form of an azo group in this study. Aromatic azo compounds are characterized by high thermal stability. During the supercritical water oxidation process of CI Reactive Orange 7 wastewater, firstly the azo group in the dye molecular structure was attacked by HOÆ and oxidized to the diazo group. The CI Reactive Orange 7 was decomposed to another aromatic compound and a diazo compound whose chemical properties were very active in the solutions. Then, in the environment of supercritical water, the diazo compound readily performed a hydrolysis reaction where the diazo groups were substituted by a hydrogen group and converted into nitrogen. As can be seen from Eq. (3.50), H^+ was generated by this reaction. As mentioned above, the reason why total nitrogen was obviously degraded, but ammonia nitrogen was scarcely produced during the supercritical water oxidation process for CI Reactive Orange 7 dye wastewater was explained.

$$\mathrm{Ar} - \overset{+}{\mathrm{N}} \equiv \mathrm{N} + \mathrm{H_2O} \xrightarrow{\Delta} \mathrm{Ar} - \mathrm{OH} + \mathrm{N_2} \uparrow + \mathrm{H^+} \tag{3.50}$$

Finally, the GC/MS (Fig. 3.18) analysis results for liquid products suggest that the main components of the supercritical water oxidation reaction for CI Reactive Orange 7 were 2,4-pentadienenitrile, 2-methylnaphthalene, 2,3-dimethylnaphthalene, fluorene, and diphenylethyne. Among these products, the 2,4-pentadienenitrile, who contains diazo, accounted for the largest proportion, which supports well the above assumptions for the reaction mechanism of nitrogen.

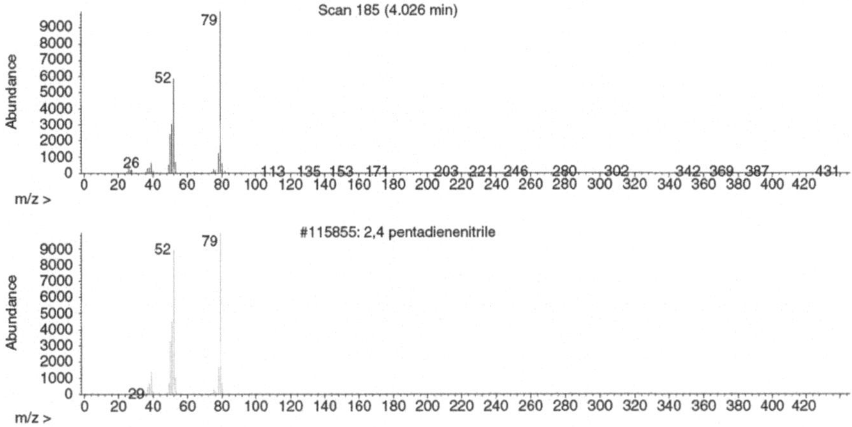

Fig. 3.18 GC/MS chromatogram showing the 2,4-pentadienenitrile obtained from the supercritical water oxidation of CI Reactive Orange 7 at 550 °C

3.9 Polyvinyl Alcohol

Polyvinyl alcohol (PVA) is a well-known water-soluble polymer [89]. It is frequently used in the textile industry as a sizing agent, with typical chemical oxygen demand (COD) values between 10,000 and 40,000 mg O_2/L. In addition, PVA is widely used in adhesive, emulsion paint, paper coating, detergent-based industries, and pharmaceutical industry as an ophthalmic lubricant. It is estimated that the global production of PVA can reach up to 650,000 tons/year [90]. Large amounts of PVA discharged from industries to natural environment cause a series of ecological problems. PVA is harmful to human health since it accumulates in human body through the food chain. Moreover, PVA can lead to environmental issues due to its ability to mobilize heavy metal from sediments in lakes and water streams [91, 92]. However, conventional biological technologies cannot effectively break down PVA, which is attributed to the restricted degradation capacity of most microorganisms for PVA. Furthermore, a great deal of foam is formed during the biological degradation of PVA wastewater, which adversely affects the stability of operation and makes it difficult to achieve the desired treatment results. Other treating PVA wastewater technologies, such as wet air oxidation, photocatalytic degradation, direct oxidation, and electro-coagulation, can only degrade the PVA into smaller molecules to improve the biodegradability of wastewater but are unable to achieve complete degradation to harmless products such as carbon dioxide and water [93]. Therefore, an effective and promising treatment method for PVA wastewater is needed.

We performed experiments on the decomposition of PVA solution under supercritical water gasification (SCWG) and SCWO conditions, respectively, with respect to various PVA solution concentrations and temperatures with or without

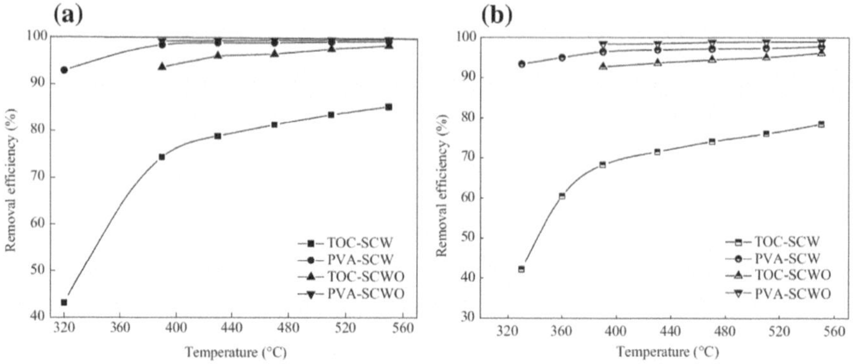

Fig. 3.19 Experiment results for TOC and PVA removals as a function of temperature under SCWG and SCWO with initial PVA concentration of 0.5 wt% (**a**) and 0.2 wt% (**b**)

NaOH [94]. The reaction path of PVA during the SCWO environment is discussed based on the experimental results. In addition, the SCWO catalytic mechanism of NaOH, as another innovation of this article, was evaluated.

Figure 3.19 shows the TOC and PVA removals as a function of temperature ranging from 320 to 550 °C with two PVA concentrations of 0.5 and 0.2 wt%, in the absence of NaOH. As seen in Fig. 3.19a, the PVA completely disappeared regardless of the addition of oxidant, and PVA removals were as high as 99%. However, the TOC was significantly lower than that of PVA, ranging from 74.3–85.1% to 93.5–98.1% under SCWG and SCWO conditions, respectively. The results indicate that in the above SCW situation, PVA can be easily decomposed to smaller molecules, which were relatively difficult to completely degrade. This consistent fact could also be obtained from Fig. 3.19b for PVA concentration of 0.2 wt%. TOC removal at SCWO conditions was obviously higher than that at SCWG, implying that the SCWO reactions can ensure the effective decomposition and complete degradation of PVA organics. Comparing Fig. 3.19a with b, we can observe that, under the same reaction conditions, the organic destruction efficiency for the 0.5 wt% PVA solution was always higher than that for the 0.2 wt% PVA solution. This fact demonstrates that the removal efficiency increased with the reactant concentration during the SCWG and SCWO reactions for PVA organic solutions, which can also be found from exponential reaction kinetic equations.

Several researchers [95, 96] have reported that the SCWO destruction of organic compounds occurs primarily through free radical pathways rather than the ionic pathways that dominate in liquid water. This is because SCW acts as a non-polar solvent with a low dielectric constant [97]. Based on the above discussion, we speculated the reaction scheme for the SCWO decomposition of PVA, as shown in the following:

$$O_2 + H_2O \rightarrow \cdot OH + HO_2\cdot$$
$$P{-}OH + OH\cdot \rightarrow P\cdot + H_2O_2$$
$$H_2O_2 \rightarrow 2\cdot OH$$
$$H_2O_2 \rightarrow H\cdot + \cdot OOH$$
$$H\cdot + H\cdot \rightarrow H_2$$
$$P\cdot + O_2 \rightarrow P{-}OO\cdot$$
$$P{-}OO\cdot + H\cdot \rightarrow PO\cdot + \cdot OH$$
$$P{=}O \rightarrow CO + Cl$$
$$P{-}OH + O_2 \rightarrow CO_2 + Cl$$
$$CO \rightarrow Cl$$
$$Cl \rightarrow CO_2$$
$$CO \rightarrow CO_2$$

During the SCWO process of the PVA solution, first, the PVA compounds were broken into smaller molecules POH. Then, most of the P–OH molecules were attacked by HO·, resulting in the loss of –OH, and the rest were oxidized by O_2 to CO_2 and CI, which represented a carbon-containing intermediate in the pathway between CO and CO_2. In a series of reaction paths, reaction $H\cdot + H\cdot \rightarrow H_2$ represents a limiting step, whose conversion controls the overall degradation. Reactions $P{-}OO\cdot + H\cdot \rightarrow PO\cdot + \cdot OH$ and $P{-}O \rightarrow CO + Cl$ stand for water-gas shift reactions, explaining the production of H_2 during the SCWO process of PVA. Reaction $P{-}OH + O_2 \rightarrow CO_2 + Cl$ represents the complete oxidation of CO to CO_2.

Figure 3.20 displays the influence of the NaOH concentration on the SCWO destruction for PVA solution at 510 °C, 25 ± 1 MPa. It can be observed that as the addition of NaOH increased from 0 up to 0.20 wt% not only the TOC removal

Fig. 3.20 Effect of NaOH on SCWO of 0.2 wt% PVA solution with an oxidation coefficient of 250% at 510 °C, 25 ± 1 MPa

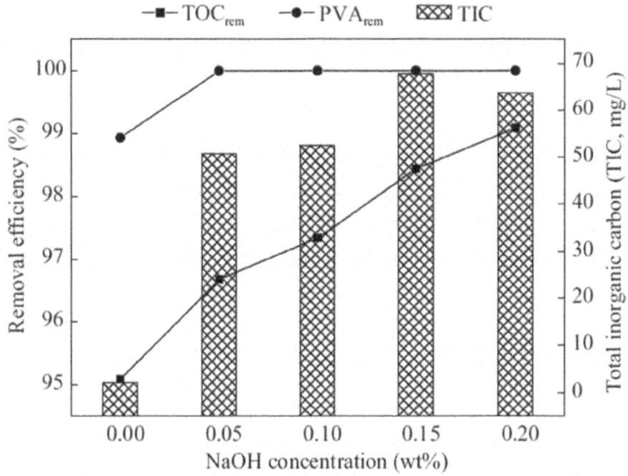

could be boosted from 95.1 to 99.1% but also the PVA could achieve complete decomposition as well. It appears that the addition of NaOH into the feed was conducive to the cleavage of PVA molecules, and meanwhile, accelerated the complete oxidation of organics.

The OH– ion helped the PVA break into smaller molecules. We assume that the catalytic reactions in the presence of OH– proceeded as follows:

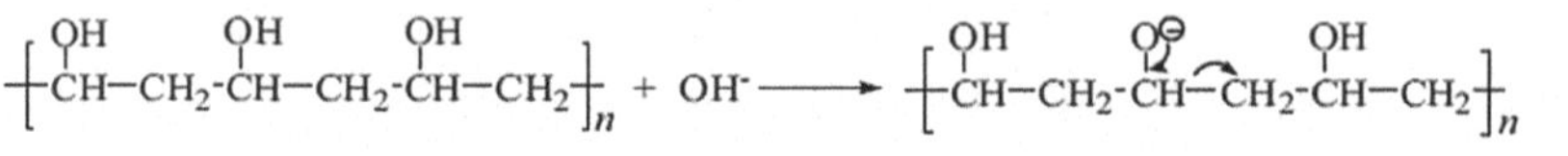

(3.51)

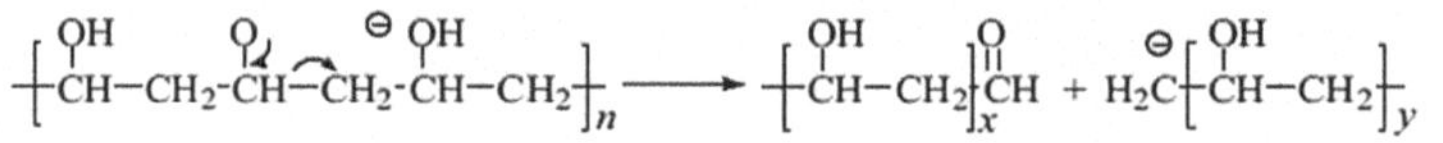

(3.52)

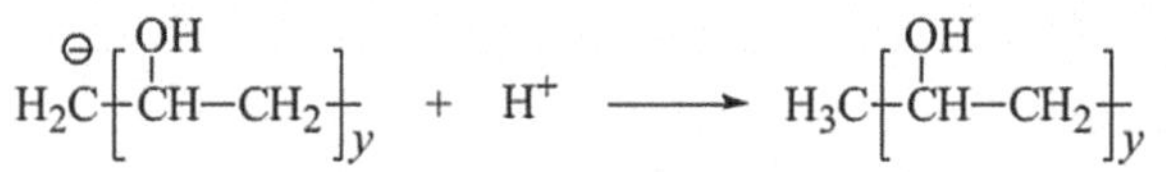

(3.53)

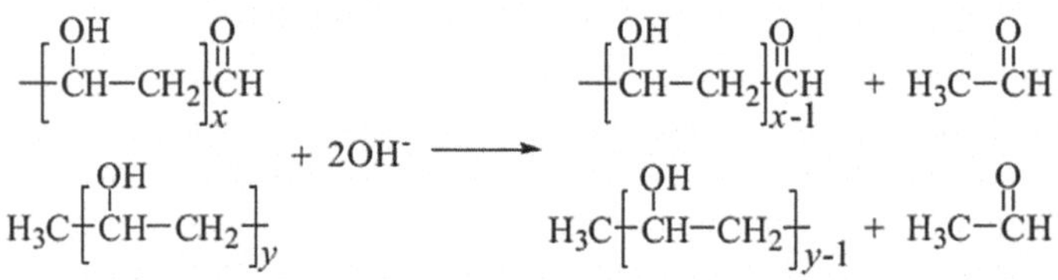

(3.54)

In Reaction (3.51), the OH· ion captured a proton from hydroxyl and converted to water. Meanwhile, an anion intermediate formed during this reaction. Then, the intermediate was broken up to P_x-CHO and another carbanion P_y-CH_2^-, as shown in Reaction (3.52), where x and y are the polymerization degree of the organics. Followed by Reaction (3.52), the carbanion attracted a proton from water, forming P_y-CH_3 (P represented the monomer CH–OH–CH_2–) and simultaneously released one OH– ion. As Reaction (3.54) displays, P_x-CHO and P_y-CH_3 combined with OH– ions again, repeating the above reaction processes, in which acetaldehydes were generated in both cases, as well as other organic compounds with lower polymerization degrees, P_{x-1}–CHO and P_{y-1}–CH_3^-. This reveals that the introduction of NaOH into the feed can promote the cleavage of the C–C bond in the PVA molecule. Based on the above discussion, acetaldehyde was a typical product. Previous studies [98, 99] have suggested that the aldehyde compounds are the primary stable organic intermediates during the SCWO of hydroxy organics. This confirms the aforementioned hypothesis for the reaction pathway.

3.10 Acetic Acid and Phenol

Methanol has higher oxidation reactivity in SCW than acetic acid and phenol. We postulated that the presence of methanol would accelerate decomposition of acetic acid or phenol during the mixture reaction. Therefore, in this investigation, we initiated SCWO experiments with different amounts of methanol added into the acetic acid and phenol aqueous feeds, respectively, aiming to examine the presence of co-oxidation effect of methanol on these two compounds. To validate the possible action mechanism, we simulated the SCWO reaction process for mixtures of methanol/acetic acid and methanol/phenol based on the detailed chemical kinetics models. This allowed us to make visual and deeper understanding of the interactions in the SCWO reaction process.

All experimental conditions are given in Table 3.18a, b for SCWO experiments of methanol/acetic acid and methanol/phenol mixtures, respectively [100].

We adopted the mechanism and kinetics parameters for methanol reported by Skodje [101] involving elementary reaction steps and 18 chemical species. By comparing this model with supercritical water oxidation reaction kinetics for methanol from Rice [102], Tester,44 and Brock8, we have validated the feasibility of this model in the condition of supercritical water oxidation. The kinetic models for phenol were obtained from the literature where DiNaro [103] developed a reduced elementary reaction mechanism for benzene oxidation in SCW. We extracted the reactions for phenol and the kinetic parameters. Gopalan et al. [104]

Table 3.18 Experimental reaction conditions for (a) SCWO of methanol/acetic acid mixtures at 24.5 MPa and (b) SCWO of methanol/phenol mixtures at 25.5 MPa

(a) SCWO of methanol/acetic acid mixtures at temperatures ranging from 673 to 753 K and 24.5 MPa				
Run ID	$[CH_3COOH]_0$ (mol/L)	$[CH_3OH]_0$ (mol/L)	Φ_I	$[O_2]_{CeO_1}$
1	0.17	0.0000	0.000	4.29
2	0.17	0.0313	0.188	4.29
3	0.17	0.0000	0.000	2.03
4	0.17	0.3125	1.875	2.03
5	0.17	0.0000	0.0000	1.24
6	0.17	0.6563	3.938	1.24
(b) SCWO of methanol/phenol mixtures at temperatures ranging from 673 to 753 K and 25.5 MPa				
Run ID	$[CH_3COOH]_0$ (mol/L)	$[CH_3OH]_0$ (mol/L)	Φ_{II}	$[O_2]_{CeO_2}$
7	0.067	0.0000	0.000	3.15
8	0.067	0.0313	0.463	3.15
9	0.067	0.0000	0.000	2.31
10	0.067	0.1563	2.317	2.31
11	0.067	0.0000	0.000	1.73
12	0.067	0.3125	4.633	1.73

Table 3.19 Supplementary reactions for phenol and acetic acid along with the kinetics parameters ($k = ATn\ \exp(-E_a/RT)$)[a]

Reaction	A	n	E_a	Ref.
$C_6H_5OH + O_2 = C_6H_5O\cdot + HO_2\cdot$	1.00×10^9	0	35,000	[32]
$C_6H_5OH = C_6H_5\cdot + H\cdot$	2.69×10^{19}	0	88,838	[32]
$C_6H_5OH + H\cdot = C_6H_5O\cdot + H_2$	1.90×10^{-10}	0	12,393.5	[46]
$CH_3OO\cdot + HO_2 = CH_3OOH + O_2$	7.69×10^{-14}	0	−2582	[48]
$CH_3OOH = CH_3O\cdot + OH\cdot$	4.00×10^{15}	0	43,000	[48]

[a]Units are cm, mol, s, and cal

analyzed the systematic paths of phenol SCWO reaction by comparing the kinetics data for several possible competing initiation steps; they recognized that the oxidation of phenol was initiated by bimolecular reactions between phenol and oxygen forming the phenoxy radical $C_6H_5O\cdot$ and peroxy radical $HO_2\cdot$. We added this initiation reaction to the detailed chemical kinetic model for phenol. The calculation results show that this reaction led to the postponement of phenol decomposition, probably due to the fact that oxygen was less reactive than the hydroxyl and peroxy radicals in attacking chemical bonds and consequently decelerated the overall reaction rate for phenol. Moreover, we complemented the other two reactions $C_6H_5OH{=}C_6H_5O\cdot + H\cdot$ and $C_6H_5OH + H\cdot = C_6H_5O\cdot + H_2$ which were obtained from the studies. The three reactions are summarized in Table 3.19.

(1) **Oxidation of Methanol/Acetic Acid**

Figures 3.21 and 3.22 illustrate the calculated temporal variations of the two key reactive intermediates, $HO_2\cdot$ and $OH\cdot$ radicals, at 733 K, 25 MPa, $[CH_3COOH]_0 = 0.17$ mol/L, and $[O_2]_{CeO_1} = 2.03$. The peak profiles of $HO_2\cdot$ concentration in Fig. 3.4 indicate that $HO_2\cdot$ radicals accumulated at first and then were consumed by intermediate organics, and the peak height showed a notable dependence on the initial methanol concentration; that is, the maximum in the $HO_2\cdot$

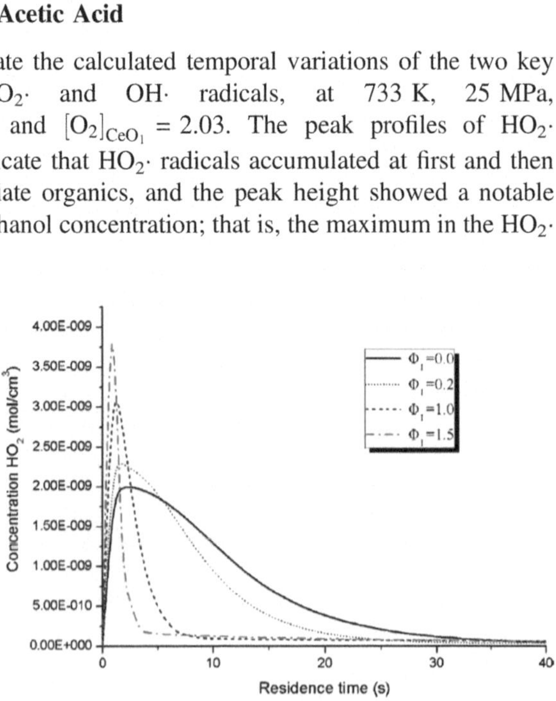

Fig. 3.21 $HO_2\cdot$ concentrations predicted for SCWO of methanol/acetic acid mixtures at $T = 733$ K, $P = 25$ MPa, $[CH_3COOH]_0 = 0.17$ mol/L, and $[O_2]_{CeO_1} = 2.03$

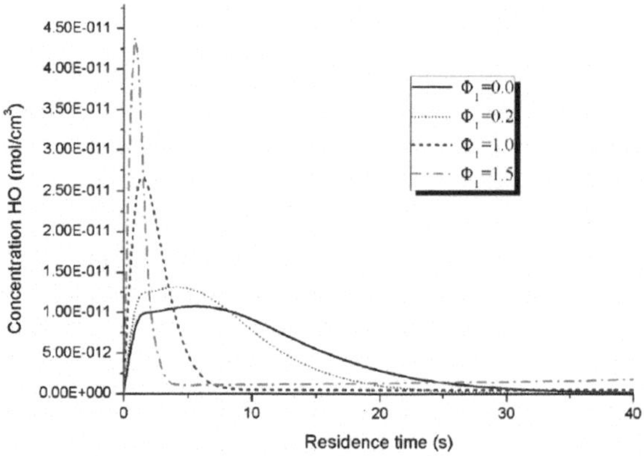

Fig. 3.22 OH· concentrations predicted for SCWO of methanol/acetic acid mixtures at T = 733 K, P = 25 MPa, $[CH_3COOH]_0$ = 0.17 mol/L, and $[O_2]_{CeO_1}$ = 2.03

concentrations shifted to higher values and appeared at shorter times as the methanol concentration increased. An increment in peak $HO_2\cdot$ concentration can be observed with Φ_I increasing from 0.0 to 0.2, while with Φ_I further increasing to 1.0 and 1.5, the maximum values of $HO_2\cdot$ concentrations were about 1.5 and 2.0 times higher than that in the absence of methanol (Φ_I = 0.0). The time profiles of the OH· concentration in Fig. 3.5 were similar to those for $HO_2\cdot$, that is, with the presence of methanol, the peak values of OH· concentrations were elevated as the methanol concentration increased. However, regardless of initial methanol concentration, the profiles of OH·concentration were always several orders of magnitude lower than those of $HO_2\cdot$. Reactions 13 and 14 were two radical producing reactions from methanol oxidation employed in the binary reaction kinetics. Clearly, an increase of methanol could facilitate the two reaction equilibriums to move in the forward direction and promoted the generation of reactive $HO_2\cdot$ and OH·. As a consequence, it was the $HO_2\cdot$ and OH· produced by methanol oxidation at lower reaction time that accelerated the acetic acid decomposition during the SCWO of methanol/acetic acid mixtures. Even with a small addition of methanol (Φ_I = 0.188), the reactive intermediates generated by CH_3OH reactions were sufficient for co-oxidation of CH_3COOH. The co-oxidative enhancement of methanol stemmed from two reasons: increasing oxidation rate constant and shortening induction time. It is reasonable to propose that both increment of oxidation rate constant and reduction of the induction time occurred during co-oxidation reactions for methanol and acetic acid. First, the time that $HO_2\cdot$ and OH· radicals reached the peak concentrations was reduced with the addition of methanol. The accumulation time for radicals was shortened, and induction time decreased. Second, the peak $HO_2\cdot$ and OH· concentrations obviously was elevated as methanol addition increased. According to the collision theory of reactions, an important factor that determines the reaction rate is the percentage of collisions between activated molecules. The obvious increment in $HO_2\cdot$ and OH· concentrations in the presence of methanol could promote the collision probabilities of acetic acid molecules and these active radicals thus enhance the reaction rate constant rate. Therefore, the two effects mentioned above accelerated the acetic acid decomposition after addition of methanol.

(2) **Oxidation of Methanol/phenol**

The temporal plots in Figs. 3.23 and 3.24 of $HO_2\cdot$ and $OH\cdot$ concentration with variable Φ_{II} ratios represent two conspicuous features. The first one is that peaks emerged with larger Φ_{II} of 2.0 and 3.0, the maximum of $HO_2\cdot$ and $OH\cdot$ concentrations shifted to higher values and appeared at a shorter time as the initial methanol concentrations increased. Specifically, increasing the Φ_{II} from 2.0 to 3.0 improved the peak values of $HO_2\cdot$ and $OH\cdot$ concentration by almost 1 order of magnitude while bringing the peak point forward temporally by 10 s. This notable characteristic confirms that, during the SCWO reaction of ethanol/phenol mixtures, methanol oxidation proceeded at much shorter times producing reactive intermediates $HO_2\cdot$ and $OH\cdot$, which remained at a later time and would participate in phenol oxidation, thereby accelerating the degradation of phenol. The second feature for HO2· and OH· profiles is that the $HO_2\cdot$ concentration with $\Phi_{II} = 0.5$ was higher than that with $\Phi_{II} = 0$; conversely, $OH\cdot$ radicals with $\Phi_{II} = 0.5$ were relatively lower. As shown in Fig. 3.8, for calculated results for phenol conversions, we

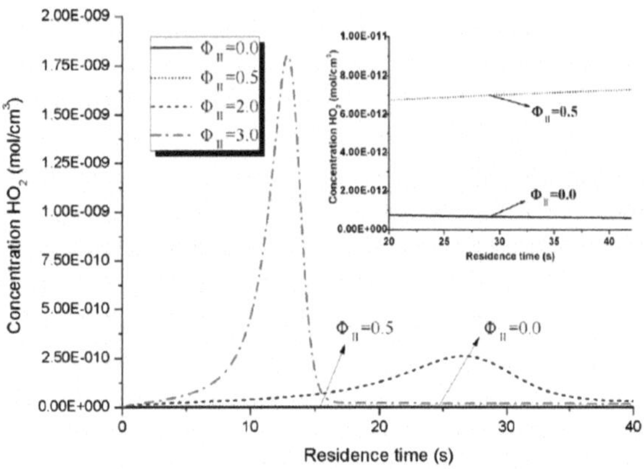

Fig. 3.23 $HO_2\cdot$ concentrations predicted for SCWO of methanol/phenol mixtures at $T = 733$ K, $P = 25$ MPa, $[C_6H_5OH]_0 = 0.067$ mol/L, and $[O_2]_{CeO_2} = 1.73$

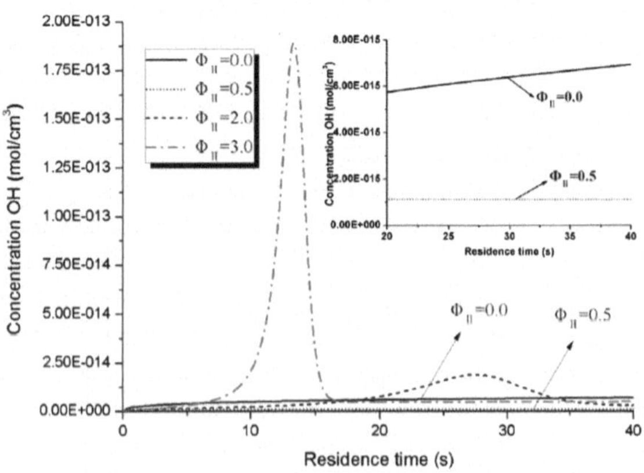

Fig. 3.24 HO· concentrations predicted for SCWO of methanol/phenol mixtures at $T = 733$ K, $P = 25$ MPa, $[C_6H_5OH]_0 = 0.067$ mol/L, and $[O_2]_{CeO_2} = 1.73$

find that phenol decomposed in Run 8 with $\Phi_{II} = 0.5$ more slowly than that in Run 7 with $\Phi_{II} = 0.0$ before 33 s, but after 33 s, the phenol conversion with $\Phi_{II} = 0.5$ was higher than that with $\Phi_{II} = 0.0$. Accordingly, this leads us to speculate whether there is a causal relationship between the observation of phenol conversions and the second feature for $HO_2\cdot$ and $OH\cdot$ profiles.

Gopalan et al. [104] have proposed a detailed reaction mechanism for supercritical water oxidation of phenol. They reported that oxidation of phenol was initiated by a bimolecular reaction between phenol and oxygen to form phenoxy radical. Following initiation, $HO_2\cdot$ and $OH\cdot$ radicals in the mixture began to attack phenol molecules. Reactions (3.55), (3.56), and (3.57) illustrate these steps. In the SCW environment, the free radical process dominated over the ionic process, and phenoxy radical, which was a stable delocalized radical, did not favor oxygen addition. Therefore, the radical–radical process became essential for the subsequent reaction of phenoxy radicals, and the ring-opening reaction for phenoxy radicals proceeded via $HO_2\cdot$ addition to the phenoxy radical. Combining the SCWO reaction mechanism for phenol, we postulate that in the case of lower methanol addition ($\Phi_{II} = 0.5$), $OH\cdot$ radicals in the system were mostly consumed by methanol, leading to less $OH\cdot$ available for phenol initiation reaction than that with pure phenol oxidation, thus restraining decomposition of phenol at earlier reaction steps before 33 s. Meanwhile, $HO_2\cdot$ radicals were important during the ring-opening of the aromatic ring at the later reaction step, and thereby, more $HO_2\cdot$ in the presence of methanol would accelerate the phenol disappearance rate after 33 s.

$$C_6H_5OH + O_2 = C_6H_5O\cdot + HO_2\cdot \tag{3.55}$$

$$C_6H_5OH + HO_2\cdot = C_6H_5O\cdot + H_2O_2 \tag{3.56}$$

$$C_6H_5OH + OH\cdot = C_6H_5O\cdot + H_2O \tag{3.57}$$

Additionally, from the sensitivity analysis for $HO_2\cdot$ and $OH\cdot$ radicals during SCWO of methanol/phenol mixtures, as shown in Table 3.20, the production of $HO_2\cdot$ at an early time was sensitive to the hydrogen abstraction reactions by O2 from C_6H_5OH, $CH_2OH\cdot$, and CH_3OH (reactions $R_{II}20$, $R_{II}94$, and $R_{III}17$). The decomposition of H_2O_2 to $OH\cdot$ was the most sensitive reaction for accumulation of $OH\cdot$ radicals, which was the same as the reactions for methanol/acetic acid mixtures, but the reactions $R_{II}22$, RII86, and $R_{II}115$ negatively affected the formation of $OH\cdot$. Figure 3.25 denotes the effect of methanol concentration in the yield of various gas constituents and total organic carbon removal for SCWO of methanol/phenol mixtures at 713 K. Apparently, the total gas yield was nearly in step with that of TOC removal efficiency. Both of them were slightly reduced with increasing ΦII from 0.0 to 0.463, which confirmed the threshold concentration of $[CH_3OH]_0$ for notable co-oxidation effect of methanol on phenol conversion from another

Table 3.20 Reaction with the largest normalized sensitivity coefficients for $HO_2\cdot$ and $OH\cdot$ during SCWO of methanol/phenol mixtures at 733 K, 25 MPa, $[C_6H_5OH]_0 = 0.067$ mol/L, $[O_2]_{CeO_2} = 1.73$, $\Phi_{II} = 3.0$, and $\tau = 2.5$ s

Reaction	Reaction number	Normalized sensitivity coefficient	
		$HO_2\cdot$	$OH\cdot$
$H\cdot + O_2(+M) = HO_2\cdot(+M)$	$R_{II}9$	0.089	
$H_2O_2(+M) = 2OH\cdot(+M)$	$R_{II}15$		0.999
$C_6H_5OH + O_2 = C_6H_5O\cdot + HO_2\cdot$	$R_{II}20$	0.266	
$C_6H_5OH + OH\cdot = C_6H_5O\cdot + H_2O$	$R_{II}22$		−0.203
$HCO\cdot + O_2 = CO + HO_2\cdot$	$R_{II}24$	0.056	
$CH_2O + OH\cdot = HCO\cdot + H_2O$	$R_{II}86$		−0.033
$CH_3O\cdot + O_2 = CH_2O\cdot + HO_2\cdot$	$R_{II}94$	0.808	
$CH_3OH + OH\cdot = CH_3O\cdot + H_2O$	$R_{II}115$		−0.366
$CH_3OH + O_2 = CH_3O\bullet + HO_2\cdot$	$R_{II}117$	0.021	
$CH_3OH\cdot + HO_2\cdot = CH_3O\cdot + H_2O_2$	$R_{II}I19$	−0.409	

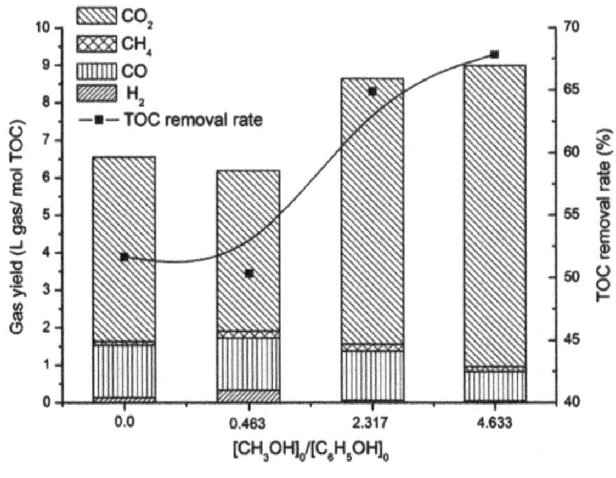

Fig. 3.25 Yields of different gaseous products and TOC removal efficiency for SCWO of methanol/phenol mixtures' solution ($T = 713$ K, $[C_6H_5OH]_0 = 0.067$ mol/L, and $[CH_3OH]_0 = 0.0$, 0.0313, 0.1563,0.3125 mol/L)

perspective. As Φ_{II} increased up to 2.317, the total gas yield and TOC removal both increased significantly. Moreover, with further increasing Φ_{II} to 4.633, the CO yield decreased, conversely leading to an obvious increment in the CO_2 yield, with the total gas yield increasing as a result. This indicates that, above the threshold of $[CH_3OH]_0$ concentration, the CO_2 yield varied positively with methanol addition, whereas the CO yield varied inversely, revealing that the increment of initial methanol concentration would obviously accelerate complete degradation of phenol and methanol mixtures during the SCWO process.

3.11 Quinazoline

As a N-containing heterocycle with a benzene ring and a pyrimidine ring, quinazoline ($C_8H_6N_2$) and its derivatives have diverse biological and pharmacological activities, such as antimalarial, antimicrobial, anti-inflammatory, anticonvulsant, antihypertensive, anti-diabetic, and anticancer activities [105, 106]. Owing to the widespread applications in a variety of synthetic drugs, bioactive natural products, pharmaceuticals and agrochemicals, quinazoline and its derivatives are often detected in treating wastewater and environmental water [107, 108]. Hence, the oxidative degradation of quinazoline and its derivatives in SCW have captured our attention. Several researchers have studied the reaction kinetics and mechanisms involved in advanced oxidation process of other N-heterocyclic organic compounds, such as quinoline [109] and pyridine [110]. The global kinetic parameters of N-heterocyclic organic compounds in SCWO in literatures are summarized in Table 3.21. Crain et al. [111] developed a simplified network of reaction pathways for pyridine oxidation in SCW on the basis of identified compounds. However, most of the previous kinetic models focused solely on global rate expression of reactant conversion but without capturing the pathways resulted in the formation and interconversion of intermediate species.

This model is based on the product distribution and reaction pathways proposed in our previous study on the SCWO mechanisms of quinazoline [114], as shown in Fig. 3.26.

Following, we defined and classified the reactions involved in this simplified model:

(1) Alternative ring-opening reaction of quinazoline

$$\begin{aligned} & C_8H_6N_2\ (\text{Quinazoline}) + 4.5O_2 \xrightarrow{k_1} C_4H_4N_2\ (\text{Pyrimidine}) \\ & \quad + 4CO_2 + H_2O \end{aligned} \tag{3.58}$$

$$\begin{aligned} & C_8H_6N_2\ (\text{Quinazoline}) + O_2 + 3H_2O \xrightarrow{k_2} C_6H_6O\ (\text{Phenol}) \\ & \quad + 2CO_2 + 2NH_3 \end{aligned} \tag{3.59}$$

The oxidation of quinazoline was triggered by two alternative ring-opening reactions. The first ring-opening reaction, as (3.58) shows, generated pyrimidine via oxidation without contribution of H_2O. The second pathway, in Eq. (3.59), yielded phenol as ring-opening intermediate by attack of H_2O and O_2 with removal of 2 molecular of NH_3.

Table 3.21 Kinetic parameters comparison of N-heterocyclic organic compounds in SCWO

Parameters	Quinoline [109]	Pyridine [111]	Pyridine [112]	Pyridine [113]		
A	$2.7 \times 10^{13\pm2}$	$10^{13.1\pm1.65}$	$10^{31.45\pm3.92}$	$10^{25.3\pm6.7}$	$10^{14.25\pm2.5}$	$10^{14.32\pm3.6}$
E_a	226 ± 19	209.5 ± 22.4	299.28 ± 40.39	287.26 ± 75.61	196.42 ± 27.8	227.61 ± 47.85
a	0.8 ± 0.1	1.0 ± 0.30	1.82 ± 0.22	2.25 ± 0.18	1.12 ± 0.13	0.42 ± 0.18
b	0.3 ± 0.1	0.20 ± 0.16	0.42 ± 0.13	0.43 ± 0.28	1.14 ± 0.23	0.73 ± 0.28
Catalyst	None	None	Pt/γ-Al_2O_3	Pt/γ-Al_2O_3	MnO_2/CeO_2	α-Al_2O_3

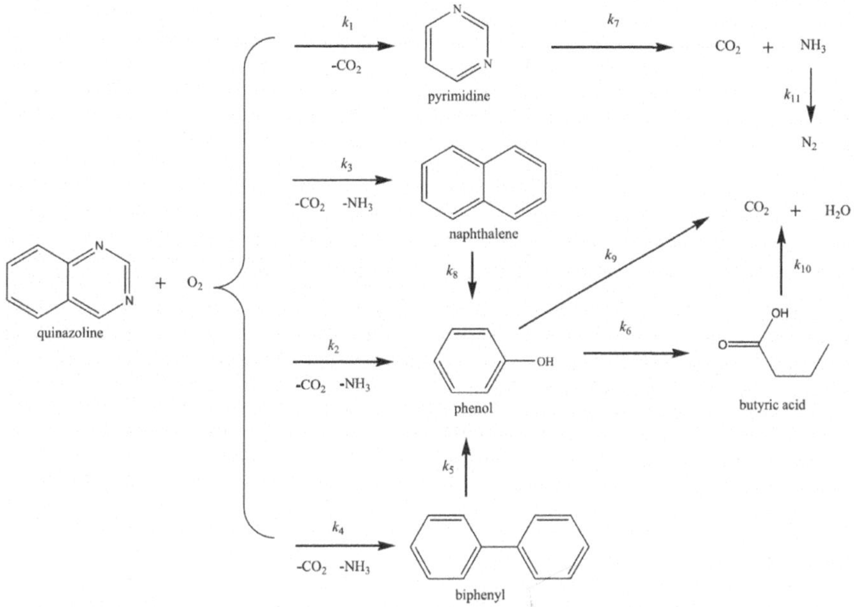

Fig. 3.26 Simplified reaction network of oxidative degradation of quinazoline in SCW

(2) Polymerization reactions

$$\begin{aligned} &2C_8H_6N_2\ (\text{Quinazoline}) + 4O_2 \\ &\quad + 4H_2O \xrightarrow{k_3} C_{10}H_8\ (\text{Naphthalene}) + 6CO_2 + 4NH_3 \end{aligned} \tag{3.60}$$

$$\begin{aligned} &2C_8H_6N_2\ (\text{Quinazoline}) + 1.5O_2 \\ &\quad + 5H_2O \xrightarrow{k_4} C_{12}H_{10}\ (\text{Biphenyl}) + 4CO_2 + 4NH_3 \end{aligned} \tag{3.61}$$

(3) Oxidation reactions of intermediates

$$C_{12}H_{10}\ (\text{Biphenyl}) + H_2O + 0.5O_2 \xrightarrow{k_5} 2C_6H_6O\ (\text{Phenol}) \tag{3.62}$$

$$C_6H_6O\ (\text{Phenol}) + 2O_2 + H_2O \xrightarrow{k_6} C_4H_8O_2\ (n\text{-Butyric acid}) + 2CO_2 \tag{3.63}$$

$$C_4H_4N_2\ (\text{Pyrimidine}) + H_2O + 3.5O_2 \xrightarrow{k_7} 4CO_2 + 2NH_3 \tag{3.64}$$

$$C_{10}H_8\ (\text{Naphthalene}\ ) + 5O_2 \xrightarrow{k_8} C_6H_6O\ (\text{Phenol}) + 4CO_2 + H_2O \quad (3.65)$$

$$C_6H_6O\ (\text{Phenol}) + 6O_2 \xrightarrow{k_9} 6CO_2 + 3H_2O \quad (3.66)$$

$$C_4H_8O_2\ (n\text{-Butyric acid}) + 5O_2 \xrightarrow{k_{10}} 4CO_2 + 4H_2O \quad (3.67)$$

Once these intermediates yielded, they could undergo a variety of oxidation reactions with or without participation of H_2O and led to the generation of lower molecular weight compounds (CO_2 and NH_3). Anitescu and Tavlarides [115] demonstrated that phenol was one of the intermediate products from SCWO degradation of biphenyl. So, we assumed that phenol was an intermediate product of oxidation reaction of biphenyl and naphthalene. In addition, we deduced that *n*-butyric acid was an intermediate product from the degradation of phenol based on the quantitative analysis data shown in Table 3.22.

Table 3.22 Data required to solve ODEs

T/K	*ρ*/ kg m^{-3}	*t*/s	[C_{Pyr}]	[C_{BA}]	[C_{Nap}]	[C_{Phe}]	[C_{Bip}]	[C_{Qu}]	[C_{CO_2}]	[C_{NH_3}]
			mmol L^{-1}							
673	166.3	0	0.00E +00	0.00E +00	0.00E +00	0.00E +00	0.00E +00	5.56E +00	0.00E +00	0.00E +00
		60	1.29E +00	2.79E −01	1.70E −01	7.45E −01	8.60E −02	1.25E +00	1.20E +01	4.82E +00
		120	2.25E +00	5.86E −01	2.64E −01	8.17E −01	9.29E −02	1.06E +00	1.33E +01	5.67E +00
		180	2.29E +00	8.21E −01	3.03E −01	6.82E −01	1.11E −01	5.72E −01	2.25E +01	6.60E +00
		240	2.54E +00	9.21E −01	2.77E −01	5.17E −01	8.41E −02	2.32E −01	2.54E +01	7.34E +00
		300	2.29E +00	1.00E +00	2.09E −01	3.77E −01	7.41E −02	1.04E −01	3.19E +01	7.40E +00
		360	2.10E +00	8.09E −01	2.08E −01	2.73E −01	7.54E −02	8.13E −02	3.07E +01	7.11E +00
		420	2.20E +00	7.49E −01	1.85E −01	1.99E −01	5.32E −02	4.02E −02	3.12E +01	7.81E +00
723	109.0	0	0.00E +00	0.00E +00	0.00E +00	0.00E +00	0.00E +00	3.64E +00	0.00E +00	0.00E +00
		60	1.71E +00	4.00E −01	4.94E −01	4.81E −01	7.30E −02	7.21E −01	1.57E +01	4.57E +00
		120	1.50E +00	6.19E −01	4.80E −01	2.66E −01	6.78E −02	1.51E −01	1.98E +01	5.04E +00
		180	1.35E +00	4.50E −01	4.07E −01	1.09E −01	3.22E −02	2.02E −02	2.10E +01	6.25E +00
		240	1.27E +00	2.47E −01	3.38E −01	8.63E −02	1.29E −02	2.10E −02	2.53E +01	5.74E +00

(continued)

Table 3.22 (continued)

T/K	ρ/ kg m^{-3}	t/s	[C_{Pyr}]	[C_{BA}]	[C_{Nap}]	[C_{Phe}]	[C_{Bip}]	[C_{Qu}]	[C_{CO_2}]	[C_{NH_3}]
			mmol L^{-1}							
			7.71E−01	2.24E−01	2.79E−01	2.65E−02	1.06E−02	1.10E−02	2.51E+01	6.26E+00
		360	7.10E−01	1.56E−01	2.31E−01	1.13E−02	8.36E−03	1.50E−03	3.01E+01	6.17E+00
		420	5.79E−01	1.07E−01	1.90E−01	1.29E−02	4.20E−03	3.58E−05	2.85E+01	6.88E+00
773	89.9	0	0.00E+00	0.00E+00	0.00E+00	0.00E+00	0.00E+00	3.01E+00	0.00E+00	0.00E+00
		60	7.92E−01	5.28E−01	2.27E−01	1.51E−01	4.89E−02	1.33E−01	1.76E+01	5.16E+00
		120	4.02E−01	2.14E−01	8.54E−02	6.09E−02	6.07E−03	5.26E−03	1.87E+01	6.59E+00
		180	6.48E−02	1.01E−01	2.89E−02	2.10E−02	2.06E−03	1.22E−03	2.46E+01	5.34E+00
		240	2.99E−02	3.79E−02	9.67E−03	4.34E−03	2.51E−04	1.92E−04	2.56E+01	6.56E+00
		300	5.13E−02	1.55E−02	3.23E−03	2.07E−03	3.12E−05	1.10E−04	2.65E+01	4.86E+00
		360	8.56E−03	8.73E−03	1.08E−03	6.51E−04	0.00E+00	0.00E+00	2.76E+01	5.06E+00
		420	2.76E−03	5.60E−04	3.60E−04	5.16E−04	0.00E+00	0.00E+00	2.64E+01	5.11E+00
823	78.6	0	0.00E+00	0.00E+00	0.00E+00	0.00E+00	0.00E+00	2.63E+00	0.00E+00	0.00E+00
		60	2.68E−01	1.27E−01	4.41E−02	2.28E−02	1.82E−03	1.99E−02	1.93E+01	4.92E+00
		120	3.91E−02	1.63E−02	2.33E−02	9.48E−03	6.46E−04	1.51E−03	2.13E+01	4.89E+00
		180	1.22E−02	3.03E−03	1.22E−03	4.86E−03	1.91E−04	1.14E−04	2.25E+01	4.31E+00
		240	2.45E−03	1.55E−04	6.35E−04	2.53E−04	0.00E+00	8.63E−05	2.24E+01	4.37E+00
		300	2.77E−04	8.07E−05	3.31E−04	1.32E−04	1.51E−05	0.00E+00	2.36E+01	3.36E+00
		360	3.13E−04	4.20E−05	1.73E−04	0.00E+00	4.23E−05	0.00E+00	2.45E+01	3.55E+00
		420	3.54E−04	0.00E+00	0.00E+00	0.00E+00	0.00E+00	0.00E+00	2.44E+01	2.64E+00
873	70.8	0	0.00E+00	0.00E+00	0.00E+00	0.00E+00	0.00E+00	2.37E+00	0.00E+00	0.00E+00
		60	2.63E−02	6.96E−03	8.21E−03	5.54E−03	2.61E−03	4.30E−02	2.10E+01	3.95E+00

(continued)

Table 3.22 (continued)

T/K	ρ/ kg m^{-3}	t/s	$[C_{Pyr}]$	$[C_{BA}]$	$[C_{Nap}]$	$[C_{Phe}]$	$[C_{Bip}]$	$[C_{Qu}]$	$[C_{CO_2}]$	$[C_{NH_3}]$
			mmol L^{-1}							
			2.14E −03	3.10E −03	4.15E −03	2.57E −03	2.01E −03	2.55E −03	2.16E +01	2.42E +00
		180	2.70E −04	4.70E −03	5.52E −03	4.11E −03	2.34E −03	8.11E −03	2.26E +01	1.85E +00
		240	7.00E −05	2.10E −03	2.28E −03	2.20E −03	1.20E −03	4.12E −03	2.36E +01	1.52E +00
		300	0.00E +00	1.60E −03	2.00E −03	0.00E +00	1.70E −03	2.08E −03	2.16E +01	1.80E +00
		360	0.00E +00	1.42E −03	1.05E −03	0.00E +00	0.00E +00	0.00E +00	2.22E +01	8.60E −01
		420	0.00E +00	7.51E −04	2.52E −03	0.00E +00	1.15E −03	1.31E −03	2.32E +01	1.17E +00

(4) Oxidation reaction of NH_3

$$NH_3 + 0.75O_2 \xrightarrow{k_{11}} 0.5N_2 + 1.5H_2O \tag{3.68}$$

There was a little NO_2^- (<1.2 mg/L) and no NO_3^- detected from liquid products, which indicates NH_3 was ultimately oxidized to N_2. On the basis of the mass action law, we described every reaction pathway in differential dynamic model, as Eqs. (3.69)–(3.77) shown. We assumed that all reactions are irreversible and first order in each reactant. Since water and oxygen were present in excess to the stoichiometric amount needed in the reaction system, we did not include them in the rate equations. Then, we obtained corresponding rate constants for the individual reactions by fitting the experimental concentrations to these differential equations, which were applied to this reaction network and the constant-volume batch reactors used in the experiments.

$$\frac{d[C_{Qu}]}{dt} = -(k_1 + k_2 + k_3 + k_4)[C_{Qu}] \tag{3.69}$$

$$\frac{d\left[C_{Pyr}\right]}{dt} = k_1[C_{Qu}] - k_7\left[C_{Pyr}\right] \tag{3.70}$$

$$\frac{d[C_{Phe}]}{dt} = k_2[C_{Qu}] + 2k_5\left[C_{Bip}\right] + k_8\left[C_{Nap}\right] - (k_6 + k_9)[C_{Phe}] \tag{3.71}$$

$$\frac{d\left[C_{Nap}\right]}{dt} = k_3[C_{Qu}] - k_8\left[C_{Nap}\right] \tag{3.72}$$

$$\frac{\mathrm{d}\left[C_{\mathrm{Bip}}\right]}{\mathrm{d}t} = k_4[C_{\mathrm{Qu}}] - k_5\left[C_{\mathrm{Bip}}\right] \tag{3.73}$$

$$\frac{\mathrm{d}[C_{\mathrm{BA}}]}{\mathrm{d}t} = k_6[C_{\mathrm{Phe}}] - k_{10}[C_{\mathrm{BA}}] \tag{3.74}$$

$$\begin{aligned}\frac{\mathrm{d}[C_{\mathrm{CO_2}}]}{\mathrm{d}t} &= (4k_1 + 2k_2 + 3k_3 + 2k_4)[C_{\mathrm{Qu}}] + (2k_6 + 6k_9)[C_{\mathrm{Phe}}] \\ &\quad + 4k_7\left[C_{\mathrm{Pyr}}\right] + 4k_8\left[C_{\mathrm{Nap}}\right] + 4k_{10}[C_{\mathrm{BA}}]\end{aligned} \tag{3.75}$$

$$\frac{\mathrm{d}[C_{\mathrm{NH_3}}]}{\mathrm{d}t} = (2k_2 + 4k_3 + 4k_4)[C_{\mathrm{Qu}}] + 2k_7\left[C_{\mathrm{Pyr}}\right] - k_{11}[C_{\mathrm{NH_3}}] \tag{3.76}$$

$$\frac{\mathrm{d}[C_{\mathrm{N_2}}]}{\mathrm{d}t} = 0.5\mathrm{k}_{11}[C_{\mathrm{NH_3}}] \tag{3.77}$$

In the equations above, $[C_{\mathrm{Qu}}]$, $[C_{\mathrm{Pyr}}]$, $[C_{\mathrm{Phe}}]$, $[C_{\mathrm{Nap}}]$, $[C_{\mathrm{Bip}}]$, $[C_{\mathrm{BA}}]$, $[C_{\mathrm{CO2}}]$, $[C_{\mathrm{NH3}}]$, and $[C_{\mathrm{N2}}]$ were concentrations (mmol/L) of quinazoline, pyrimidine, phenol, naphthalene, biphenyl, *n*-butyric acid, CO_2, NH_3, and N_2 in the SCWO reactor, respectively.

Oxidation coefficient, OC, is defined as:

$$\mathrm{OC} - \frac{[\mathrm{O_2}]_{\mathrm{add}}}{[\mathrm{O_2}]_{\mathrm{need}}} \tag{3.78}$$

where, $[O_2]$ add is the oxygen (mg) formed from the decomposition of H_2O_2. $[O_2]$ need is the stoichiometric oxygen (mg) needed to completely oxidize the feed to carbon dioxide, nitrogen, and water. In this experiment, the oxidation coefficient was controlled to the value of 2.0.

After determining optimized values for the rate constants at each temperature, the Arrhenius parameters were determined by using linear regression. Table 3.23 lists the rate constants, pre-exponential factors, and activation energy data under different testing temperatures.

The value of k_1 was always higher than that of k_2, k_3, and k_4 in all testing temperature conditions, suggesting that the alternative ring-opening of quinazoline for pyrimidine yielding was the dominant pathway. When the temperature reached 823 K, k_7 increased rapidly until close to k_1. In this case, the ring-opening reaction of pyrimidine was accelerated obviously. High values of k_5 and k_6 at 873 K indicate high oxidation rates of biphenyl to phenol and further to *n*-butyric acid at the

Table 3.23 Chemical reaction rate constant and kinetic parameters

Reaction rate $k/(s^{-1})$	Temperature T/K					lnA	Activation energy E_a/kJ mol^{-1}	R^2
	673	723	773	823	873			
k_1	6.44E−03	1.87E−02	2.59E−02	7.13E−02	7.29E−02	6.01 ± 1.24	61.15 ± 7.87	0.937
k_2	4.12E−03	9.05E−03	1.64E−02	1.34E−01	6.81E−02	9.01 ± 3.11	81.62 ± 19.75	0.801
k_3	8.44E−04	2.57E−03	8.21E−03	4.46E−02	4.87E−02	12.10 ± 1.80	107.64 ± 11.41	0.957
k_4	3.34E−04	1.17E−03	3.55E−03	2.09E−02	1.06E−02	9.51 ± 3.02	97.30 ± 19.14	0.861
k_5	2.77E−03	1.14E−02	3.79E−02	2.44E−01	2.29E−01	15.08 ± 2.16	117.17 ± 13.68	0.948
k_6	9.49E−03	2.54E−02	6.50E−02	4.71E−01	3.13E−01	12.70 ± 2.69	97.50 ± 17.05	0.888
k_7	9.57E−04	3.41E−03	1.60E−02	1.04E−02	7.23E−02	10.2 ± 2.88	95.52 ± 18.3	0.868
k_8	1.84E−03	7.69E−03	1.93E−02	6.33E−03	1.19E−01	7.92 ± 4.93	78.90 ± 31.28	0.573
k_9	1.93E−04	7.28E−04	1.74E−03	1.24E−02	1.48E−02	11.55 ± 1.93	112.80 ± 12.25	0.954
k_{10}	4.56E−03	1.33E−02	3.76E−02	5.60E−02	1.88E−01	10.10 ± 1.05	86.69 ± 6.64	0.977
k_{11}	1.59E−05	1.54E−04	5.46E−04	2.10E−03	8.60E−03	15.78 ± 0.83	149.23 ± 5.27	0.995

participation of water. Comparing with k_9 and k_{10}, without participation of water, oxidative degradation of phenol was relatively slower than that of n-butyric acid due to its high molecular structural stability, which was in accordance with the previous literature.

As demonstrated, E_a for reaction (3.58) and reaction (3.59) was relatively lower than the others and the values were 61.15 ± 7.87 kJ mol/L and 81.62 ± 19.75 kJ mol/L, respectively. It indicates that ring-opening reaction of quinazoline was relatively easy. However, activation energy for reactions (3.64) and (3.66) was high, illustrating that pyrimidine and phenol were difficult to be decomposed. Given the minimum values of rate constants and the highest E_a value observed for reaction (3.68), the degradation of NH_3 was suggested to be the most difficult pathway in the network, which was consistent with previous literatures.

Figure 3.27 provides a detailed assessment by showing the experimental temporal variations of the concentrations of the major liquid and gaseous products ($[C_{Qu}]$, $[C_{Pyr}]$, $[C_{Phe}]$, $[C_{NH3}]$, $[C_{Bip}]$, $[C_{Nap}]$, $[C_{BA}]$, $[C_{CO2}]$) along with the model calculations at testing temperatures. As expected, the model fits experimental concentration changes of major compounds reasonably well in the process. Reaction rate analysis shows that reactions (3.58–3.61) reached their thermodynamic equilibrium within 100 s and their contributions to quinazoline decomposition follow the order: reaction (3.58) > reaction (3.59) > reaction (3.60) > reaction (3.61). The model results suggested the alternative ring-opening of quinazoline for pyrimidine yielding (reaction 3.58) was the dominant pathway, and the ring-opening reaction of pyrimidine (reaction 3.64) was accelerated obviously when the temperature was above 823 K. Reaction (3.59) was the dominate reaction for the generation of phenol and NH_3. Degradation of NH_3 was quite difficult with the low value of conversion rate constant and the high E_a value. As confirmed by sensitivity analysis results, reaction (3.58) played a key role in the generation of pyrimidine. Nitrogen content in liquid products decreased, whereas that in gaseous phase increases as reaction time prolonged. The nitrogen predicted by the model in gaseous phase combined with experimental nitrogen in liquid products gave an accurate nitrogen balance, which was in the range of 81.80–100.96%.

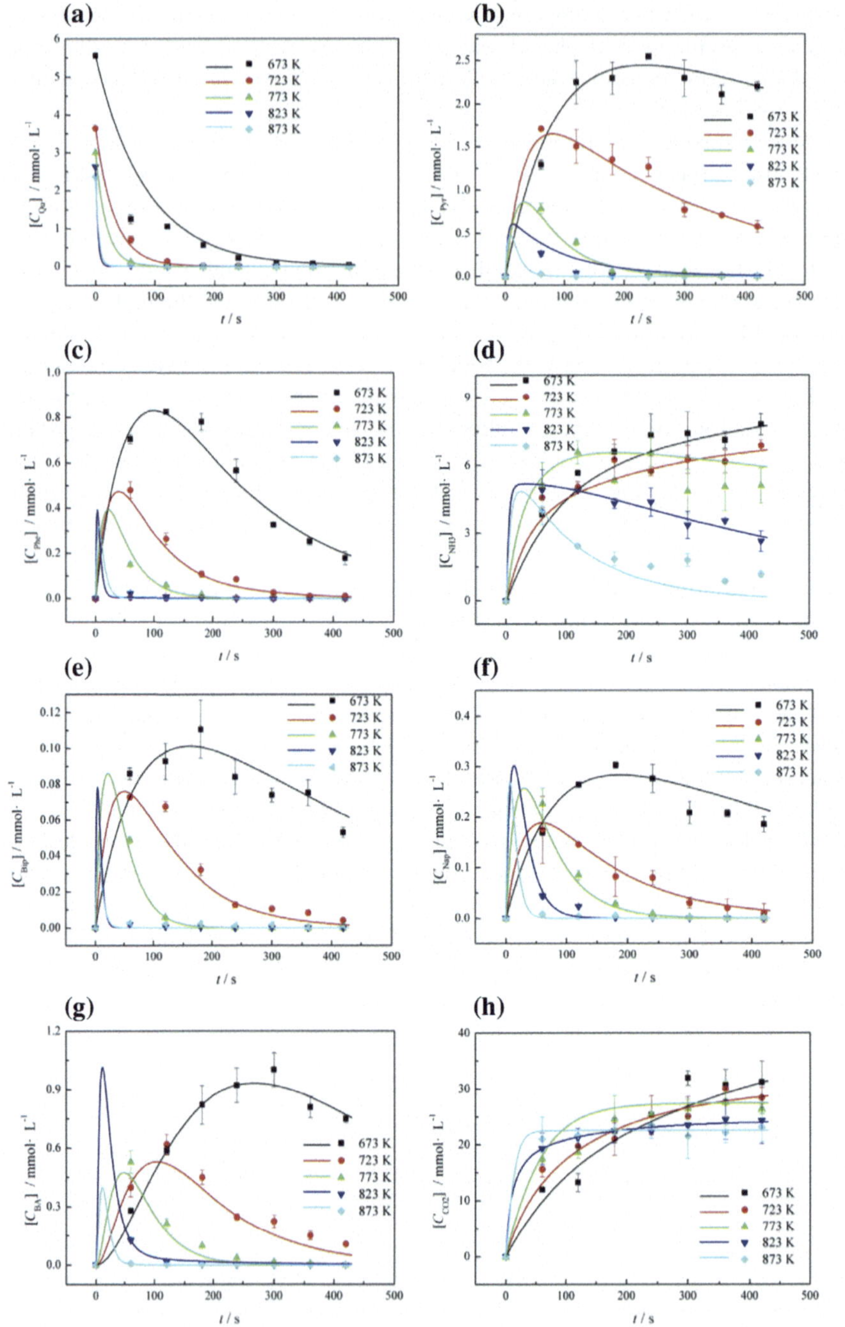

Fig. 3.27 Comparison between experimental and model calculation results (**a** [C_{Qu}], **b** [C_{Pyr}], **c** [C_{Phe}], **d** [C_{NH3}], **e** [C_{Bip}], **f** [C_{Nap}], **g** [C_{BA}], **h** [C_{CO2}]) at 673, 723, 773, 823 and 873 K. (scattered points: experiment data; solid lines: model predictions)

References

1. M. Goto, T. Nada, S. Kawajiri, A. Kodama, T. Hirose, Decomposition of municipal sludge by supercritical water oxidation. J Chem Eng Japan, 1997, 30: 813–818.
2. M. Goto, T. Nada, A. Kodama, T. Hirose, Kinetic Analysis for Destruction of Municipal Sewage Sludge and Alcohol Distillery Wastewater by Supercritical Water Oxidation. Ind Eng Chem Res, 1999, 38: 1863–1865.
3. K. M. Benjamin, P. E. Savage, Hydrothermal reactions of methylamine. J Supercrit Fluids, 2004, 31: 301–311.
4. O. M. Ogunsola, Decomposition of isoquinoline and quinoline by supercritical water. J Hazard Mater, 2000, 74: 187–195.
5. P. A. Webley, J. W. Tester, H. R. Holgate, Oxidation-kinetics of Ammonla and Ammonla-methanol Mixtures in Supercritical WAater in the Temperature-range 530-Degrees-C 700-Degrees-C at 246 BAR. Ind. Eng. Chem. Res., 1991, 30: 1745–1754.
6. N. Segond, Y. Matsumura, K. Yamamoto, Determination of ammonia oxidation rate in sub- and supercritical water. Ind. Eng. Chem. Res., 2002, 41: 6020–6027.
7. H.-C. Lee, J.-H. In, K.-Y. Hwang, C.-H. Lee, Decomposition of ethylenediaminetetraacetic acid by supercritical water oxidation. Ind Eng Chem Res, 2004, 43: 3223–3227.
8. X. Du, R. Zhang, Z. Gan, J. Bi, Treatment of high strength coking wastewater by supercritical water oxidation. Fuel, 2013, 104: 77–82.
9. F. M. Jin, T. Moriya, H. Enomoto. Oxidation reaction of high molecular weight carboxylic acids in supercritical water. Environmental Science & Technology, 2003, 37 (14): 3220–3231.
10. K. Okuda, M. Umetsu, S. Takami, T. Adschiri, Disassembly of lignin and chemical recovery - rapid depolymerization of lignin without char formation in water-phenol mixtures. Fuel Process Technol, 2004, 85: 803–813.
11. Wahyudiono, M. Sasaki, M. Goto, Recovery of phenolic compounds through the decomposition of lignin in near and supercritical water. Chem Eng Process, 2008, 47: 1609–1619.
12. A. Sinag, A. Kruse, V. Schwarzkopf, Key compounds of the hydropyrolysis of glucose in supercritical water in the presence of K2CO3. Ind Eng Chem Res, 2003, 42: 3516–3521.
13. A. Kruse, P. Maniam, F. Spieler, Influence of Proteins on the Hydrothermal Gasification and Liquefaction of Biomass. 2. Model Compounds. Ind Eng Chem Res, 2006, 46: 87–96.
14. J. R. Heberle, C. F. Edwards, Coal energy conversion with carbon sequestration via combustion in supercritical saline aquifer water. International Journal of Greenhouse Gas Control, 2009, 3: 568–576.
15. N. Akiya, P. E. Savage, Roles of Water for Chemical Reactions in High-Temperature Water. Chemical Reviews, 2002, 102: 2725–2750.
16. Y. Guo, S. Z. Wang, D. H. Xu, Y. M. Gong, H. H. Ma, X. Y. Tang, Review of catalytic supercritical water gasification for hydrogen production from biomass. Renewable and Sustainable Energy Reviews, 2010, 14: 334–343.
17. T. Oe, H. Suzugaki, I. Naruse, A. T. Quitain, H. Daimon, K. Fujie, Role of Methanol in Supercritical Water Oxidation of Ammonia. Industrial & Engineering Chemistry Research, 2007, 46: 3566–3573.
18. M. W. Haenel, Recent progress in coal structure research. Fuel, 1992, 71: 1211–1223.
19. P. Dagaut, M. Cathonnet, J.-C. Boettner, Chemical kinetic modeling of the supercritical-water oxidation of methanol. The Journal of Supercritical Fluids, 1996, 9: 33–42.
20. P. E. Savage, J. Yu, N. Stylski, E. E. Brock, Kinetics and mechanism of methane oxidation in supercritical water. The Journal of Supercritical Fluids, 1998, 12: 141–153.
21. P. Kjeldsen, M. A. Barlaz, A. P. Rooker, A. Baun, A. Ledin, T. H. Christensen, Present and long-term composition of MSW landfill leachate: A review. Critical Reviews in Environmental Science and Technology, 2002, 32: 297–336.

22. J. Wiszniowski, D. Robert, J. Surmacz-Gorska, K. Miksch, J. V. Weber, Landfill leachate treatment methods: A review. Environmental Chemistry Letters, 2006, 4: 51–61.
23. Y. M. Gong, S. Z. Wang, X. Y. Tang, D. H. Xu, H. H. Ma, Supercritical water oxidation of acrylic acid production wastewater. Environ Technol, 2014, 35: 907–916.
24. Y. W. Lee, B. Veriansyah, T. J. Park, J. S. Lim, Supercritical water oxidation of wastewater from LCD manufacturing process: kinetic and formation of chromium oxide nanoparticles. Journal of Supercritical Fluids, 2005, 34: 51–61.
25. H. Erkonak, O. O. Sogut, M. Akgun, Treatment of olive mill wastewater by supercritical water oxidation. J. Supercrit. Fluid., 2008, 46: 142–148.
26. F. Vogel, J. L. D. Blanchard, P. A. Marrone, S. F. Rice, P. A. Webley, W. A. Peters, K. A. Smith, J. W. Tester, Critical review of kinetic data for the oxidation of methanol in supercritical water. J. Supercrit. Fluid., 2005, 34: 249–286.
27. W. J. Gong, X. J. Duan, Degradation of landfill leachate using transpiring-wall supercritical water oxidation (SCWO) reactor. Waste Management, 2010, 30: 2103–2107.
28. M. D. Bermejo, M. J. Cocero, Supercritical water oxidation: A technical review. Aiche J., 2006, 52: 3933–3951.
29. Z. B. Wu, R. B. Jin, Y. Liu, H. Q. Wang, Ceria modified MnO_x/TiO_2 as a superior catalyst for NO reduction with NH_3 at low-temperature. Catalysis Communications, 2008, 9: 2217–2220.
30. M. Machida, M. Uto, D. Kurogi, T. Kijima, MnO_x-CeO_2 binary oxides for catalytic NO_x sorption at low temperatures. Sorptive removal of NO_x. Chemistry of Materials, 2000, 12: 3158–3164.
31. P. E. Savage, J. B. Dunn, J. L. Yu, Recent advances in catalytic oxidation in supercritical water. Combust. Sci. Technol., 2006, 178: 443–465.
32. Z. Y. Ding, L. X. Li, D. Wade, E. F. Gloyna, Supercritical water oxidation of NH_3 over a MnO_2/CeO_2 catalyst. Ind. Eng. Chem. Res., 1998, 37: 1707–1716.
33. X. Q. Dong, Y. N. Zhang, Y. W. Xu, M. H. Zhang, Catalytic mechanism study on manganese oxide in the catalytic supercritical water oxidation of nitrobenzene. Rsc Advances, 2015, 5: 47488–47497.
34. E. Shoko, B. McLellan, A. L. Dicks, Hydrogen from coal: Production and utilisation technologies. Int. J. Coal Geol., 2006, 65: 213.
35. D. Feng, Z. Yu, Y. Chen, Y. Qian, Novel Single Stripper with Side-Draw to Remove Ammonia and Sour Gas Simultaneously for Coal-Gasification Wastewater Treatment and the Industrial Implementation. Ind Eng Chem Res, 2009, 48: 5816.
36. H. Erkonak, O. O. Sogut, M. Akgun, Treatment of olive mill wastewater by supercritical water oxidation. J. Supercrit. Fluid., 2008, 46: 142.
37. R. Sridhar, V. Sivakumar, V. P. Immanuel, J. P. Maran, Development of model for treatment of pulp and paper industry bleaching effluent using response surface methodology. Environmental Progress & Sustainable Energy, 2012, 31: 558–565.
38. M. D. Bermejo, F. Cantero, M. J. Cocero, Supercritical water oxidation of feeds with high ammonia concentrations: Pilot plant experimental results and modeling. Chem Eng J, 2008, 137: 542–549.
39. M. Goto, D. Shiramizu, A. Kodama, T. Hirose, Kinetic analysis for ammonia decomposition in supercritical water oxidation of sewage sludge. Ind Eng Chem Res, 1999, 38: 4500.
40. T. Mizuno, M. Goto, A. Kodama, T. Hirose, Supercritical water oxidation of a model municipal solidwaste. Ind Eng Chem Res, 2000, 39: 2807.
41. N. Segond, Y. Matsumura, K. Yamamoto, Determination of ammonia oxidation rate in sub- and supercritical water. Ind Eng Chem Res, 2002, 41: 6020.
42. K. M. Benjamin, P. E. Savage, Supercritical water oxidation of methylamine. Ind Eng Chem Res, 2005, 44: 5318.
43. M. Krajnc, J. Levec, Oxidation of phenol over a transition-metal oxide catalyst in supercritical water. Ind Eng Chem Res, 1997, 36: 3439.
44. Z. Y. Ding, L. Li, D. Wade, E. F. Gloyna, Supercritical water oxidation of NH3 over a MnO2/CeO2 catalyst. Ind Eng Chem Res, 1998, 37: 1707.

45. M. J. Angeles-Hernandez, G. A. Leeke, R. C. D. Santos, Catalytic supercritical water oxidation for the destruction of quinoline over MnO2/CuO mixed catalyst. Ind Eng Chem Res, 2009, 48: 1208.
46. Z. Y. Ding, S. N. V. K. Aki, M. A. Abraham, K. W. Hutchenson, N. R. Foster, Innovations in Supercritical Fluids Science and Technology. Vol. 608. 1995. 232.
47. R. Hayashi, K. Ohkuma, K. Tonokura, Y. Oshima, Contribution of ionic reactions to sub- and supercritical water oxidation of phenol. J. Chem. Eng. Jpn., 2007, 40: 556.
48. C. J. Martino, P. E. Savage, Supercritical water oxidation kinetics and pathways for ethylphenols, hydroxyacetophenones, and other monosubstituted phenols. Ind Eng Chem Res, 1999, 38: 1775.
49. J. L. Yu, P. E. Savage, Catalytic oxidation of phenol over MnO2 in supercritical water. Ind Eng Chem Res, 1999, 38: 3793.
50. W. J. Gong, L. Fang, D. L. Xi, Supercritical water oxidation CI disperse red 60 dyeing wastewater using transpiring water reactor. Prog Env Sci Tec, 2007, 1: 836.
51. E. Kipcak, M. Akgün, In situ gas fuel production during the treatment of textile wastewater at supercritical conditions. Water Sci Technol, 2013, 67: 1058.
52. O. Söğüt, M. Akgün, Treatment of textile wastewater by SCWO in a tube reactor. J Supercrit Fluid, 2007, 43: 106.
53. O. Söğüt, M. Akgün, Treatment of dyehouse waste-water by supercritical water oxidation: a case study. J Chem Technol Biot, 2010, 85: 640.
54. F. H. Hussein, Effect of Photocatalytic Treatments on Physical and Biological Properties of Textile Dyeing Wastewater. ASIAN JOURNAL OF CHEMISTRY, 2013, 25: 9387–9392.
55. R. M. Jain, K. H. Mody, J. Keshri, B. Jha, Biological neutralization and biosorption of dyes of alkaline textile industry wastewater. Marine Pollution Bulletin, 2014, 84: 83–89.
56. M. Goto, T. Nada, A. Kodama, T. Hirose, Kinetic analysis for destruction of municipal sewage sludge and alcohol distillery wastewater by supercritical water oxdation. Ind. Eng. Chem. Res., 1999, 38: 1863–1865.
57. James W. Griffith, D. H. Raymond, The first commercial supercritical water oxidation sludge processing plant. Waste Management, 2002, 22: 453–459.
58. Jing GL, Qin SP, Cui BC, Li M, Xing LJ, L. SL., Oxidation of Oilfield Sludge in Supercritical Water. Research Journal of Chemistry and Environment, 2009, 13: 18–22.
59. B.-c. Cui, S.-z. Liu, F.-y. Cui, G.-l. Jing, X.-j. Liu, Lumped kinetics for supercritical water oxidation of oily sludge. Process. Saf. Environ., 2011, 89: 198–203.
60. J. Zhang, S. Wang, Y. Li, J. Lu, S. Chen, X. Luo, Supercritical water oxidation treatment of textile sludge. Environ Technol, 2017, 38: 1949–1960.
61. N. Boukis, V. Diem, W. Habicht, E. Dinjus, Methanol reforming in supercritical water. Ind. Eng. Chem. Res., 2003, 42: 728–735.
62. P. T. Williams, J. Onwudili, Composition of Products from the Supercritical Water Gasification of Glucose: A Model Biomass Compound. Industrial & Engineering Chemistry Research, 2005, 44: 8739–8749.
63. P. T. Williams, J. Onwudili, Subcritical and Supercritical Water Gasification of Cellulose, Starch, Glucose, and Biomass Waste. Energy & Fuels, 2006, 20: 1259–1265.
64. R. Muangrat, J. A. Onwudili, P. T. Williams, Reaction products from the subcritical water gasification of food wastes and glucose with NaOH and H2O2. Bioresource. Technol., 2010, 101: 6812–6821.
65. Y. Guo, S. Wang, Y. Wang, J. Zhang, D. Xu, Y. Gong, Gasification of two and three-components mixture in supercritical water: Influence of NaOH and initial reactants of acetic acid and phenol. Int J Hydrogen Energy, 2012, 37: 2278–2286.
66. W.-J. Gong, F. Li, D.-L. Xi, Supercritical Water Oxidation of Acrylic Acid Production Wastewater in Transpiring Wall Reactor. Environ. Eng. Sci., 2009, 26: 131–136.
67. V. M. Bednarik V, Removal of formaldehyde from acrylic acid production wastewater. Environ Eng Sci., 2003, 20: 703–707.

68. B. Veriansyah, T. J. Park, J. S. Lim, Y. W. Lee, Supercritical water oxidation of wastewater from LCD manufacturing process: kinetic and formation of chromium oxide nanoparticles. J. Supercrit. Fluid., 2005, 34: 51–61.
69. W. Gong, X. Duan, Degradation of landfill leachate using transpiring-wall supercritical water oxidation (SCWO) reactor. Waste Manage., 2010, 30: 2103–2107.
70. P. C. Mandal, Wahyudiono, M. Sasaki, M. Goto, Reduction of total acid number (TAN) of naphthenic acid (NA) using supercritical water for reducing corrosion problems of oil refineries. Fuel, 2012, 94: 620–623.
71. S. Wang, Y. Guo, L. Wang, Y. Wang, D. Xu, H. Ma, Supercritical water oxidation of coal: Investigation of operating parameters' effects, reaction kinetics and mechanism. Fuel Process. Technol, 2011, 92: 291–297.
72. S. Koda, N. Kanno, H. Fujiwara, Kinetics of supercritical water oxidation of methanol studied in a CSTR by means of Raman spectroscopy. Ind. Eng. Chem. Res., 2001, 40: 3861–3868.
73. T. B. H. Steven F. Rice, Åsa C. Rydén, and Russell G. Hanush, Raman spectroscopic measurement of oxidation in supercritical water. 1. Conversion of methanol to formaldehyde. Industrial & engineering chemistry research 1996, 35: 2161–2171.
74. H. S. S. Baura, A.Krämerb, J. Gerberb, The destruction of industrial aqueous waste containing biocides in supercritical water—development of the SUWOX process for the technical application[The Journal of supercritical fluids, 2005, 33.
75. D. Xu, S. Wang, J. Zhang, X. Tang, Y. Guo, C. Huang, Supercritical water oxidation of a pesticide wastewater. Chem. Eng. Res. Des., 2015, 94: 396–406.
76. K. P., Corrosion in high-temperature and supercritical water and aqueous solutions: a review. The Journal of Supercritical Fluids, 2004, 29: 1–29.
77. M. B. Garcia Jarana, J. Sanchez-Oneto, J. R. Portela, E. Nebot Sanz, E. J. Martinez de la Ossa, Supercritical water gasification of industrial organic wastes. J. Supercrit. Fluid., 2008, 46: 329–334.
78. M. C. Zhang J, Reaction mechanism and influence factors of ethylenediamine wastewater treated by supercritical water oxidation. 5th International Conference on Bioinformatics and Biomedical Engineering, 2011.
79. V. I. Anikeev, A. Yermakova, Technique for complete oxidation of organic compounds in supercritical water. Russ. J. Appl. Chem + . 2011, 84: 88–94.
80. O. Ö. Sögüt, Ekin, K., Akgün, M.,, Treatment of whey wastewater by supercritical water oxidation. Water Sci. Technol., 2011a, 63: 908–916.
81. S. Wang, Y. Guo, C. Chen, J. Zhang, Y. Gong, Y. Wang, Supercritical water oxidation of landfill leachate. Waste Manage., 2011, 31: 2027–2035.
82. E. D. A. Kruse, Hot compressed water as reaction medium and reactant: properties and synthesis reactions. J. Supercrit. Fluids, 2007, 39: 362–380.
83. T.-N. Wang, L. Lu, G.-F. Li, J. Li, T.-F. Xu, M. Zhao, Decolorization of the azo dye reactive black 5 using laccase mediator system. African Journal of Biotechnology, 2011, 10: 17186–17191.
84. O. O. Sogut, M. Akgun, Treatment of dyehouse waste-water by supercritical water oxidation: a case study. J. Chem. Technol. Biot., 2010, 85: 640–647.
85. J. Zhang, S. Wang, Y. Guo, D. Xu, Y. Gong, X. Tang, Experimental study on supercritical water oxidation of CI Reactive Orange 7 dye wastewater using response surface methodology. Color. Technol., 2012, 128: 323–330.
86. O. Y. J. Li H, Elementary reaction mechanism of methylamine oxidation in supercritical water. Ind. Eng. Chem. Res., 2005, 44: 8756–8764.
87. M. D. Bermejo, F. Cantero, M. J. Cocero, Supercritical water oxidation of feeds with high ammonia concentrations Pilot plant experimental results and modeling. Chem. Eng. J., 2008, 137: 542–549.
88. a. P. E. S. Kenneth M. Benjamin, Supercritical Water Oxidation of Methylamine. Ind. Eng. Chem. Res., 2005, 44: 5318–5324.

89. W.-L. Chou, C.-T. Wang, K.-Y. Huang, Investigation of process parameters for the removal of polyvinyl alcohol from aqueous solution by iron electrocoagulation. Desalination., 2010, 251: 12–19.
90. Y. H. Q. Zhang S J, Ge X W, Zhu R F, Optimization of radiolytic degradation of poly(vinyl alcohol). Engineering Chemistry Research, 2005, 44: 1995–2001.
91. W.-L. Chou, C.-T. Wang, C.-W. Hsu, K.-Y. HuangA, T.-C. Liu, Removal of total organic carbon from aqueous solution containing polyvinyl alcohol by electrocoagulation technology. Desalination., 2010, 259: 103–110.
92. O. M. Hayashi R, Sugiyama M, Koda S, Oshima Y, Kinetic analysis on alcohol concentration and mixture effect in supercritical water oxidation of methanol and ethanol by elementary reaction model. The Journal of Supercritical Fluids, 2007, 40: 74–83.
93. L. L. T. Hsu L J, Lin C C, Adsorption and photocatalytic degradation of polyvinyl alcohol in aqueous solutions using P-25 TiO2. Chem. Eng. J., 2011, 173: 698–705.
94. J. Zhang, S. Wang, Y. Guo, D. Xu, Y. Gong, X. Tang, Supercritical water oxidation of polyvinyl alcohol and desizing wastewater: Influence of NaOH on the organic decomposition. Journal of Environmental Sciences-China, 2013, 25: 1583–1591.
95. R. Hayashi, M. Onishi, M. Sugiyama, S. Koda, Y. Oshima, Kinetic analysis on alcohol concentration and mixture effect in supercritical water oxidation of methanol and ethanol by elementary reaction model. J. Supercrit. Fluid., 2007, 40: 74–83.
96. B. P. A. Ploeger J M, Lachance R P, Tester J W, Co-oxidation of methylphosphonic acid and ethanol in supercritical water ¢: Experimental results. The Journal of Supercritical Fluids, 2006, 39: 233–238.
97. A. Loppinet-Serani, C. Aymonier, F. Cansell, Current and Foreseeable Applications of Supercritical Water for Energy and the Environment. Chemsuschem, 2008, 1: 486–503.
98. C. E. Rice S F, Oxidation of simple alcohols in supercritical water III. Formation of intermediates from ethanol. In: Ind. Eng. Chem. Res., 2000, American Chemical Society.: 86–93.
99. W. P. T. Onwudili J A, Reaction mechanisms for the decomposition of phenanthrene and naphthalene under hydrothermal conditions. The Journal of Supercritical Fluids, 2007, 39: 399–408.
100. J. Zhang, S. Wang, Y. Guo, D. Xu, X. Li, X. Tang, Co-Oxidation Effects of Methanol on Acetic Acid and Phenol in Supercritical Water. Ind. Eng. Chem. Res., 2013, 52: 10609–10618.
101. R. T. T. Skodje, A. S.; Klippenstein, S. J.; Harding, L. B.; Davis, M. J., Theoretical validation of chemical kinetic mechanisms: Combustion of methanol. J. Phys. Chem. A, 2010, 114.
102. S. H. Rice, T.; Ryden, A.; Hanush, R., Raman spectroscopic measurement of oxidation in supercritical water. 1. Conversion of methanol to formaldehyde. Ind. Eng. Chem. Res, 1996, 35: 2161–2171.
103. J. L. H. DiNaro, J. B.; Green, W. H.; Tester, J. W.; Bozzelli, J. W., Elementary reaction mechanism for benzene oxidatio in supercritical water. J. Phys. Chem. A, 2000, 104: 10576–10586.
104. S. Gopalan, P. E. Savage, Reaction mechanism for phenol oxidation in supercritical water. J. Phys. Chem., 1994, 98: 12646–12652.
105. D. Cao, X. Wang, L. Lei, L. Ma, F. Wang, C. Wang, M. Tang, W. Xiang, T. Wang, H. Li, L. Chen, Synthesis, in vitro and in vivo evaluation of novel substituted N-(4-(2-(4-benzylpiperazin-1-yl)ethoxy)phenyl)-N-methylquinazolin-4-amin es as potent antitumor agents. Bioorg. Med. Chem. Lett, 2016, 26: 1931–1935.
106. I. Khan, A. Ibrar, N. Abbas, A. Saeed, Recent advances in the structural library of functionalized quinazoline and quinazolinone scaffolds: Synthetic approaches and multifarious applications. Eur. J. Med. Chem., 2014, 76: 193–244.
107. E. Kaiser, C. Prasse, M. Wagner, K. Broeder, T. A. Ternes, Transformation of Oxcarbazepine and Human Metabolites of Carbamazepine and Oxcarbazepine in Wastewater Treatment and Sand Filters. Environ. Sci. Technol., 2014, 48: 10208–10216.

108. H. R. A. Tina Kosjek, Boris Kompare, Anna Ledin and Ester Heath, Fate of Carbamazepine during Water Treatment. Environ. Sci. Technol., 2009, 43: 6256–6261.
109. L. D. S. Pinto, L. M. F. dos Santos, R. C. D. Santos, B. Al-Duri, Supercritical water oxidation of quinoline in a continuous plug flow reactor - part 2: kinetics. J. Chem. Technol. Biot., 2006, 81: 919–926.
110. S. Aki, M. A. Abraham, Catalytic supercritical water oxidation of pyridine: kinetics and mass transfer. Chem. Eng. Sci., 1999, 54: 3533–3542.
111. N. Crain, S. Tebbal, L. X. Li, E. F. Gloyna, KINETICS AND REACTION PATHWAYS OF PYRIDINE OXIDATION IN SUPERCRITICAL WATER. Ind. Eng. Chem. Res., 1993, 32: 2259–2268.
112. S. Aki, M.A. Abraham. Catalytic supercritical water oxidation of pyridine: comparison of catalysts. Industrial & Engineering Chemistry Research, 1999a, 38 (2), 358–367.
113. S. Aki, M.A. Abraham. Catalytic supercritical water oxidation of pyridine: kinetics and mass transfer. Chemical Engineering Science, 1999b, 54 (15-16), 3533–3542.
114. M. Gong, W. Zhu, Y. Fan, H. Zhang, Y. Su, Influence of the reactant carbon-hydrogen-oxygen composition on the key products of the direct gasification of dewatered sewage sludge in supercritical water. Bioresource. Technol., 2016, 208: 81–86.
115. G. Anitescu, L. L. Tavlarides, Oxidation of biphenyl in supercritical water: Reaction kinetics, key pathways, and main products. Ind. Eng. Chem. Res., 2005, 44: 1226–1232.

Chapter 4
Hydrothermal Combustion of Coal in Supercritical Water

Today's commercial operation of power systems, such as Supercritical Generating Unit and IGCC, presents some congenital problems [1]. First, both the investment and operation cost of the desulfurization and denitration equipment are very expensive. Second, the carbon capture is not only inefficient, but also leads to the power generation efficiency reduce exceeding 8%. Third, because only radiation heat transfer works, it is very difficult to improve the parameters of the steam in heating surface. Fourth, the dust contained in the flue gas needs to be removed, while it is hard to realize the zero produces and zero emissions for PM2.5. As the reasons mentioned above, the power generation efficiency could not exceed 44%, and it will greatly decrease if the carbon capture equipment is installed [2].

Therefore, seeking a revolutionary breakthrough combustion technology is the goal of the scientists. The new technology should produce zero pollutants, capture carbon dioxide cost-effectively, break through the heat transfer limitations of only radiation, and improve the power generation efficiency. Is there such a combustion technology nowadays? Supercritical water hydrothermal combustion of coal is a novel efficient and clean fuel applied technology with many advantages. First, a coal particle with the diameter of 90 μm can burn out in 1.8 s. Second, not only radiation but also heat conduction and convection work. Third, there is no sulfur oxide, nitric oxide, carbon dioxide, or PM2.5 emission. Why the combustion technology has so many advantages over others? This is closely connected with the special properties of supercritical water.

4.1 Characteristics and Difficulties of SWHC

Using the special thermophysical properties of SCW, scientists put the fuel and oxygen into SCW to form a homogeneous reaction medium and explore SCWO technology. It typically implies temperatures and pressures varying between 400–650 °C and 25–35 MPa, respectively. In a few minutes, organics are oxidized to

S. Wang et al., *Supercritical Water Processing Technologies for Environment, Energy and Nanomaterial Applications*,
https://doi.org/10.1007/978-981-13-9326-6_4

lower molecular compounds, ultimately to carbon dioxide and water. Heteroatoms, such as chlorine, sulfur, and phosphorus, are converted into their corresponding acids (e.g., HCl, H_2SO_4, and H_3PO_5). The presence of cations results in the formation of inorganic salts or oxides.

However, SCWO has encountered a number of difficult problems: easily precipitate carbon [3]; the reaction time is long and results in poor economy [4, 5]; solid fuel is difficult to completely react [6]. Can we figure out these problems from the SCW technology itself? Supercritical water hydrothermal combustion (SWHC) is the answer. Two conditions are required for SWHC: a high fuel and oxidant concentrations, and the fuel and oxidant streams preheated to a high enough temperature. Then auto-ignition occurs and forms flames compatibility between fire and water. As the combustion proceeds, hydrothermal flame sustainedly spreads in the reactor [7]. Compared with the flameless of SCWO, SWHC shows significant technological advantages, which can be used to solve the obstacles confronting of SCWO [8] A low preheated temperature is enough for the fuel stream to catch fire, and thus carbon deposition and coking will not take place [9]. Only 10–100 ms is needed for complete combustion of the fuel [10, 11]. The hydrothermal flame temperature is excess of 1000 °C, which is conducive to recover heat [12, 13].

When coal is applied as the fuel, SCWO and SWHC of coal are explored one after another. Generally speaking, SCWO is a relatively slow oxidation process without flame, while SWHC is characterized by an extremely fast combustion process with hydrothermal flame formed. SCWO of coal is characterized as the little temperature difference between the coal partial and the SCW fluid. However, for SWHC, the temperature difference usually exceeds 500 K. In fact, if they cannot observe the hydrothermal flames, the scientists often determine that SWHC of coal occurs according to the temperature difference between the coal particle and SCW.

So far, only SCWO of coal has been investigated. Fujie et al. [14] studied the carbon-, nitrogen-, and sulfur-distribution in products during SCWO of coal. The research results proved the well environmental performance of this technology, which provides high reference value for the research on the reaction pathway of elements. Wang and Zhu [15, 16] investigated the reaction kinetics of SCWO of coal applying in the first-order homogeneous model ($T < 420$ °C). Vostrikov et al. [17–19] investigated kinetics of the organic matter of coal. Wang et al. [20] adopted chemical oxygen demand of water coal slurry as the research object to investigate the reaction kinetics. Moreover, the power generation system based on SCWO was explored by Bermejo et al. [21] and Donatini et al. [22], respectively.

All the studies above offer the great reference value for the research on SWHC of coal. However, compared with SCWO, there are some special problems for SWHC of coal required to be solved. (1) Ignition characteristics of coal in SCW. This is the key fundamental problem. (2) Burning rate of coal in SCW. The SWHC of coal is a complex heterogeneous reaction among coal, SCW, and oxygen, which coupling the heat transfer and mass transfer process. Therefore, the reaction rate predicted by homogeneous model cannot describe the real reaction process. (3) Cleanliness of SWHC of coal. The carbon, sulfur, and nitrogen in coal are in the form of inorganic carbon and organic carbon, thioalcohol, thioether and pyrite, pyridine, pyrrole, and amino. If considering coal as the research object only, we are not sure whether SWHC of coal is always clean when another coal is applied as the fuel. Therefore, different forms of carbon, sulfur, and nitrogen need to be investigated. (4) Could the high power generation efficiency, desulfurization, denitrification, and carbon emissions reductions be integrated for a power generation system based on the SWHC?

4.2 Ignition Characteristics

During the auto-ignition of coal, on the one hand, the coal temperature is elevated because of the heterogeneous oxidation reaction [23]; on the other hand, the heat transfers from the coal particle to the surrounding SCW [24]. The particle size decreases, and the edge of the initial particle is removed, keeping the shape, density, and composition nearly unchanged. The auto-ignition of the coal particle takes place when the coal temperature suddenly increases rapidly. The main factors affecting the coal auto-ignition are the SCW temperature, oxygen concentration, and particle size, etc. The effect of SCW temperature on the ignition characteristics of Coal is shown in Fig. 4.1 (taking Chinese Shenmu coal as the case). When SCW temperature is 797 K, the coal temperature is always equal to that of SCW. When the value is 798 K, the coal temperature suddenly rapidly increases to much higher than that of SCW. The auto-ignition induction time is 2700 ms. The results indicate that 798 K is the critical temperature for the coal ignition. The higher the SCW temperature is, the easier the coal ignition will be. By the similar method, the critical value of oxygen concentration and coal particle size is 7.97 wt% and 168 μm, respectively. Increasing the oxygen concentration or decreasing the particle size enhances the auto-ignition of coal. The induction time for auto-ignition of coal is approximate 100 ms at the conditions of SCW temperature of 798 K, oxygen concentration of 10.0 wt%, particle size of 100 μm.

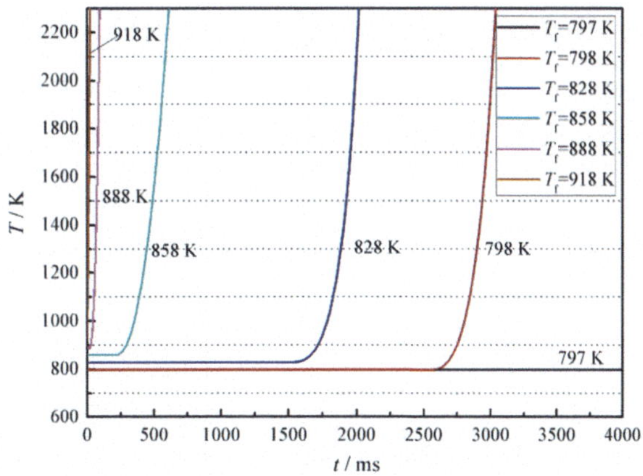

Fig. 4.1 Effect of SCW temperature on the auto-ignition of coal (case in Shenmu coal)

4.3 Combustion Rate of SWHC of Coal

Because of the oxidation reaction on the coal surface, the coal temperature is higher than SCW. It is assumed that the temperature is homogeneous in the particle and that the different between the coal and SCW is hundreds of K according to the film theory estimation. Figure 4.2 shows the influence of SCW temperature and oxygen concentration on the combustion rate. If the limiting step is the surface reaction, the combustion rate will generally increase with temperature with Arrhenius-type dependence. Such temperature dependence is observed when SCW temperature is below 753 K, suggesting that the combustion rate is limited by the surface reaction. On the other hand, if the limiting step is the mass transfer of oxygen, the rate will be nearly independent of temperature, and a high oxygen concentration will enhance the mass transfer and thus the combustion rate. Therefore, the mass transfer limits the combustion rate at SCW temperature above 793 K. Moreover, the surface reaction and the mass transfer play a comparable role for the combustion rate at the SCW temperature range of 753–793 K.

Thus, integrating all the kinetics information discussed above, we can obtain the reaction rate at different temperature. Furthermore, the combustion rate observed in the experiment and calculated by the kinetics information is compared, and it shows a good agreement to the calculated values and experimental data. Therefore, the kinetics information obtained is quite satisfactory. The coal particle with different size is chosen to compare with the traditional combustion method in the air environment. As shown in Fig. 4.3, when surface reaction, or surface reaction and mass transfer colimit the combustion rate, the coal particle burns slowly and is difficult to burn out. When mass transfer is the rate-limited step in the SCW of 820 K, the complete combustion time for 45, 90, 120 μm, coal particles is 2, 5, 10 s, respectively. If the SCW temperature is 860 K, the coal with 90 μm requires only 1.8 s to burn out. This combustion rate is comparable to that of the 1200 °C in the air.

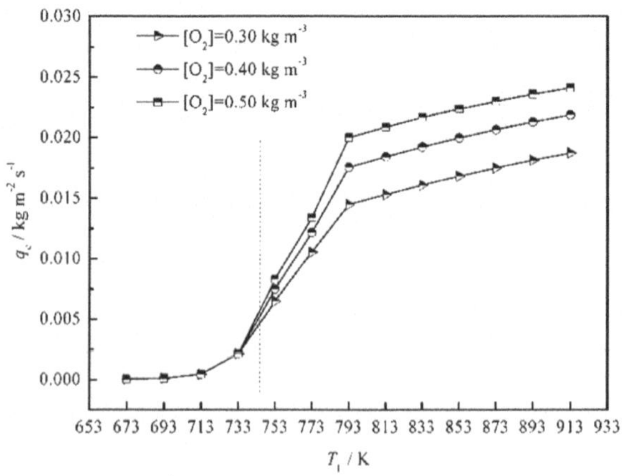

Fig. 4.2 Influence of reaction parameters on the combustion rate of coal

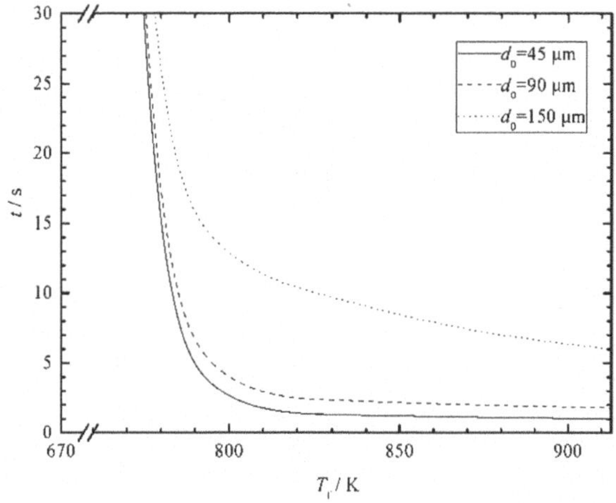

Fig. 4.3 Influence of SCW temperature and size on the burnout characteristics of coal

4.4 Reaction Pathway of Major Elements

Carbon, nitrogen, and sulfur are presented in many forms in coal. For example, carbon, nitrogen, and sulfur are divided into organic carbon and fixed carbon, pyridine, pyrrole and amino nitrogen, thioalcohol, thioether, and pyrite. Therefore, the previous conclusions are limited, which cannot provide adequate theoretical support for the green performance and for controlling pollutant emissions of SWHC of coal. The previous studies show that organic carbon of coal is firstly converted into BTEX like aniline. Furthermore, aniline is also the main intermediate product during SCWO of pyridine or pyrrole. Therefore, aniline is selected to investigate the reaction rule of nitrogen. In addition, the homemade fixed carbon, methanthiol, thiirane, and iron sulfide are chosen as the object to study the pathway of fixed carbon, thioalcohol, thioether, and pyrite, respectively.

The results indicate that the organic carbon is firstly converted into aniline, and then phenol, and then oxidized to organic acids, and finally, all oxidized to carbon dioxide and a small amount of carbon monoxide. Carbon monoxide is a secondary intermediate product on the pathway of carbon dioxide, and no carbon monoxide is formed above 833 K. Both pyridine and pyrrole are primarily degradations of aniline, and then ammonia, and further, oxidation of nitrogen and a trace amount of nitrate. In all cases, no nitric oxide is produced. All ammonia will be oxidized into nitrogen and nitrate above 863 K. Fixed carbon is oxidized into carbon dioxide besides minute quality carbon monoxide, which will not be formed above 82 K. Likewise, the reaction pathway of methanthiol, thiirane, and iron sulfide can be described as: methanthiol → sulfion → sulfite → sulfate, thiirane → thiosulfate → sulfite → sulfate, iron sulfide → sulfion → sulfite → sulfate, respectively. The formation of sulfite is the rate-limited step. Moreover, the heavy metal in the form of carbonates or iron and manganese oxides, such as copper and mercury, is very unstable and will cause serious heavy metal pollution. However, they will be transformed into stable and harmless residual after SWHC.

4.5 Power Generating Efficiency of SWHC of Coal

According to the research findings above, a power generating system is theoretically built based on SWHC of Shenmu coal. This system reasonably couples the steam power cycle, gas expansion work circulation cycle, and hydroelectric power generation technology, achieves the cascade utilization of resources, and realizes the desulfurization, denitrification, and dusting. The combustion product carbon dioxide passes through the expansion work, the gas–liquid separation and the liquid oxygen cooling processes in sequence, and all are trapped in the liquid state with zero emission of carbon dioxide. The net power generation efficiency of this system can be promoted to 48.2% through the optimization of the operating parameters.

Currently, the net power generation efficiency of ultra-supercritical unit with the operating parameters of 31 MPa/593 °C/593 °C/593 °C is up to 44%, and this value is 42–44% for the power generation efficiency of the latest large-scale IGCC. However, if the carbon capture system installed, the power generation efficiency of the ultra-supercritical unit and IGCC will decrease by 7–9% and 4–6%, respectively. Therefore, the power generation efficiency of SWHC system is 10–13% higher than that of ultra-supercritical unit and IGCC. Why does the SWHC of coal have such a high power generation efficiency? We can attribute it to three reasons. (1) Not only the hydrothermal flame heats the working fluid through the radiation, but also the hydrothermal flame has a strong ability of conduction and convection because of the specific heat of SCW far greater than the fuel gas. Therefore, SWHC is much more efficient than that of the conventional pulverized coal combustion. (2) After combustion, the fluid can be cooled to the room temperature, thus avoiding the heat loss of the fuel gas. (3) The combustion products are only water and carbon dioxide, and only carbon dioxide is presented after condensation. So, it is easy to completely capture carbon dioxide.

4.6 Conclusions

SWHC of coal can produce hydrothermal flames compatible of fire and water, which sustainedly transmits in SCW. The critical conditions for Sehnmu coal auto-ignition are the SCW temperature of 798 K, the oxygen concentration of 7.97 wt%, the volatile matter content of 10.0 wt%, and the particle size of 168 μm. Increasing SCW temperature, oxygen concentration, volatile matter content, or reducing the particle size, all promote the coal ignition. The shortest induction time for coal ignition is 80–100 ms.

The coal particle with 90 μm requires only 1.8 s for burning out when SCW temperature above 860 K. The combustion rate is comparable to that of the 1200 ° C in the air. Carbon dioxide is the exclusive carbon-containing product without carbon monoxide formed above 833 K. Pyridine, pyrrole, and amino nitrogen are all oxidized into nitrogen (95%) and nitrate (5%). All forms of sulfur contained in coal are oxidized into sulfate without any sulfur oxide produced. The metal elements are formed into stable state oxidation without any dust emission. The power generation system based on SWHC of coal can realize integration of desulfurization, denitration, dedusting, and carbon dioxide zero emissions. The power generation efficiency of this system is up to 48.2%, which is 10–13% higher than that of ultra-supercritical unit and IGCC.

References

1. M D Bermejo, M J Cocero, F Fernández-Polanco. A process for generating power from the oxidation of coal in supercritical water. Fuel, 2004, 83(2):195–204.
2. F Marias, F Mancini, F Cansell, et al. Energy recovery in supercritical water oxydation process. Environmental Engineering Science, 2008, 25(1):123–130.
3. D Xu, S Wang, X Tang, et al. Design of the first pilot scale plant of China for supercritical water oxidation of sewage sludge. Chemical Engineering Research & Design, 2012, 90 (2):288–297.
4. B Wellig, K Lieball, P R Rohr. Operating characteristics of a transpiring-wall SCWO reactor with a hydrothermal flame as internal heat source. Journal of Supercritical Fluids, 2005, 34 (1):35–50.
5. B Wellig, M Weber, K Lieball, et al. Hydrothermal methanol diffusion flame as internal heat source in a SCWO reactor. Journal of Supercritical Fluids, 2009, 49(1):59–70.
6. L Han, R Zhang, J Bi. Experimental investigation of high-temperature coal tar upgrading in supercritical water. Fuel Processing Technology, 2009, 90(2):292–300.
7. H Sato, S Endo, M Sugiyama, et al. Baddeleyite-Type High-Pressure Phase of TiO_2. Science, 1991, 251(4995):786–788.
8. R M Serikawa, T Usui, T Nishimura, et al. Hydrothermal flames in supercritical water oxidation: investigation in a pilot scale continuous reactor. Fuel, 2002, 81(9):1147–1159.
9. M D Bermejo, P Cabeza , J P S Queiroz, et al. Analysis of the scale up of a transpiring wall reactor with a hydrothermal flame as a heat source for the supercritical water oxidation. Journal of Supercritical Fluids, 2011, 56(1):21–32.
10. A Sobhy, I S Butler, J A Kozinski. Selected profiles of high-pressure methanol–air flames in supercritical water. Proceedings of the Combustion Institute, 2007, 31(2):3369–3376.
11. K Hirosaka, M Fukayama, K Wakamatsu, et al. Combustion of ethanol by hydrothermal oxidation. Proceedings of the Combustion Institute, 2007, 31(2):3361–3367.
12. K Koido, K Hirosaka, T Kubo, et al. Numerical study on premixed hydrothermal combustion in tube reactor. Combustion Theory & Modelling, 2009, 13(2):295–318.
13. P Cabeza, M D Bermejo, C Jiménez, et al. Experimental study of the supercritical water oxidation of recalcitrant compounds under hydrothermal flames using tubular reactors. Water Research, 2011, 45(8):2485–2495.
14. J Abeln, M Kluth, G Petrich, et al. Supercritical water oxidation (SCWO): a process for the treatment of industrial waste effluents. High Pressure Research, 2001, 20(1–6):537-547.
15. T Wang, X Zhu. Sulfur transformations during supercritical water oxidation of a Chinese coal. Fuel, 2003, 82(18):2267–2272.

16. T Wang, X Zhu. Conversion and Kinetics of the Oxidation of Coal in Supercritical Water. Energy & Fuels, 2004, 18(5): 1569–1572.
17. A A Vostrikov, O N Fedyaeva, S A Psarov, et al. Brown coal conversion under the action of supercritical water. Solid Fuel Chemistry, 2007, 41(5):280–289.
18. A A Vostrikov, D Y Dubov, S A Psarov, et al. Oxidation of a coal particle in flow of a supercritical aqueous fluid. Combustion Explosion & Shock Waves, 2008, 44(2):141–149.
19. A A Vostrikov, O N Fedyaeva, D Y Dubov, et al. Conversion of brown coal in supercritical water without and with addition of oxygen at continuous supply of coal–water slurry. Energy, 2011, 36(4):1948–1955.
20. S Wang, Y Guo, L Wang, et al. Supercritical water oxidation of coal: Investigation of operating parameters' effects, reaction kinetics and mechanism. Fuel Processing Technology, 2011, 92(3):291–297.
21. M D Bermejo, M J Cocero, F Fernández-Polanco. A process for generating power from the oxidation of coal in supercritical water. Fuel, 2004, 83(2):195–204.
22. F Donatini, G Gigliucci, J Riccardi, et al. Supercritical water oxidation of coal in power plants with low CO_2 emissions. Energy, 2009, 34(12):2144–2150.
23. Y Yoneyama, M Okamura, K Morinaga, et al. Role of Water in Hydrogenation of Coal without Catalyst Addition. Energy & Fuels, 2001, 16(1): 48-53.
24. S Murata, K Murata, K Kidena, et al. A Novel Oxidative Desulfurization System for Diesel Fuels with Molecular Oxygen in the Presence of Cobalt Catalysts and Aldehydes. Energy & Fuels An American Chemical Society Journal, 2004, 18(1): 116–121.

Chapter 5
Supercritical Hydrothermal Synthesis of Inorganic Nanomaterials

Nanomaterials are always at the leading emerging area of material and nanotechnology. Especially in this age, the ever-increasing population in the world makes energy and resource efficiency more important in sustaining concept [1]. The significant reduction in size (1–100 nm) often improves their electronic and other properties due to quantum confinement effects [2]. In the size range of 1–100 nm, nanomaterials can also acquire highly increased surface area to volume ratios [3]. These tunable and unique properties make nanomaterials applicable in many modern technological applications, such as optics [4, 5], pigments [6, 7], biomedicine [8, 9], antimicrobials [10], and semiconductors [11, 12]. Nanomaterials' technologies begin to enter commercial marketplace because of their unique performance in all aspects of life. More recently, as one of the most promising, readily large-scale, and green inorganic nanomaterial synthesis method, supercritical continuous hydrothermal flow synthesis process attracts more and more attention from researches [13–15].

The supercritical state is defined as the thermodynamic state above the critical temperature and critical pressure. As for water, the critical temperature and pressure are ca. 374 °C and 22.1 MPa, respectively. A significant change in properties of water occurs when approaching this critical point, such as extremely low density of molecules, high diffusion coefficient, low viscosity, low dielectric constant, and high thermal conductivity [16]. The dielectric constant of water decreases significantly with increasing temperature [17], as shown in Fig. 5.1. Around the critical point of water, this value is nearly equal to the dielectric constant of polar organic solvents. The decreasing dielectric constant results in the improvement of reaction rate, based on the theory of Born, $\ln k = \ln k_0 - \frac{\omega}{RT}\left(\frac{1}{\varepsilon} - \frac{1}{\varepsilon_0}\right)$, in which k is reaction rate, ε is the dielectric constant, k_0 is the reaction rate at ε_0, ω is a constant decided by a reaction system, R is the gas constant, T is the absolute temperature. In the continuous hydrothermal synthesis process, once metal salts under ambient conditions instantaneously mix with supercritical water, the conditions favor rapid hydrolysis of metal salts and the subsequent hydrolysis of reaction intermediate to

S. Wang et al., *Supercritical Water Processing Technologies for Environment, Energy and Nanomaterial Applications*,
https://doi.org/10.1007/978-981-13-9326-6_5

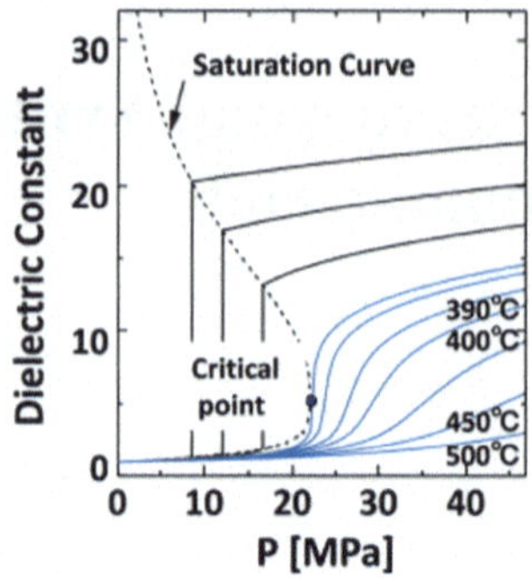

Fig. 5.1 Dielectric constant of water [13]

hydroxides or oxides. The low dielectric constant also results in low solubility of inorganic oxides or reaction intermediate. According to the classical nucleation theory [18], low solubility makes an enhancement in supersaturation. In supercritical state, the extremely high supersaturation plays a vital role in producing smaller particles because high supersaturation increases the rate of nucleation [19].

A commonly seen continuous hydrothermal flow synthesis process consists of two flows; one flow of metal salt solution mixes with another flow of preheated fluid in a mixer and then reacts in a reactor, followed by fast cooling after reaction. The schematic diagram can be seen in Fig. 5.2. Two high-pressure pumps are applied to supply solvent and precursor separately. Before mixing with precursor solution, solvent is preheated to reach a set temperature. In the mixer, room-temperature precursor is rapidly heated to high temperature, thus high

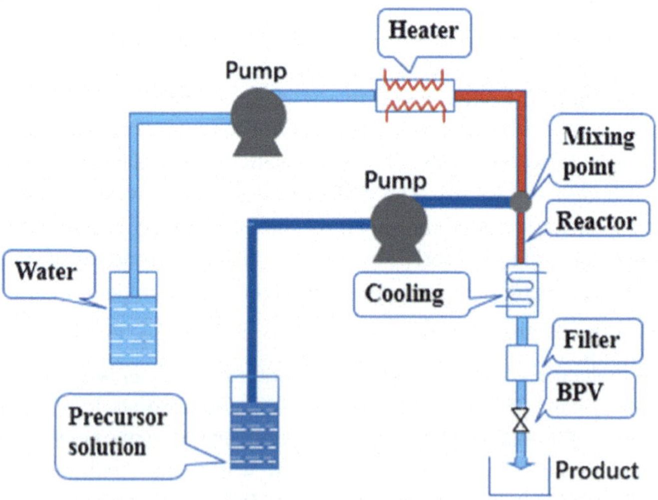

Fig. 5.2 Schematic diagram of continuous hydrothermal flow synthesis process

supersaturation is generated. A cooling water jacket is set to cool effluent after reaction. A back pressure valve is used to maintain the system pressure.

Adschiri et al. [20] reported the first try of synthesizing anatase at contributed of supercritical hydrothermal synthesis technology (30 MPa and 400–450 °C) in 1992, attracting extensive attention from other scholars. In recent decades, various materials including metal oxides and zero-valent metals are prepared by this method, such as metal oxides ZnO [21], CuO [22], Co_3O_4 [23], TiO_2 [20, 24–30], metals Ag [31, 32], Cu [31, 32], Ni [31, 32], and Co [33]. Besides, a potential battery material $LiFePO_4$ is also synthesized taking advantage of this method [34]. In our research group, TiO_2, CuO, and Cu were studied completely.

5.1 Synthesis of TiO_2 Nanoparticles

5.1.1 *Effect of Reaction Parameters*

Nanoscale titanium dioxide (TiO_2) is a new type of inorganic semiconductor material with diverse applications, photocatalysts [35], lithium batteries [36], wastewater treatment processes [37], dye-sensitized solar cells [38], and so on. These are attributed to their unique properties such as photocatalytic function, better thermal and chemical stability, strong absorbing UV ability, non-toxicity, and self-cleaning function [39, 40].

The effect of reaction temperature on synthesized products was studied using batch reactor for 3 min reaction time. 0.04 mol/L titanium (IV) sulfate ($Ti(SO_4)_2$) solution was used as precursor with the addition of 0.16 mol/L KOH. Influence of different temperature on reaction was taken at 30 MPa. The XRD results are shown in Fig. 5.3. We can see that the diffraction peaks became much sharper and narrower with increasing temperature. This evolution trend indicates that higher reaction temperature facilitated the formation of higher crystallinity product. The TEM measurement can also be another evidence of this, as shown in Fig. 5.4. TiO_2 synthesized at 300 °C was aggregated without clear boundary; however, when temperature increased to 400 °C, product's regular polygon was aggregated with clear boundary. This is because high temperature improved the reaction rate. The average particle size measured from TEM pictures at 350 °C and 400 °C was 8.8 nm and 19.8 nm, respectively. According to nucleation theory, increasing temperature from subcritical to supercritical, the size of crystal should decrease. However, at subcritical temperature, the products are amorphous-like or low degree of crystallinity product; this may explain the inverse trend of size.

In order to explore the effect of reaction pressure on products, 0.04 mol/L Ti $(SO_4)_2$, 0.16 mol/L KOH solution was used as precursor. Reaction temperature and time were set to be 400 °C and 3 min. The results show that reaction pressure had influence on conversion of Ti^{4+}, as shown in Fig. 5.5. The conversion of Ti^{4+} increased with increasing pressure. Higher pressure promoted the hydroxide ion concentration, thus improved the hydrolysis.

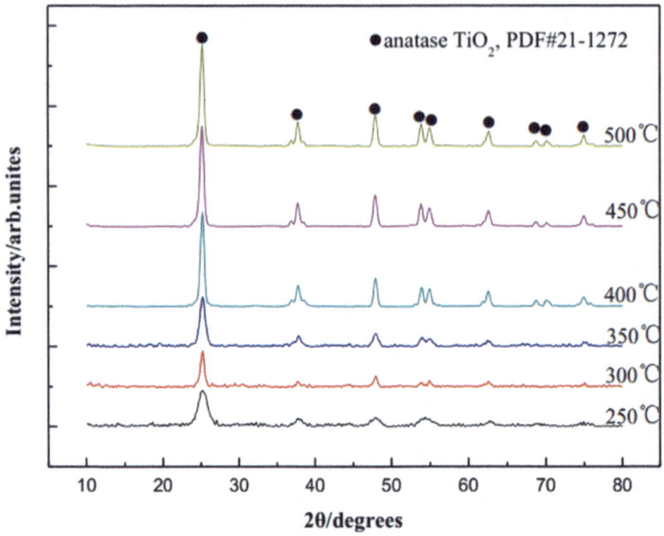

Fig. 5.3 XRD results of products synthesized at different temperatures and 30 MPa

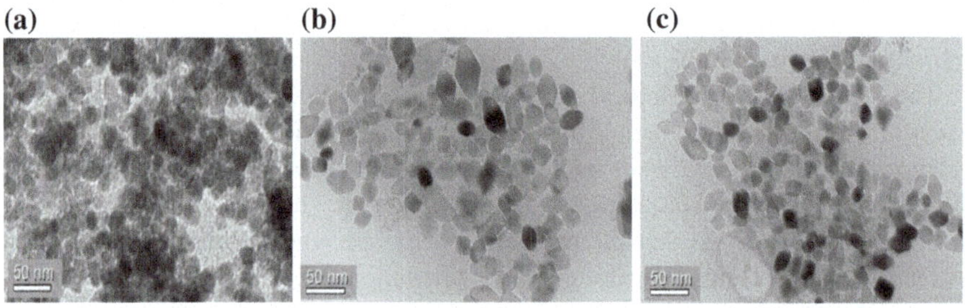

Fig. 5.4 TEM pictures of TiO_2 synthesized at 30 MPa and different temperatures: **a** 300 °C, **b** 400 °C, and **c** 500 °C as 0.04 M $Ti(SO_4)_2$ being precursor and 0.16 M KOH was added

5.1.2 Effect of Reaction PH and Precursor Species

For inorganic reaction, pH value plays an important role. Different ratios of KOH were added to $Ti(SO_4)_2$ to study the pH effect. Reaction temperature and pressure were set at 400 °C and 30 MPa. The TEM results are shown in Fig. 5.6. As mentioned before, supercritical hydrothermal process consisted of hydrolysis and dehydration procedures. The existence of OH^- promoted the hydrolysis of Ti^{4+}. Increased ratio between OH^- and Ti^{4+} favored the formation of uniform and high-crystallite products. Beside this, the addition of KOH could also increase the conversion of titanium ion.

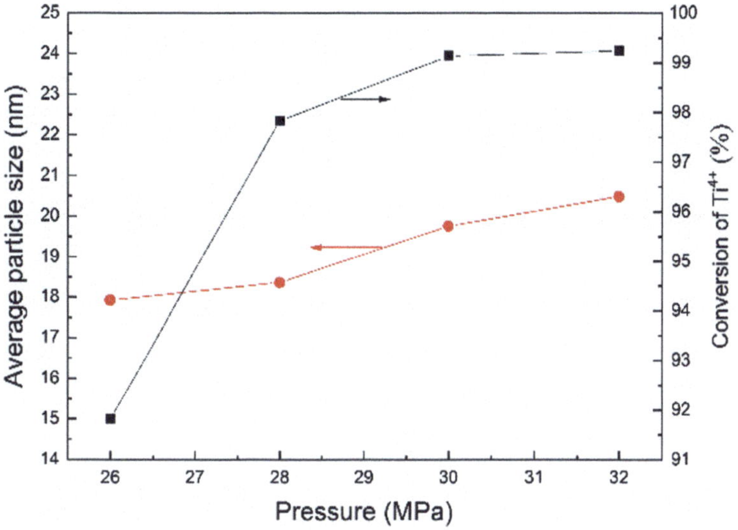

Fig. 5.5 Evolution of average size and conversion of Ti^{4+} with reaction pressure at 400 °C

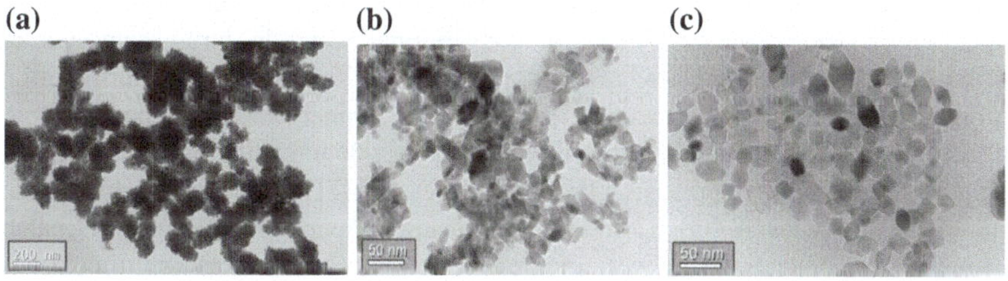

Fig. 5.6 TEM images of TiO_2 collected from different ratios between KOH and $Ti(SO_4)_2$. **a** $[OH^-]/[Ti^{4+}] = 0$, **b** $[OH^-]/[Ti^{4+}] = 2$, and **c** $[OH^-]/[Ti^{4+}] = 4$

Titanium(IV) chloride ($TiCl_4$) was also used as Ti source, and a compare was made with the products from $Ti(SO_4)_2$. Reaction conditions were 400 °C and 30 MPa. Concentration of precursor was 0.04 mol/L, and 0.16 mol/L KOH was added. After reaction for 3 min, products were collected and TEM images are shown in Fig. 5.7. The products obtained from $TiCl_4$ were bigger (average size: 52 nm) and had a wider size distribution. The morphology was not uniform, which consisted of cubic and elliptical ball-shaped particles.

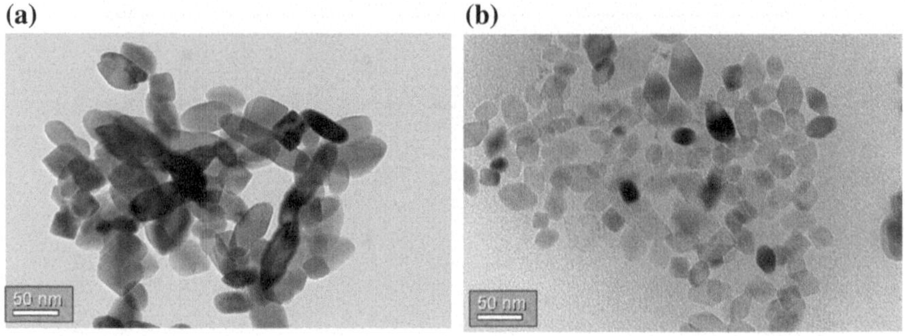

Fig. 5.7 TEM pictures of products collected from different Ti sources. **a** $TiCl_4$, **b** $Ti(SO_4)_2$

5.2 Synthesis of CuO Nanoparticles

Numerous methods have been explored to synthesize various copper oxide nanoparticles, such as chemical precipitation [41], electrochemical processes [42], spray pyrolysis [43, 44], thermal oxidation [45], and sonochemical synthesis [46]. However, the processes of these methods are usually complicated and require a long period of reaction time and not suitable for large-scale production of submicron copper oxide. On the whole, for conventional methods, large-scale synthesizing of high-quality and low-cost nanoparticles is difficult and even turns into a critical issue in the development of manufacturing copper oxide nanoparticles.

A number of works about synthesizing copper oxide ultra-fine particles in SCW have been done by some research groups. Kim et al. [47] synthesized copper oxide micrometer particles at 400 °C and 30 MPa using copper nitrate with 23 ml batch stainless steel reactor. However, the reaction time (include heating time) was 10 min, which was too long for the quick SHS process. Sue et al. [48] studied the synthesizing of copper oxide nanoparticles taking advantage of a continuous hydrothermal synthesis system with T-type micro mixer and the effect of residence time (0.002–2 s) was demonstrated. Outokesh et al. [49] made a research about manufacturing CuO nanoparticles in 200 ml batch stainless steel reactor, and the residence time was 1–3 h. Detailed information of the reaction parameters on the copper oxide particles prepared using supercritical water as solvent was studied by our research group.

Copper(II) nitrate ($Cu(NO_3)_2$) was used to supply copper ions. The experiments were listed in Table 5.1, including the average particle size and conversion of copper ions. These reactions were taken place inside batch reactor, which had a volume of 5.2 mL (inner diameter and length are 8.7 and 80 mm).

The products are all monoclinic CuO. According to the reaction mechanism proposed by Adschiri [20], the production of CuO can be expressed by Eqs. (5.1) and (5.2):

Table 5.1 Summary of experimental conditions and obtained results

No.	Cu $(NO_3)_2$ (mol/L)	T (°C)	P (MPa)	NaOH (mol/L)	Conversion (%)	Average particle size (nm)	Standard deviation (nm)
1	0.1	400	30	0	24.3	–	–
2	0.1	400	30	0.05	33.6	112.1	52.6
3	0.1	400	30	0.1	53.1	107.7	48.8
4	0.1	400	30	0.2	96.0	96.5	28.8
5	0.1	400	30	0.3	96.1	97.2	29.8
6	0.1	250	30	0.2	97.0	–	15.6
7	0.1	300	30	0.2	96.0	–	13.4
8	0.1	350	30	0.2	97.4	100.2	44.1
9	0.1	450	30	0.2	97.8	104.6	29.3
10	0.1	400	22	0.2	91.6	83.9	44.5
11	0.1	400	24	0.2	92.3	86.2	34.6
12	0.1	400	26	0.2	97.1	97.9	24.9
13	0.1	400	28	0.2	96.7	107.0	27.6
14	0.01	400	30	0.02	99.8	161.6	43.1
15	0.02	400	30	0.04	98.3	119.5	46.1
16	0.05	400	30	0.1	97.3	112.1	48.0
17	0.2	400	30	0.4	96.0	168.6	51.4

$$\mathrm{Cu(NO_3)_2 + 2OH^- = Cu(OH)_2 + 2NO_3^{2-}} \tag{5.1}$$

$$\mathrm{Cu(OH)_2 = CuO + H_2O} \tag{5.2}$$

5.2.1 *Effects of NaOH Concentrations*

The relationship between NaOH concentration and the conversion of copper ions are shown in Fig. 5.8, a significant change happened once base was added. The conversion degree of copper ions increased with an increase of additive NaOH concentrations when the concentration was below 0.2 M. XRD and FESEM micrographs' results of corresponding experiments are shown in Figs. 5.9 and 5.10. From the XRD results, we can see that all of the products are monoclinic CuO. Micron-order sized products were obtained in the absence of NaOH. With the increase of NaOH concentration, the average particle size of products decreased. When the ratio between NaOH and $Cu(NO_3)_2$ was 2, the products' size reached 96.5 nm. Similar results were also observed by Sue [21], whose research was fabricating zinc oxide. In the research, the conversion of zinc ions increased with the increased molar ratio of base to zinc ions from 2:1 to 4:1. The hydrolysis of metal salt increased due to the increased OH^- concentration at near-critical conditions [34]. However, in the supercritical hydrolysis of $Cu(NO_3)_2$ reaction, the acceleration of reaction conditions was not enough, thus base was added to further improve the hydrolysis of copper nitrate. This can also be explained by that higher concentration of OH^- promotes the reaction to right direction.

The NaOH concentration also has an influence on particle size. The addition of NaOH favors the formation of smaller and more uniform particles, as shown in Fig. 5.10. This can be also from the higher supersaturation after the addition of base [50]. The addition of hydroxyl ion leads to more copper oxide nucleus in the SCW. According to the definition of supersaturation, the supersaturation of copper oxide becomes higher. The higher supersaturation is, the more copper ions consume in nucleate, then smaller and uniform particles are produced.

5.2.2 *Effects of Temperature and Pressure*

The influence of temperature was studied at different reaction temperatures from 250 to 450 °C with an interval of 50 °C. The XRD detection demonstrates that all the products belonged to tenorite without the appearance of any impurities, as shown in Fig. 5.11. The percent conversions of copper ions listed in Table 5.1 demonstrate that the reaction temperature had a little effect on the percent conversion of copper ions in SCW with the addition of NaOH. The FESEM pictures of

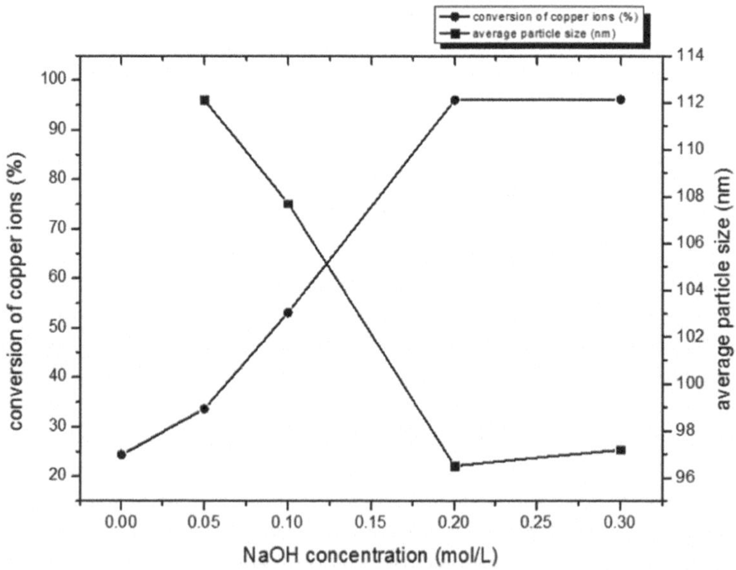

Fig. 5.8 Relationship between conversion of copper ions (filled circle represents), average particle size (filled square represents), and NaOH concentrations at 400 °C and 30 MPa with a precursor concentration of 0.1 mol/L

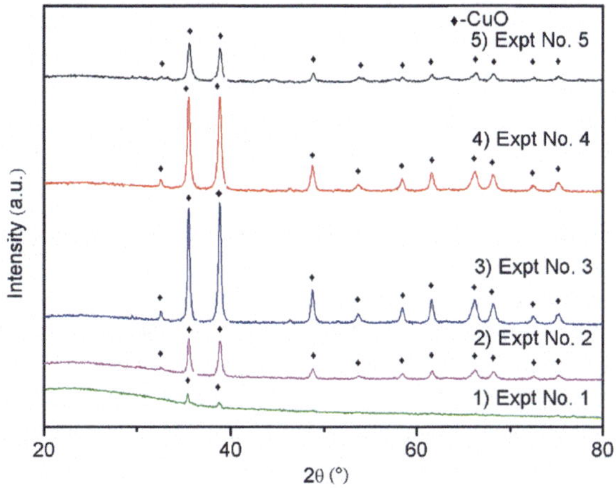

Fig. 5.9 XRD results with different NaOH concentrations: (1) 0 mol/L, (2) 0.05 mol/L, (3) 0.1 mol/L, (4) 0.2 mol/L, (5) 0.3 mol/L at 400 °C and 30 MPa with precursor concentration 0.1 mol/L

Fig. 5.10 FESEM images of products with the addition of different NaOH concentrations: **a** 0.05 mol/L, **b** 0.1 mol/L, **c** 0.2 mol/L, and **d** 0.3 mol/L at 400 °C and 30 MPa

the products obtained at different reaction temperatures and 30 MPa are shown in Fig. 5.12. The morphologies of obtained products at different reaction temperatures had a dramatic change. The average particles sizing 100.2 nm of the products obtained at subcritical conditions (350 °C) was a little bigger than that 96.5 nm produced at supercritical conditions (400 °C) and increased to 104.6 nm when the reaction temperature reached to 450 °C. However, the average particle size did not change significantly when the reaction temperature was above critical point of water. The standard deviation of particle size was bigger at subcritical conditions than that at supercritical conditions, namely the homogeneity of the products synthesized at subcritical conditions was not satisfactory. In addition, the products synthesized at low temperatures were inclined to aggregate.

This result is consistent with some researches [51, 52]. The dielectric constant of water decreased with the increase of temperature, thus the dissolution power of inorganic substances in water reduced, and the reaction rate increased. The solubility of copper oxide increased with an increase of temperature from 250 to 350 °C and then immediately decreased around the critical point of water [53] in a great speed. Accordingly, as the temperature increased from 350 to 400 °C, the supersaturation of dehydration reaction intermediate products increased, giving rise to a larger quantity of crystallization nuclei. Based on the classical nucleation theory, when the nucleation nuclei number was greater, the particle size and distribution width tended to be smaller. That is the reason for the variation of average particle size with different water conditions.

Up to now, for effects of reaction pressure on the products, there are still numerous disputes having not reached to unified understandings. For different metal oxides, the influence is various [54, 55]. In order to explore the effects of reaction pressure on synthesizing copper oxide nanoparticles in SHS, we studied the process of synthesizing copper oxide materials at different reaction pressures from 22 to 30 MPa. XRD detection demonstrates that all the products were monoclinic CuO. With the increase of reaction pressure from 22 to 26 MPa, the average particle size

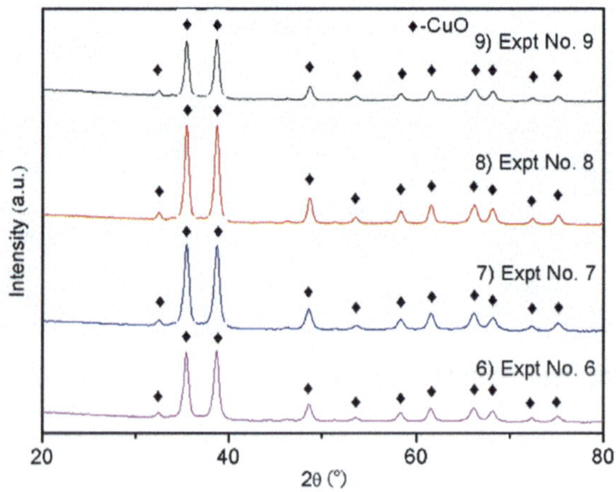

Fig. 5.11 XRD results with different reaction temperatures: (6) 250 °C, (7) 300 °C, (8) 350 °C, and (9) 450 °C at 30 MPa

Fig. 5.12 FESEM pictures of copper oxide produced at different reaction temperatures, i.e., **a** 250 °C, **b** 300 °C, **c** 350 °C, and **d** 450 °C and 30 MPa

grew from 83.9 to 97.9 nm. The results showed tell us that the reaction pressure mainly affected the uniformity of the copper oxide products fabricated from copper nitrate using supercritical hydrothermal synthesis. This might be due to the solubility change of copper oxide with the reaction pressure. Water density increased with pressure under supercritical conditions, resulting in an increase in copper oxide solubility, thus some bigger particles could be gained.

5.2.3 *Effects of Precursor Concentrations*

Precursor concentration always plays a vital role in controlling the size of products. Five different copper ion concentrations, 0.01, 0.02, 0.05, 0.1, and 0.2 M, were investigated in the SHS reaction, and reaction temperature and pressure maintained at 400 °C and 30 MPa, respectively. The products were all CuO. Figure 5.13 depicts the corresponding FESEM pictures of copper oxide products obtained from different copper ion concentrations in precursor. The average particle size of the products was 161.6 nm when the precursor concentration was 0.01 M. Increasing the precursor concentration to 0.02 and 0.05 M, their average particle size decreased to 119.5 and 112.1 nm, respectively. The morphologies of the products

Fig. 5.13 Micrographs of copper oxide for precursors with different copper ion concentrations at 400 °C and 30 MPa: **a** 0.01 M, **b** 0.02 M, **c** 0.05 M, and **d** 0.2 M

changed going on with the copper ion concentrations in precursors, as we can see obviously. The average particle size reached to 96.5 nm when the precursor concentration was 0.1 M as above-mentioned, it further rose to 168.6 nm when increasing the precursor concentration to 0.2 M.

According to classical nucleation theory, the supersaturation of reaction intermediates changed with the precursor concentrations. For lower concentration range (<0.1 M), the supersaturation increased with precursor concentrations, more copper ions would be inclined to nucleate rather than be consumed for the further growth of nucleation particles. Thus, smaller particles could be formed with increasing precursor concentrations. As the concentration increased to 0.1 M, the above-mentioned low concentration effects disappeared, i.e., the particle size did not decrease with precursor concentrations any more but presented an increasing trend. These results are consistent with the research of Sue [21], in which as the $Zn(NO_3)_2$ concentrations changed from 5×10^{-4} mol/kg to 5×10^{-3} mol/kg and 5×10^{-2} mol/kg, at first the particle size decreased from 49 nm to 28 nm and then increased to 44 nm. The nucleation rate and the rate of subsequent growth had a competitive impact on the ultimate particle size of formed products. At higher concentrations, there might be many copper ions failing to be nucleated, which would promote the growth of initially formed nucleuses in later reaction, resulting in the formation of many bigger product particles.

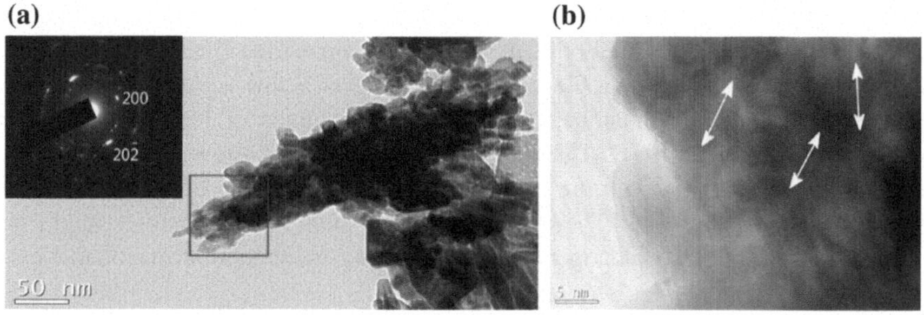

Fig. 5.14 TEM and SAED examination of the materials obtained at 250 °C and 30 MPa with precursor concentration being 0.01 M

5.2.4 Structure and Morphology of CuO Crystals

It can be seen that the morphologies of products are different at different SHS reaction conditions. The products were spindle-like at lower reaction temperatures (250 and 300 °C), whereas at higher temperatures (350 and 450 °C), the morphology of products changed to be hexagon flake-like. From the micrographs Fig. 5.13a–c, it can be seen that for precursor concentration of 0.01 to 0.05 M, the products were all rod-like. The morphologies of products were primary hexagon flake-like when the precursor concentrations were 0.1 and 0.2 M. In Fig. 5.14, the TEM detection results of the as-prepared products at 250 °C and 30 MPa are showed. The TEM image of individual spindle-like CuO crystal is presented in Fig. 5.14a. It is found that each spindle-like copper oxide crystal is consisted of many small particles. The elongated spots in the selected area electron diffraction (SAED) image (left side of Fig. 5.14a) indicate that the CuO crystal has a defective single-crystal-like feature. This suggests that the crystallographic orientations among particle within CuO architecture are nearly the same as a single-crystal-like

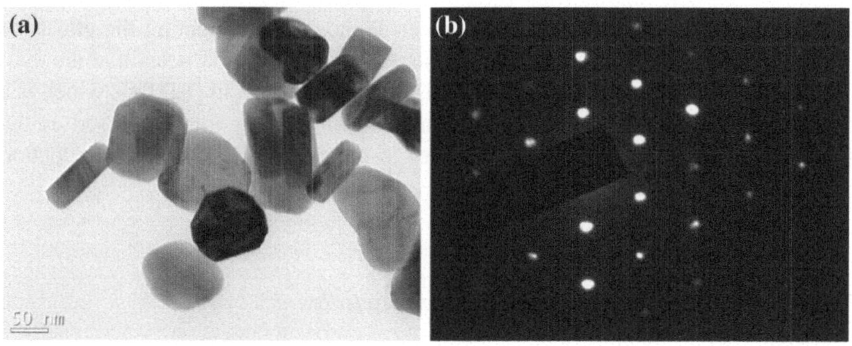

Fig. 5.15 TEM and SAED examination of the materials obtained at 400 °C and 30 MPa with precursor concentration being 0.01 M

structure, which might be attributed to the oriented attachment growth mechanism determined by the self-assembly of the smaller nanoparticles. The HRTEM image showed in Fig. 5.14b reveals the crystalline nature of a few CuO submicron particles with straight (200) lattice fringes. The TEM image of hexagon flake-like products is showed in Fig. 5.15a, and the SAED image is showed in Fig. 5.15b. The clear diffraction spots in the pattern demonstrate that each hexagon flake-like product was single crystal. These flake-like CuO submicron structures might be formed through the Ostwald ripening by consuming rather smaller nanoparticles.

5.3 Synthesis of Cu Nanoparticles

Copper nanoparticles are widely used in conductive coatings, metal surface's wear-resistant coatings, microelectronics, and lubricant additives [56]. In the area of inkjet printing, copper nanoparticles are more competitive than silver due to its excellent conducting properties and low cost [57–60]. As far as we know, there are few researches about copper nanoparticles synthesis using supercritical water [61]. Various metal (e.g., Ag, Ni, and Pd) nanoparticles have been synthesized using hydrogen as reducing agent at supercritical reaction conditions. Thus, it is interesting to explore the reaction process of zero-valent copper synthesizing in supercritical water. In the field of supercritical hydrothermal synthesis, controlling the morphological structure and particle size of the nanoparticles is one of the research emphases. Particle size control of metal oxide nanoparticles during the supercritical hydrothermal synthesis has been extensively studied. However, only a limited number of investigations deal with the particle size control of the metal nanoparticles during the reductive supercritical hydrothermal synthesis.

Sue et al. [52] synthesized Ni fine particles in near-critical and supercritical water. During the reaction, 1,10-phenanthroline was used as complexing agent that can complex with nickel ion to form stable chelate compounds at lower temperatures, which may contribute to the control of the reduction rate of nickel ion and thus to the particle size of final products. To the best of author's knowledge, no systematic and experimental study has been found for investigating the effects of the synthetic parameters on the particle size of copper nanoparticles that are synthesized in supercritical water. Ethylenediaminetetraacetic acid (EDTA), which is a carboxylate forming relatively stable complex with copper ion, was used as the complexing agent. Copper sulfate was used as precursor. CH_2O acted as reduction agent.

5.3.1 Production of Copper Nanoparticles

First of all, the amount of reducing agent required to produce zero-valent copper was determined. The effect of reducing agent was investigated by varying the molar

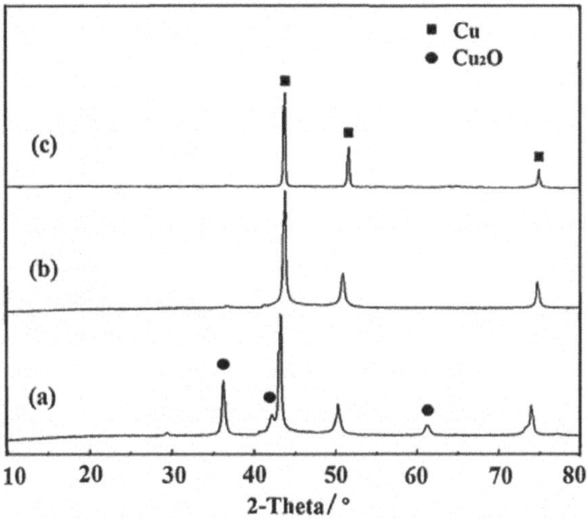

Fig. 5.16 XRD profiles of the nanoparticles synthesized at different molar ratios of HCOH to Cu^{2+}([HCHO]/[Cu^{2+}]): **a** 1:2, **b** 1:1, and **c** 3:1

ratio of formaldehyde to copper ion from 1:2 to 3:1. Figure 5.16 shows the XRD profiles of the samples. When the molar ratio of formaldehyde to copper ion was 1:2, in addition to zero-valent Cu, small amounts of Cu_2O were also identified in the product. When the molar ratio increased to 1:1 or higher, pure zero-valent Cu was obtained. This result shows that the formaldehyde required to produce Cu nanoparticles by the supercritical hydrothermal synthesis method was much lower than that of the traditional chemical reduction method [62].

5.3.2 *Control of Particle Size*

Figure 5.17 shows the TEM images of the synthesized nanoparticles under different copper sulfate concentrations. When the concentration of copper sulfate increased from 0.05 to 0.5 mol/L, the average particle size of the nanoparticles increased from about 14 to 50 nm. It seems that the nanoparticles may obtain sufficient growth at higher copper sulfate concentrations. This result is consistent with the findings of prior studies on the preparation of γ-AlO(OH) [54] and Fe_2O_3 [48] nanoparticles by supercritical hydrothermal synthesis method. In addition, the particle-size distribution of products became wider as the concentration of copper sulfate increased. When the copper sulfate concentration increased to 0.5 mol/L, the particle size of the products showed obvious heterogeneous distribution. It suggests that the copper sulfate concentration may have a significant impact on the crystallization kinetics of copper nanoparticles. At low copper sulfate concentration, primary nuclei were rapidly generated at supercritical water, which greatly reduced the concentration of copper ion and thus suppressed the formation of secondary nucleus. In this case, the growth time of nanoparticles was limited, and small particles with good uniformity

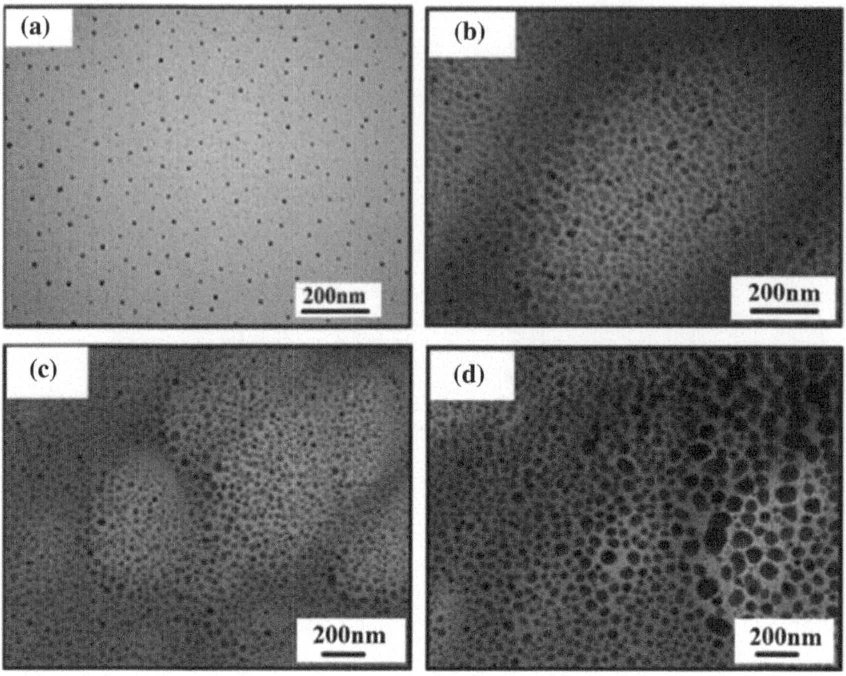

Fig. 5.17 TEM micrographs of copper nanoparticles synthesized at different copper sulfate concentrations: **a** 0.05, **b** 0.1, **c** 0.2, and **d** 0.5 mol/L

were readily produced. Whereas, at high copper sulfate concentration, the nucleation period was prolonged. Repeated heterogeneous phase nucleation thus occurred, which ultimately resulted in heterogeneous distribution of nanoparticles.

5.3.3 Effect of EDTA Concentration

The effect of EDTA concentration on the copper particle size was studied. The TEM pictures of obtained products are shown in Fig. 5.18. Dispersible ellipsoidal copper nanoparticles were observed. The particle size range of products was obtained from different ratios between EDTA and Cu^{2+} were 24–405 nm, 17–114 nm, and 22–88 nm, respectively, corresponding to EDTA: Cu^{2+} 1:1, 2:1, and 4:1. The concentration of EDTA plays an important role in controlling the homogeneity and dispersity of nanoparticles. This is because the increasing concentration of EDTA provided enough atoms to coordinate with copper ions to form stable chelate compounds during the heating process, which prevented premature nucleation of the nanoparticles.

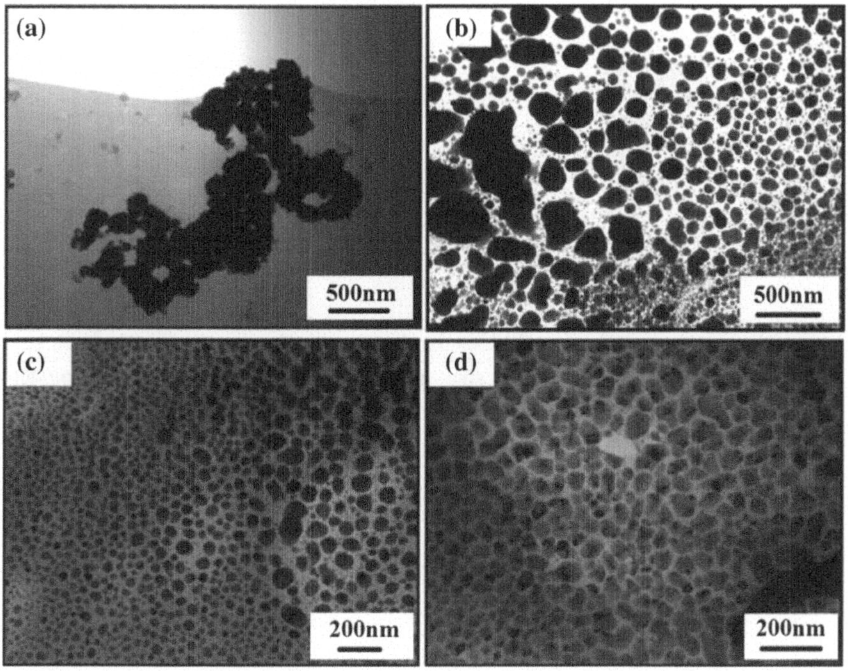

Fig. 5.18 TEM images of copper nanoparticles synthesized at different EDTA concentrations: **a** Without EDTA, **b** [EDTA]/[Cu^{2+}] = 1:1, **c** [EDTA]/[Cu^{2+}] = 2:1, and **d** [EDTA]/[Cu^{2+}] = 4:1

5.3.4 Effect of Sodium Hydroxide Concentration

The effect of sodium hydroxide concentration was also explored in synthesizing copper nanoparticles. When the molar ratio of NaOH to copper ion increased from 0:1 to 2:1, the particle size range of the products significantly decreased from 21–210 nm to 5–19 nm, and the corresponding average particle size decreased from about 85 to 14 nm. This finding has the same trend with copper oxide synthesizing, the same reason increasing alkalinity of the reactant solution that leads to the increase of the degree of the supersaturation.

5.3.5 Mechanism Analysis

To investigate the chemical bonding between EDTA and copper nanoparticles, FT-IR analysis was conducted. Figure 5.19 shows the spectra results of different EDTA concentrations. Broad hydroxyl stretch peaks at 3400 cm^{-1} were observed even for the nanoparticles synthesized without NaOH. It indicates that the unsaturated bonds on the surfaces of nanoparticles tended to be combined with hydroxy

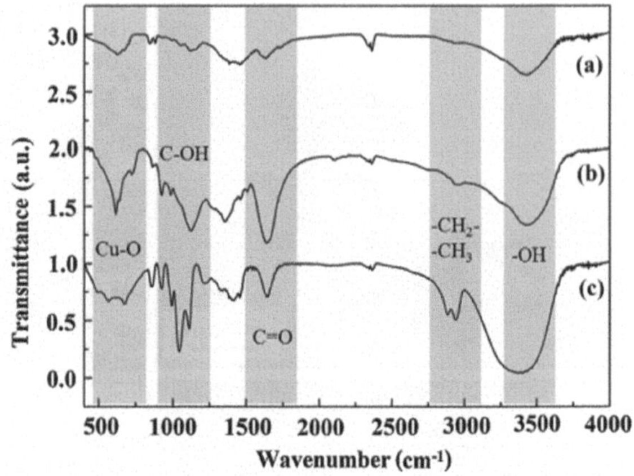

Fig. 5.19 FT-IR spectra of copper nanoparticles synthesized at different NaOH additions: **a** Without NaOH, **b** $[NaOH]/[Cu^{2+}]$ = 1:1, and **c** $[NaOH]/[Cu^{2+}]$ = 2:1

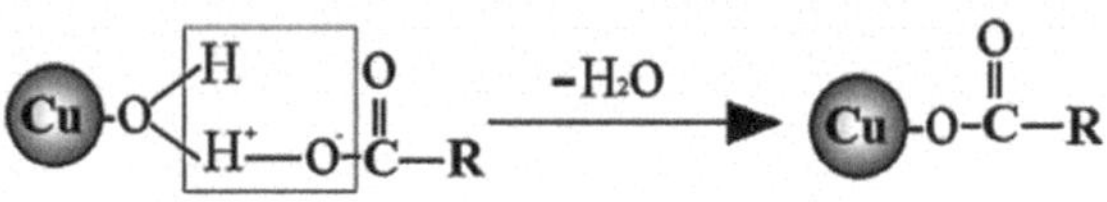

Fig. 5.20 Adsorption mechanism of EDTA on the surfaces of copper nanoparticles

which was hydrolyzed from water. The absorption bands, corresponding to $–CH_2$ and $–CH_3$ stretching vibration on the particles, were observed at 2885 and 2947 cm^{-1}. The bands at 1047 and 1640 cm^{-1} were the vibration bands of C–OH and C–O stretching vibration, which attributed to the adsorption of carboxyl group (–COOH). The bands at 620 cm^{-1} corresponded to the stretching vibrations of Cu–O.

Based on above analysis, the adsorption mechanism of EDTA on the surfaces of copper nanoparticles was proposed. As shown in Fig. 5.20, the complexing agent was adsorbed on the surface of nanoparticles by the dehydration reaction of carboxyl groups in the molecule of EDTA and hydroxyl groups on surface of copper nanoparticles (Cu–OH).

5.4 Optimization and Design of Mixer

In the continuous supercritical hydrothermal synthesis process, proper design of the mixer is a key factor in obtaining products with decreased particle sizes and narrow particle-size distributions [63]. The common process is that metal salt solution mixes directly with supercritical water and is rapidly heated to supercritical state. The process has advantages including high heating rate, short residence time, and easy industrial scale-up. However, in the continuous supercritical hydrothermal reaction system, particle plugging and hard-to-control particle size frequently encounter problems. The source of the problems is unreasonable design of the mixer, resulting in slow and non-uniform mixing of the metal salt solution and supercritical water.

To date, various types of mixers have been developed using this method for preparation of nanoparticles. The most commonly used mixer is traditional tee mixer, besides, some new mixers have also been reported, such as Y-type [64], nozzle-type [65], central collisiontype [66], and swirl-type [34] mixers. From the engineering perspective, to optimize the design of the mixer, it is necessary to acknowledge the fluid dynamics inside the mixer and the crystallization kinetics on the mixing time scale of tens of milliseconds [67].

At present, the computational fluid dynamics (CFD) method has been employed as an effective means in studying the mixing and crystallization behavior inside SWHS mixers. However, because of the complexity of the interaction of mixing and the crystallization reaction, the quantitative study of this effect is still inadequate. Our purpose is to develop a mathematical model capable of quantitatively studying the effect of mixing efficiency on the crystallization in SWHS mixers to produce inorganic nanoparticles. The mixing efficiency was evaluated in terms of the time to achieve micro-mixing. The micro-mixing is molecular-scale mixing and has been confirmed to have decisive effects on the PSD of the products in the crystallization reaction. The flow fields inside the mixer under different operating parameters were simulated. The effects of fluid dynamic factors on the mixing efficiency and crystallization kinetics were investigated. In this way, the effect of fluid dynamics on the formation of nanoparticles was elucidated, and the corresponding design rules were established.

5.4.1 Mixing Model

Fox simplified the simulation of turbulent reactions using mixture fraction probability [68], which relates the concentration of each component to the mixture fraction. The mixture fraction $\langle\xi\rangle$ is defined as:

$$\langle\xi\rangle = \frac{m_2}{m_1 + m_2} \tag{5.3}$$

where m_1 and m_2 are the local mass flow rates from the two mixer inlets. $\langle\xi\rangle$ represents the local mean mixture fraction. The squared difference of $\langle\xi\rangle$ and ξ' (where ξ' is the mean mixture fraction when mixing is completed) is defined as the macro-mixing variance $\langle\xi^2\rangle_{\mathrm{Lss}}$:

$$\langle\xi^2\rangle_{\mathrm{Lss}} = (\langle\xi\rangle - \xi')^2 \tag{5.4}$$

ξ' can be computed as $m_{10}/(m_{10}+m_{20})$, where m_{10} and m_{20} are the mass flow rates at each mixer inlet. The micro-mixing variance $\langle\xi\rangle^2$ is defined as follows:

$$\langle\xi^2\rangle = (\xi - \langle\xi\rangle)^2 \tag{5.5}$$

where ξ is the local mixture fraction. $\langle\xi\rangle^2$ is the measure of the extent of mixing at the microscale. Its value equals zero when mixing is complete at the molecular scale.

$\langle\xi\rangle$, $\langle\xi^2\rangle_{\mathrm{Lss}}$, and $\langle\xi\rangle^2$ were calculated using the user-defined scalars in Fluent software. The transport equations of these scalars were established as follows:

$$\frac{\partial\rho\langle\xi\rangle}{\partial t} + \nabla(\rho\langle U\rangle\langle\xi\rangle) = \nabla(\rho\Gamma_{\mathrm{T}}\nabla\langle\xi\rangle) \tag{5.6}$$

$$\frac{\partial\rho\langle\xi^2\rangle_{\mathrm{Lss}}}{\partial t} + \nabla\left(\rho\langle U\rangle\langle\xi^2\rangle_{\mathrm{Lss}}\right) = \nabla\left(\rho\Gamma_{\mathrm{T}}\nabla\langle\xi^2\rangle_{\mathrm{Lss}}\right) - 2\rho\Gamma_{\mathrm{T}}|\nabla\langle\xi\rangle|^2 \tag{5.7}$$

And

$$\frac{\partial\rho\langle\xi^2\rangle}{\partial t} + \nabla\left(\rho U\langle\xi\rangle^2\right) = \nabla\left(\rho\Gamma_{\mathrm{T}}\nabla\langle\xi\rangle^2\right) - 2\rho\Gamma_{\mathrm{T}}|\nabla\langle\xi\rangle|^2 - 2\rho\gamma\langle\xi\rangle \tag{5.8}$$

where ρ is the fluid density, γ is the micro-mixing rate, Γ_{T} is the turbulent diffusivity, and U is the fluid velocity

$$\Gamma_{\mathrm{T}} = \frac{C_\mu}{S_{\mathrm{CT}}}\frac{k_{\mathrm{t}}^2}{\varepsilon_{\mathrm{t}}} \tag{5.9}$$

where C_μ and turbulent Schmidt number S_{CT} take empirical material constants of 0.09 and 0.7, k_{t} is the dissipation of turbulent kinetic energy, and ε_{t} is the turbulent kinetic energy

$$\gamma = C_\varphi\frac{\varepsilon_{\mathrm{t}}}{2k_{\mathrm{t}}} \tag{5.10}$$

where C_φ is calculated as a function of Reynolds number Re and Schmidt number Sc [68].

5.4.2 *Reaction Model*

The experiments to determine the copper ion conversions were carried out at 200–350 °C and 30 MPa. The copper sulfate concentration was 0.05 M. From the decrease of the copper ion concentration in the effluent, the conversion rate of copper ions, x, was determined. $-\ln(1 - x)$ was in a linear relationship with the reaction time. The first-order reaction rate constant k at 200–350 °C was calculated. The Arrhenius plot provided estimates of the global activation energy E_{a0} of 41.135 ± 4.986 kJ mol^{-1} and a frequency factor a_0 of 77.092 ± 3.062 s^{-1}. The conversions of copper ion at 400 °C, however, reached almost 100% for all experimental runs. In fact, the supercritical reaction proceeded very rapidly, so it was difficult to obtain reliable reaction rates at supercritical conditions experimentally. In this study, the reaction rates at supercritical temperatures were estimated based on the Hakuta equation [69].

In fact, quantitatively describing the SWHS reaction kinetics is a difficult task because of the lack of sufficient experimental k data. The mixing rate and the reaction rate are comparable at high Reynolds numbers. Thus, the mixing-controlled reaction rate model is not appropriate in this study. According to previous reports, the Hakuta equation has been well used in research on the SWHS reaction kinetics. The term $(\varepsilon^{-1} - \varepsilon_0^{-1})$ in equation is approximately zero at subcritical conditions, and thus ln k is approximately linearly related to T^{-1}. In contrast, under supercritical conditions, ln k increases rapidly because of the significant decrease in ε.

5.4.3 Population Balance Equation

The population balance equation (PBE) coupled with the Eulerian multiphase flow model is used to describe the interphase mass-transfer rate of the heterogeneous reaction. The PBE is effective in describing the properties of the PSD in detail. Within this model, the number density function $n(L; x, t)$ (i.e., the total number of particles per unit volume at time t in the size range from L to L + dL) is expressed as a function of the variables "external coordinate" x and "internal coordinate" L, which denote the spatial positions of particles and the particle size, respectively.

5.4.4 Nucleation Rate and Growth Rate

The nucleation rate J is defined as the number of crystal nuclei formed per unit time and unit volume. For crystallization in supercritical water, primary nucleation is the predominant mechanism controlling the nucleation rate because of the high-system supersaturation. Therefore, the rate of homogeneous nucleation is used to simulate the nucleation reaction based on classical nucleation theory.

The growth rate G is given by:

$$G = k_g(S - 1)^{N_g} \tag{5.11}$$

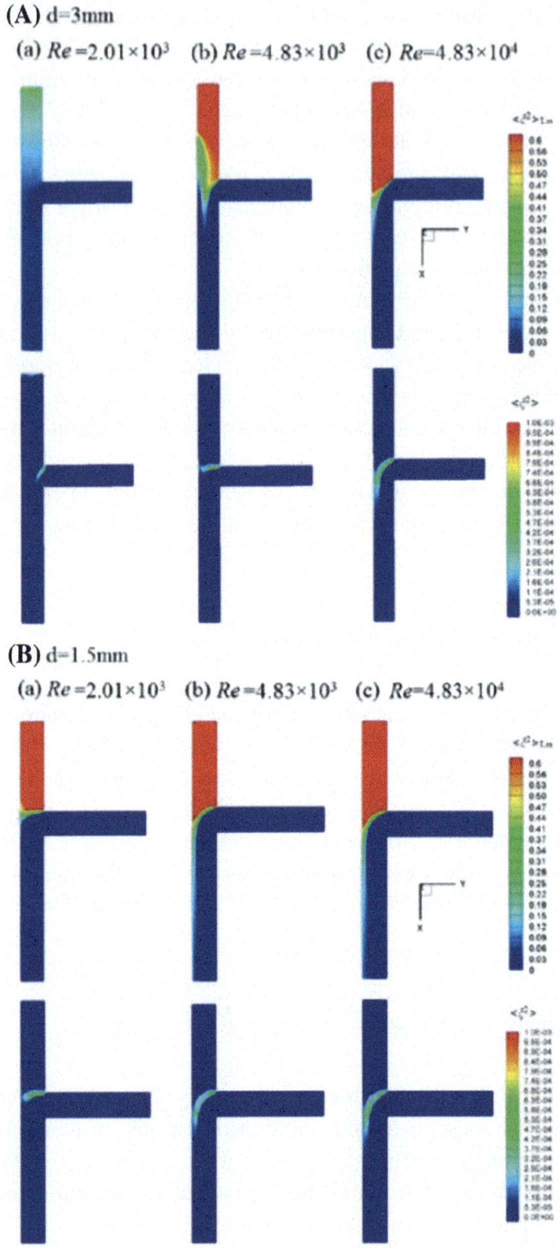

Fig. 5.21 Contours of $\langle \xi^2 \rangle_{\mathrm{Lss}}$ and $\langle \xi \rangle^2$ at different Reynolds numbers and mixer diameters d: d = (**A**) 3, (**B**) 1.5, and (**C**) 0.3 mm; Re = (a) 2.01×10^3, (b) 4.83×10^3, and (c) 4.83×10^4

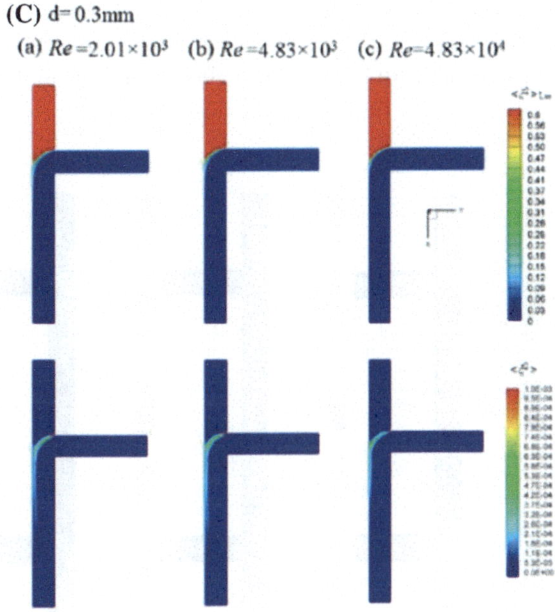

Fig. 5.21 (continued)

where $k_g = 3 \times 10^{-10}$ m s^{-1} is the growth rate constant, N_g is the exponent with a value of 2.0 (usual value range of 1.0–2.0), and S is the supersaturation.

5.4.5 Influence of Reynolds Number and Mixer

The influence of the mixer diameter *d* and the mixture Reynolds number on the mixing efficiency of the tee mixer and the crystallization of nanoparticles was studied. The selected diameters were 3, 1.5, and 0.3 mm. The Reynolds number was varied within the range from 2.01×10^3 to 2.09×10^5 by changing the total flow rate. The mass flow ratio of metal salt solution to supercritical water was 1:4. Figure 5.21 shows the mixture fraction variances $\langle \xi^2 \rangle_{\text{Lss}}$, and $\langle \xi \rangle^2$, for various Reynolds numbers and mixer diameters *d*. From the distributions of $\langle \xi^2 \rangle_{\text{Lss}}$, and $\langle \xi \rangle^2$, the macro- and micro-mixing processes were observed. The maximum value of $\langle \xi \rangle^2$ was generated in the mixing zone, indicating that the micro-mixing was the most inhomogenous in this region. Then, it was destroyed by the scalar dissipation rate and reached complete mixing (i.e., $\langle \xi \rangle^2 = 0$). For the 3 mm mixer, it was observed that the supercritical water flowed countercurrently into the other mixer inlet and thus resulted in the premature heating of the metal salt solution at Reynolds numbers of 2.01×10^3 and 4.83×10^3. This bad mixing phenomenon is known as "back-mixing." It was reported that back-mixing occurred when the buoyancy lift of supercritical water was greater than the inertial force of metal salt solution, where the buoyancy lift became the main driven force of mixing. It can be

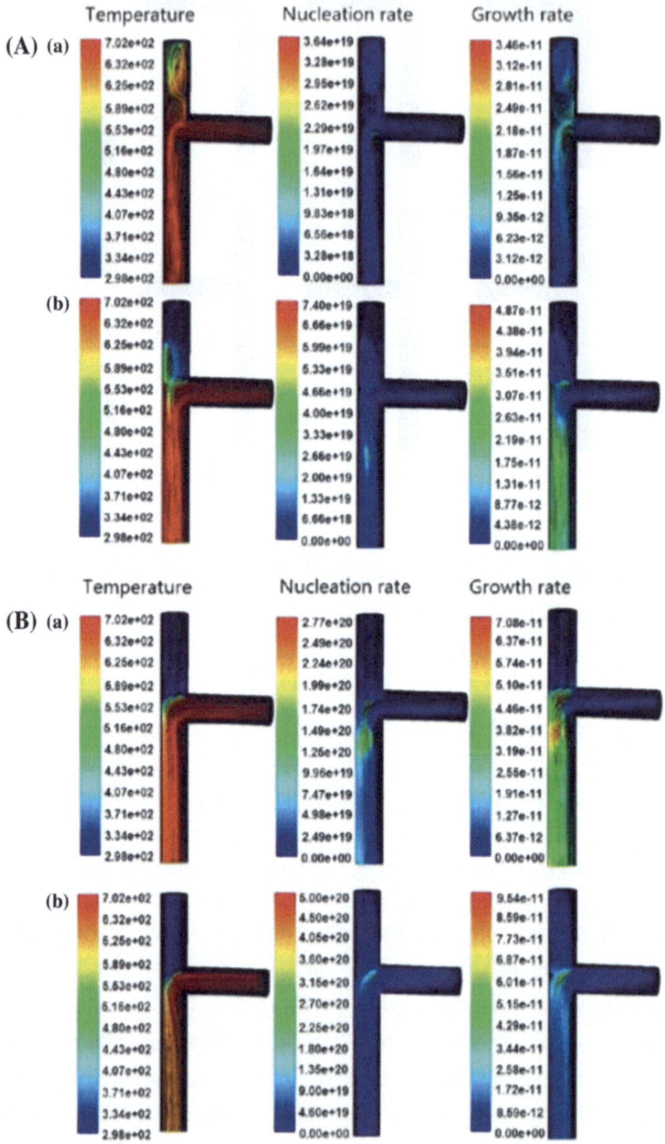

Fig. 5.22 Flowcharts of (left) temperature, (middle) nucleation rate, and (right) growth rate inside the mixer at different Re and *d* values: *d* = (**A**) 3, (**B**) 1.5, and (**C**) 0.3 mm; Re = (a) 2.01×10^3 and (b) 4.83×10^3

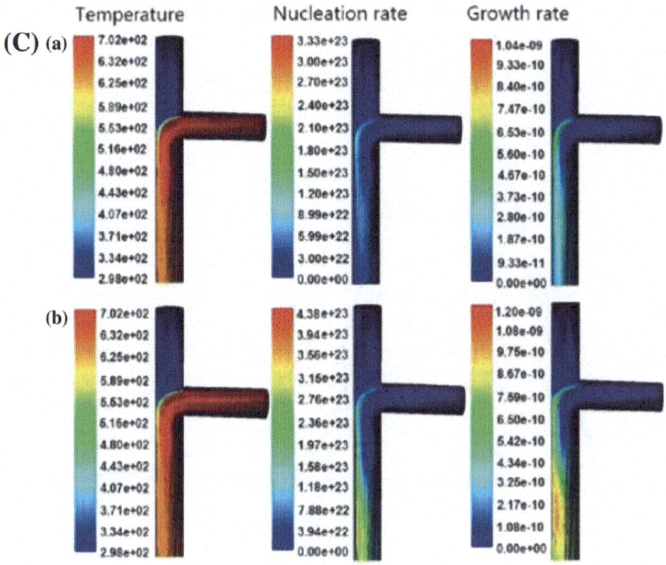

Fig. 5.22 (continued)

seen in Fig. 5.22. Flowcharts of (left) temperature, (middle) nucleation rate, and (right) growth rate inside the mixer at different Re and d values: d = (A) 3, (B) 1.5, and (C) 0.3 mm; Re = (a) 2.01×10^3 and (b) 4.83×10^3. Figure 5.23 predicted (a) PSDs and (b) APSs at different mixing times.

Figure 5.21 shows that the back-mixing was suppressed with the increase in Reynolds number. In addition, it was found that an increase in Reynolds number or a decrease in mixer diameter d can lead to a more confined mixing zone [67]. Takami et al. [70] performed neutron radiography on a T-junction to visualize the mixing behavior of supercritical water and room-temperature water. The supercritical water flowed in from the top of the mixer (4.5 mm in diameter). It was observed that a density-stratified layer was formed near the mixing point. It can be concluded that a low flow rate should be avoided for the large mixer diameter in the SWHS process.

Figure 5.24 shows the mass-average crystallization rate (nucleation rate and growth rate) at various Re and d values. As shown in the figure, both the nucleation rate and growth rate increased with increasing Re and decreasing d. This is because better mixing resulted in higher supersaturation and, thus, higher nucleation and growth rates.

Figure 5.22 shows the examples of the predicted distributions of temperature, nucleation rate, and growth rate within the mixers. It was observed that the maximum nucleation and growth rates also increased when Re increased and d decreased. For the cases in Fig. 5.22a, b, the metal salt solution was prematurely heated, which resulted in crystallization of the metal salt solution at the inlet arm.

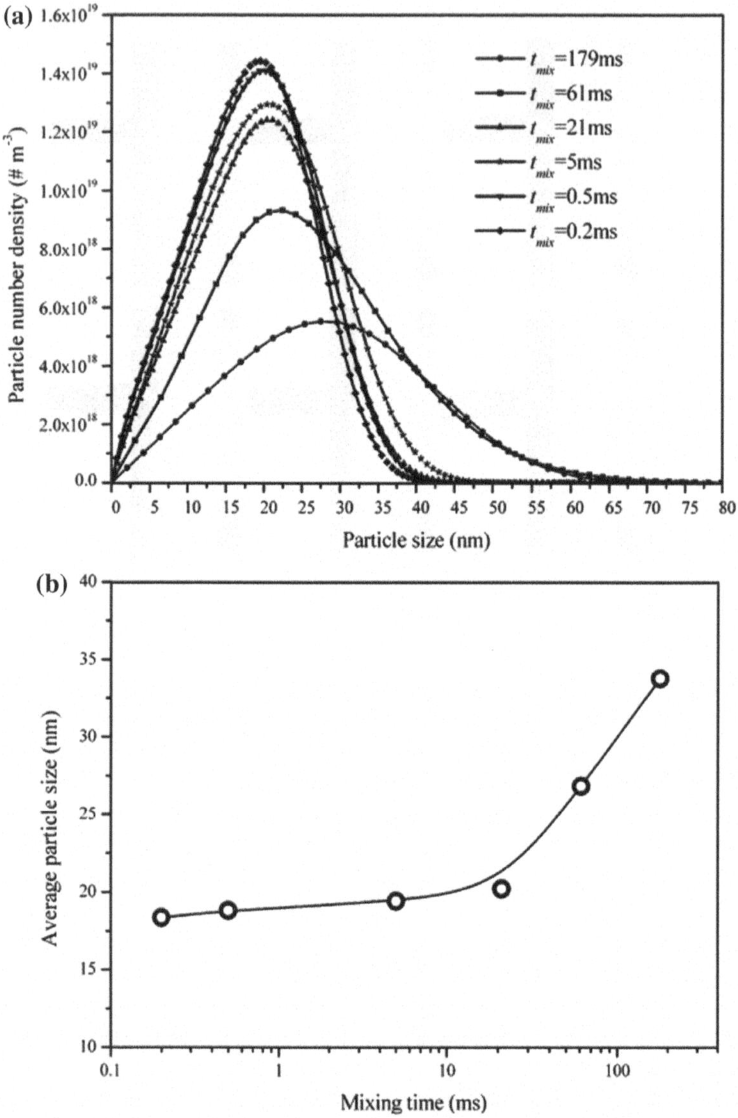

Fig. 5.23 Predicted **a** PSDs and **b** APSs at different mixing times

This might give rise to the danger of blockage for the mixer and should be prevented during the SWHS process. These results further evidence that mixing plays a key role in the crystallization behavior.

Figure 5.23 displays the APSs and PSDs predicted at various mixing times. As the mixing time decreased, narrower PSDs and smaller particle sizes were resulted. This indicates that the nucleation rate affects the particle size more than the growth

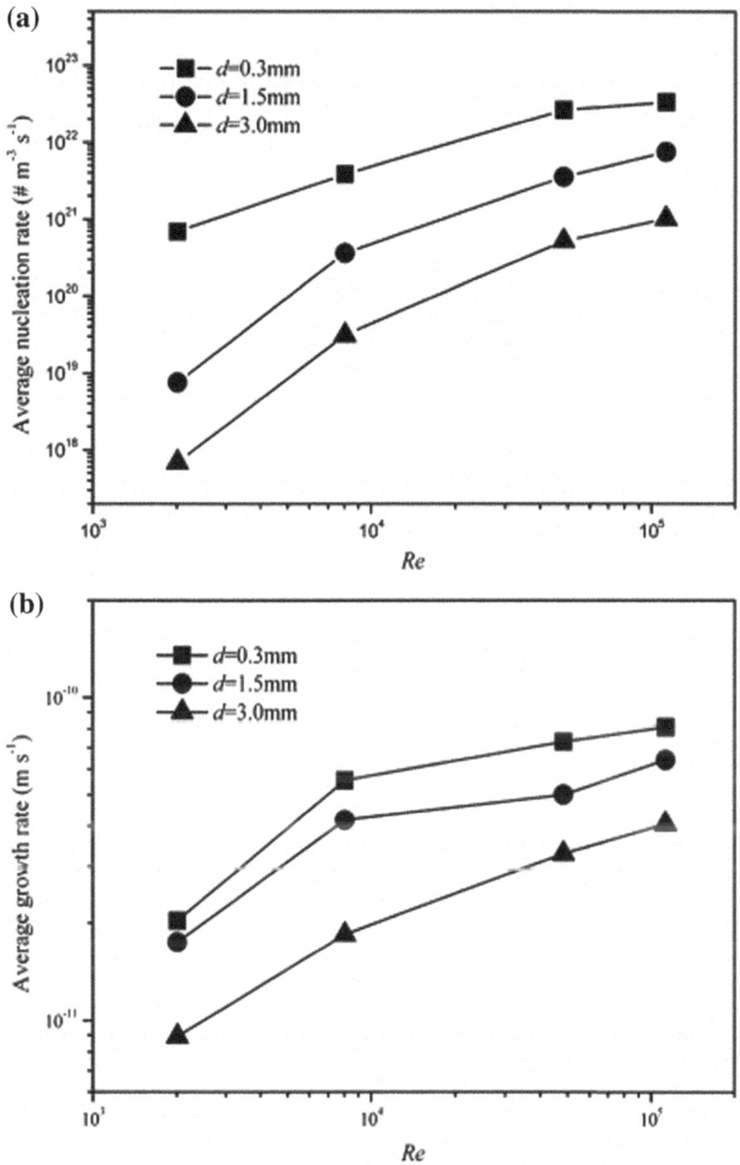

Fig. 5.24 Mass-average crystallization rates of the tee mixer at different Re and d values: **a** nucleation rate and **b** growth rate

rate does. In contrast, for mixing times of less than 20 ms, no significant difference was observed with decreasing mixing time. This suggests that the mixing process is no longer the rate-determining step of the crystallization reaction in this range.

References

1. S. RAI, A. RAI, Nanotechnology-The secret of fifth industrial revolution and the future of next generation. JURNAL NASIONAL, 2017, 7: 61–66.
2. J. A. Scher, J. M. Elward, A. Chakraborty, Shape Matters: Effect of 1D, 2D, and 3D Isovolumetric Quantum Confinement in Semiconductor Nanoparticles. The Journal of Physical Chemistry C, 2016, 120: 24999–25009.
3. E. Roduner, Size matters: why nanomaterials are different. Chem Soc Rev, 2006, 35: 583–592.
4. J.-H. In, H.-C. Lee, M.-J. Yoon, K.-K. Lee, J.-W. Lee, C.-H. Lee, Synthesis of nano-sized YAG: Eu3 + phosphor in continuous supercritical water system. The Journal of supercritical fluids, 2007, 40: 389–396.
5. W. Van der Weg, T. J. Popma, A. Vink, Concentration dependence of UV and electron-excited Tb3 + luminescence in Y3Al5O12. J Appl Phys, 1985, 57: 5450–5456.
6. M. Jansen, H. Letschert, Inorganic yellow-red pigments without toxic metals. Nature, 2000, 404: 980.
7. R. Ricceri, S. Ardizzone, G. Baldi, P. Matteazzi, Ceramic pigments obtained by sol-gel techniques and by mechanochemical insertion of color centers in Al2O3 host matrix. J Eur Ceram Soc, 2002, 22: 629–637.
8. K. Byrappa, S. Ohara, T. Adschiri, Nanoparticles synthesis using supercritical fluid technology - towards biomedical applications. Adv Drug Del Rev, 2008, 60: 299–327.
9. N. T. Thanh, L. A. Green, Functionalisation of nanoparticles for biomedical applications. Nano Today, 2010, 5: 213–230.
10. S. Noimark, K. Page, J. C. Bear, C. Sotelo-Vazquez, R. Quesada-Cabrera, Y. Lu, E. Allan, J. A. Darr, I. P. Parkin, Functionalised gold and titania nanoparticles and surfaces for use as antimicrobial coatings. Faraday Discuss, 2015, 175: 273–287.
11. S. Anandan, S. H. Yang, Emergent methods to synthesize and characterize semiconductor CuO nanoparticles with various morphologies - an overview. J Exp Nanosci, 2007, 2: 23–56.
12. K. Huang, M. Lei, Y. J. Wang, C. Liang, C. X. Ye, X. S. Zhao, Y. F. Li, R. Zhang, D. Y. Fan, Y. G. Wang, Green hydrothermal synthesis of Ce0(2) NWs-reduced graphene oxide hybrid with enhanced photocatalytic activity. Powder Diffr, 2014, 29: 8–13.
13. T. Adschiri, Y. W. Lee, M. Goto, S. Takami, Green materials synthesis with supercritical water. Green Chemistry, 2011, 13: 1380–1390.
14. C. Aymonier, A. Loppinet-Serani, H. Reveron, Y. Garrabos, F. Cansell, Review of supercritical fluids in inorganic materials science. J Supercrit Fluids, 2006, 38: 242–251.
15. J. A. Darr, J. Zhang, N. M. Makwana, X. Weng, Continuous Hydrothermal Synthesis of Inorganic Nanoparticles: Applications and Future Directions. Chem Rev, 2017.
16. J. A. Darr, M. Poliakoff, New directions in inorganic and metal-organic coordination chemistry in supercritical fluids. Chem Rev, 1999, 99: 495–542.
17. T. Adschiri, Y. Hakuta, K. Arai, Hydrothermal synthesis of metal oxide fine particles at supercritical conditions. Ind Eng Chem Res, 2000, 39: 4901–4907.
18. H. Vehkamäki, Classical nucleation theory in multicomponent systems. 2006: Springer Science & Business Media.
19. D. Kashchiev, On the relation between nucleation work, nucleus size, and nucleation rate. The Journal of Chemical Physics, 1982, 76: 5098–5102.

20. T. Adschiri, K. Kanazawa, K. Arai, Rapid and Continuous Hydrothermal Crystallization of Metal-oxide Particles in Supercritical Water. J Am Ceram Soc, 1992, 75: 1019–1022.
21. K. Sue, K. Murata, K. Kimura, K. Arai, Continuous synthesis of zinc oxide nanoparticles in supercritical water. Green Chemistry, 2003, 5: 659–662.
22. P. P. Sun, S. Z. Wang, T. Zhang, Y. H. Li, Y. Guo, Supercritical Hydrothermal Synthesis of Submicrometer Copper(II) Oxide: Effect of Reaction Conditions. Ind Eng Chem Res, 2017, 56: 6286–6294.
23. Y. L. Hao, A. S. Teja, Continuous hydrothermal crystallization of alpha-Fe2O3 and Co3O4 nanoparticles. J Mater Res, 2003, 18: 415–422.
24. T. Mousavand, J. Zhang, S. Ohara, M. Umetsu, T. Naka, T. Adschiri, Organic-ligand-assisted supercritical hydrothermal synthesis of titanium oxide nanocrystals leading to perfectly dispersed titanium oxide nanoparticle in organic phase. J Nanopart Res, 2007, 9: 1067–1071.
25. J. R. Eltzholtz, C. Tyrsted, K. M. O. Jensen, M. Bremholm, M. Christensen, J. Becker-Christensen, B. B. Iversen, Pulsed supercritical synthesis of anatase TiO2 nanoparticles in a water-isopropanol mixture studied by in situ powder X-ray diffraction. Nanoscale, 2013, 5: 2372–2378.
26. N. M. Makwana, C. J. Tighe, R. I. Gruar, P. F. McMillan, J. A. Darr, Pilot plant scale continuous hydrothermal synthesis of nano-titania; effect of size on photocatalytic activity. Mater Sci Semicond Process, 2016, 42: 131–137.
27. J. L. Mi, C. Clausen, M. Bremholm, N. Lock, K. M. O. Jensen, M. Christensen, B. B. Iversen, Rapid Hydrothermal Preparation of Rutile TiO_2 Nanoparticles by Simultaneous Transformation of Primary Brookite and Anatase: An in Situ Synchrotron PXRD Study. Crystal Growth & Design, 2012, 12: 6092–6097.
28. J. L. Mi, S. Johnsen, C. Clausen, P. Hald, N. Lock, L. So, B. B. Iversen, Highly controlled crystallite size and crystallinity of pure and iron-doped anatase-TiO2 nanocrystals by continuous flow supercritical synthesis. J Mater Res, 2013, 28: 333–339.
29. S. Kawasaki, Y. Xiuyi, K. Sue, Y. Hakuta, A. Suzuki, K. Arai, Continuous supercritical hydrothermal synthesis of controlled size and highly crystalline anatase TiO2 nanoparticles. J Supercrit Fluids, 2009, 50: 276–282.
30. H. Hayashi, K. Torii, Hydrothermal synthesis of titania photocatalyst under subcritical and supercritical water conditions. J Mater Chem, 2002, 12: 3671–3676.
31. H. Choi, B. Veriansyah, J. Kim, J. D. Kim, J. W. Kang, Continuous synthesis of metal nanoparticles in supercritical methanol. J Supercrit Fluids, 2010, 52: 285–291.
32. K. Jaehoon, K. Daewoo, B. Veriansyah, K. Jeong Won, K. Jae-Duck, Metal nanoparticle synthesis using supercritical alcohol. Mater Lett, 2009, 63: 1880–1882.
33. N. C. Shin, Y.-H. Lee, Y. H. Shin, J. Kim, Y.-W. Lee, Synthesis of cobalt nanoparticles in supercritical methanol. Mater Chem Phys, 2010, 124: 140–144.
34. S. A. Hong, S. J. Kim, K. Y. Chung, M. S. Chun, B. G. Lee, J. Kim, Continuous synthesis of lithium iron phosphate (LiFePO4) nanoparticles in supercritical water: Effect of mixing tee. J Supercrit Fluids, 2013, 73: 70–79.
35. C. Xue, T. X. Zhang, S. J. Ding, J. J. Wei, G. D. Yang, Anchoring Tailored Low-Index Faceted BiOBr Nanoplates onto TiO2 Nanorods to Enhance the Stability and Visible-Light-Driven Catalytic Activity. Acs Applied Materials & Interfaces, 2017, 9: 16091–16102.
36. M. Wagemaker, R. van de Krol, A. P. M. Kentgens, A. A. van Well, F. M. Mulder, Two phase morphology limits lithium diffusion in TiO2 (anatase): A Li-7 MAS NMR study. J Am Chem Soc, 2001, 123: 11454–11461.
37. S. Fukahori, H. Ichiura, T. Kitaoka, H. Tanaka, Photocatalytic decomposition of bisphenol A in water using composite TiO2-zeolite sheets prepared by a papermaking technique. Environ Sci Technol, 2003, 37: 1048–1051.
38. A. Hagfeldt, M. Gratzel, Molecular photovoltaics. Acc Chem Res, 2000, 33: 269–277.
39. M. Dahl, Y. Liu, Y. Yin, Composite Titanium Dioxide Nanomaterials. Chem Rev, 2014, 114: 9853.

40. X. Chen, A. Selloni, Introduction: Titanium Dioxide (TiO 2) Nanomaterials. Chem Rev, 2014, 114: 9281.
41. J. W. Zhu, H. P. Bi, Y. P. Wang, X. Wang, X. J. Yang, L. Lu, CuO nanocrystals with controllable shapes grown from solution without any surfactants. Mater Chem Phys, 2008, 109: 34–38.
42. G. Q. Yuan, H. F. Jiang, C. Lin, S. J. Liao, Shape- and size-controlled electrochemical synthesis of cupric oxide nanocrystals. J Cryst Growth, 2007, 303: 400–406.
43. J. Morales, L. Sanchez, F. Martin, J. R. Ramos-Barrado, M. Sanchez, Nanostructured CuO thin film electrodes prepared by spray pyrolysis: a simple method for enhancing the electrochemical performance of CuO in lithium cells. Electrochim Acta, 2004, 49: 4589–4597.
44. G. Q. Jian, L. Liu, M. R. Zachariah, Facile Aerosol Route to Hollow CuO Spheres and its Superior Performance as an Oxidizer in Nanoenergetic Gas Generators. Adv Funct Mater, 2013, 23: 1341–1346.
45. J. T. Chen, F. Zhang, J. Wang, G. A. Zhang, B. B. Miao, X. Y. Fan, D. Yan, P. X. Yan, CuO nanowires synthesized by thermal oxidation route. J Alloys Compd, 2008, 454: 268–273.
46. S. Anandan, G. J. Lee, J. J. Wu, Sonochemical synthesis of CuO nanostructures with different morphology. Ultrason Sonochem, 2012, 19: 682–686.
47. M. Kim, W. S. Son, K. H. Ahn, D. S. Kim, H. S. Lee, Y. W. Lee, Hydrothermal synthesis of metal nanoparticles using glycerol as a reducing agent. J Supercrit Fluids, 2014, 90: 53–59.
48. K. Sue, S. Kawasaki, M. Suzuki, Y. Hakuta, H. Hayashi, K. Arai, Y. Takebayashi, S. Yoda, T. Furuya, Continuous hydrothermal synthesis of Fe2O3, NiO, and CuO nanoparticles by superrapid heating using a T-type micro mixer at 673 K and 30 MPa. Chem Eng J, 2011, 166: 947–953.
49. M. Outokesh, M. Hosseinpour, S. J. Ahmadi, T. Mousavand, S. Sadjadi, W. Soltanian, Hydrothermal Synthesis of CuO Nanoparticles: Study on Effects of Operational Conditions on Yield, Purity, and Size of the Nanoparticles. Ind Eng Chem Res, 2011, 50: 3540–3554.
50. K. Sue, M. Suzuki, K. Arai, T. Ohashi, H. Ura, K. Matsui, Y. Hakuta, H. Hayashi, M. Watanabe, T. Hiaki, Size-controlled synthesis of metal oxide nanoparticles with a flow-through supercritical water method. Green Chemistry, 2006, 8: 634–638.
51. P. Y. Wang, K. Ueno, H. Takigawa, K. Kobiro, Versatility of one-pot, single-step synthetic approach for spherical porous (metal) oxide nanoparticles using supercritical alcohols. J Supercrit Fluids, 2013, 78: 124–131.
52. K. Sue, N. Kakinuma, T. Adschiri, K. Arai, Continuous production of nickel fine particles by hydrogen reduction in near-critical water. Ind Eng Chem Res, 2004, 43: 2073–2078.
53. K. Sue, Y. Hakuta, R. L. Smith, T. Adschiri, K. Arai, Solubility of lead(II) oxide and copper (II) oxide in subcritical and supercritical water. J Chem Eng Data, 1999, 44: 1422–1426.
54. Y. Hakuta, H. Ura, H. Hayashi, K. Arai, Effects of hydrothermal synthetic conditions on the particle size of gamma-AlO(OH) in sub and supercritical water using a flow reaction system. Mater Chem Phys, 2005, 93: 466–472.
55. H. Hayashi, A. Suino, K. Shimoyama, M. Takesue, S. Tooyama, R. L. Smith, Jr., Continuous hydrothermal synthesis of ZnGa2O4:Mn2 + nanoparticles at temperatures of 300–500 degrees C and pressures of 25-35 MPa. J Supercrit Fluids, 2013, 77: 1-6.
56. Y. Xuan, Q. Li, Heat transfer enhancement of nanofluids. Int J Heat Fluid Flow, 2000, 21: 58–64.
57. L.-Y. Shao, J. P. Coyle, S. T. Barry, J. Albert, Anomalous permittivity and plasmon resonances of copper nanoparticle conformal coatings on optical fibers [Invited]. Optical Materials Express, 2011, 1: 128–137.
58. J. Toyir, P. R. r. de la Piscina, J. L. G. Fierro, N. s. Homs, Highly effective conversion of CO2 to methanol over supported and promoted copper-based catalysts: influence of support and promoter. Applied Catalysis B: Environmental, 2001, 29: 207–215.
59. E. K. Athanassiou, R. N. Grass, W. J. Stark, Large-scale production of carbon-coated copper nanoparticles for sensor applications. Nanotechnology, 2006, 17: 1668–1673.

60. H. L. Yu, Y. Xu, P. J. Shi, B. S. Xu, X. L. Wang, Q. Liu, Tribological properties and lubricating mechanisms of Cu nanoparticles in lubricant. Transactions of Nonferrous Metals Society of China, 2008, 18: 636–641.
61. S. Kubota, T. Morioka, M. Takesue, H. Hayashi, M. Watanabe, R. L. Smith, Continuous supercritical hydrothermal synthesis of dispersible zero-valent copper nanoparticles for ink applications in printed electronics. J Supercrit Fluids, 2014, 86: 33–40.
62. C. G. Wen, K. M. Wang, X. Q. Li, S. S. Gao, W. L. Huan, Y. Zhao, Preparation of copper nanoparticles using liquid phase precipitation method. J. Anshan Univ. Sci. Technol., 2003, 26: 176–178.
63. S. I. Kawasaki, K. Sue, R. Ookawara, Y. Wakashima, A. Suzuki, Y. Hakuta, K. Arai, Engineering study of continuous supercritical hydrothermal method using a T-shaped mixer: Experimental synthesis of NiO nanoparticles and CFD simulation. J Supercrit Fluids, 2010, 54: 96–102.
64. A. Leybros, R. Piolet, M. Ariane, H. Muhr, F. Bernard, F. Demoisson, CFD simulation of ZnO nanoparticle precipitation in a supercritical water synthesis reactor. Journal of Supercritical Fluids the, 2012, 70: 17–26.
65. C. Y. Ma, M. Chen, X. Z. Wang, Modelling and simulation of counter-current and confined jet reactors for hydrothermal synthesis of nano-materials. Chem Eng Sci, 2014, 109: 26–37.
66. K. Sue, T. Sato, S. Kawasaki, Y. Takebayashi, S. Yoda, T. Furuya, T. Hiaki, Continuous Hydrothermal Synthesis of Fe2O3 Nanoparticles Using a Central Collision-Type Micromixer for Rapid and Homogeneous Nucleation at 673 K and 30 MPa. Ind Eng Chem Res, 2010, 49: 8841–8846.
67. J. Sierra-Pallares, D. L. Marchisio, E. Alonso, M. Teresa Parra-Santos, F. Castro, M. Jose Cocero, Quantification of mixing efficiency in turbulent supercritical water hydrothermal reactors. Chem Eng Sci, 2011, 66: 1576–1589.
68. R. O. Fox, Computational Models for Turbulent Reacting Flows. Cambridge University Press: New York, 2003.
69. Y. Hakuta, H. Hayashi, K. Arai, Fine particle formation using supercritical fluids. Current Opinion in Solid State & Materials Science, 2003, 7: 341–351.
70. S. Takami, K.-i. Sugioka, T. Tsukada, T. Adschiri, K. Sugimoto, N. Takenaka, Y. Saito, Neutron radiography on tubular flow reactor for hydrothermal synthesis: In situ monitoring of mixing behavior of supercritical water and room-temperature water. The Journal of Supercritical Fluids, 2012, 63: 46–51.

Chapter 6
Corrosion Behavior of Alloy Steels in Supercritical Water Environments

Energy, especially electricity, plays a vital role in our modern industrial society and in our current civilization. In order to increase the power generation efficiency of fossil fuels, supercritical and ultra-supercritical power plants, using supercritical water (SCW) as the working medium, have been developed over the past century [1]. In the early part of this current, twenty-first century, the supercritical, water-cooled (nuclear) reactor (SCWR) has been envisioned as one of the six Generation IV nuclear power plant designs, because of its simplified core and higher thermal efficiency (45 vs. 33% in current light water reactors) [2]. Due to the highly corrosive SCW environments, the corrosion resistance of candidate structural materials becomes a key requirement, especially when the presence of additional aggressive species such as salts and/or oxygen in a series of chemical systems with SCW as the reaction medium (SCWO, SCWG, SCHS, SCHC, etc.). Generally, the SCWO systems result in much severer corrosion issues than those under near-pure SCW. Actually, the former is likely the most corrosive at least among the known SCW processes [3, 4]. Supercritical water oxidation (SCWO) provides a potential alternative technology for incineration technology to treat high-concentration organic wastewater without secondary pollution. However, there is serious corrosion for materials under the harsh working conditions of high temperature, high pressure, and strong corrosive anions. The corrosion problem is the main obstacle to commercialization and large-scale application of SCWO. A volume of work has been carried out on solving the material corrosion problem in SCWO systems, especially at condition of hazardous waste oxidation. In order to find the appropriate material, the corrosion behaviors of candidate materials under different SCW conditions were investigated. Ni-based alloys, iron-based alloys, ceramics, titanium-based alloys, zirconium-based alloys, noble metals, tantalum, and niobium were selected for corrosion tests as the candidate materials. The purpose of this chapter is to review and summarize our reported related studies, and to share new understanding on corrosion mechanisms in SCW and some ideas with potential values to improve the corrosion resistance of objective alloys or to decrease the corrosivity of SCW systems of interest.

S. Wang et al., *Supercritical Water Processing Technologies for Environment, Energy and Nanomaterial Applications*,
https://doi.org/10.1007/978-981-13-9326-6_6

6.1 Research Status and Progress in Supercritical Water Environment

Corrosion tests of typical materials in SCW were stainless steels, nickel-based alloys, titanium, ceramic, zirconium, niobium, tantalum, and gold. In consideration of economy and corrosion resistance, stainless steels and nickel-based alloys have drawn more research attention than other materials.

As Fig. 6.1a shows, occurrence numbers of different test alloys were presented in the statistical analysis of 114 articles from 1994 to 2014 in SCWO corrosion research. The nickel-based alloys hold the largest proportion among all tested materials in previous corrosion researches, indicating the nickel-based alloys are excellent candidate materials for fabricating SCWO system. As shown in Fig. 6.1b, Inconel 625 makes up the highest proportion of the occurrence numbers of nickel-based alloys (accounted for 34.2%), followed by Hastelloy C-276 (accounted for 23.61%). Both Inconel 625 and Hastelloy C-276 own good corrosion resistance and mechanical strength under the hashing operated condition, so these two alloys are usually applied in experimental and commercial plant (Table 6.1).

6.1.1 Effect of Materials Class

(1) **Iron-based alloy**

316 SS, 316L SS, 304 SS, and 304L SS were four main iron-based alloys which were selected for constructing SCWO systems by many researchers. Chromium and TiO_2 were used as coatings on 316 SS for improving the resistant to corrosion under the SCWO condition [12, 20]. The outer and inner layers of the tested

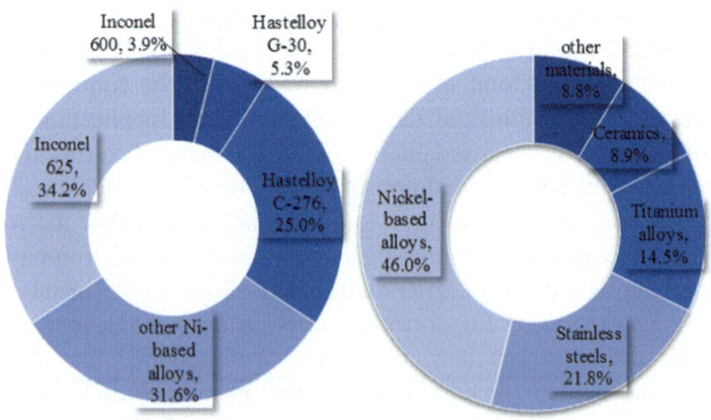

Fig. 6.1 Percentage of alloys occurrence numbers in 114 articles from 1994 to 2014, **a** different kinds of alloys and **b** different Ni-based alloys

Table 6.1 Preliminary summary on corrosion studies in SCW environments

Material	Material	Oxidant/acid/alkali/salt	*T* (°C)	*P* (MPa)	Corrosion results
Iron-based	304L	Deionized water	550	24	304L was susceptible to stress corrosion cracking [5]
	304	0.01 mol L^{-1} HCl	380	24	Electrochemical mechanism was the dominant corrosion mechanism [6]
	304L/316L	Deionized water	400–500	24	Outer layer was porous magnetite, and the inner layer was a denser iron–chromium spinel [7]
	316L	2.0% H_2O_2	500–550	24	Loose outer layer, interface, and inner layer were Fe, Ni, and Cr and Fe enrichment, respectively [8]
	316L	4.8 wt.% H_2O_2/2 wt.% H_2SO_4/2.4 wt% NH_3	380–390	24	Corrosion rate was 39 $mmyr^{-1}$ at this condition [9]
	316	2.0% H_2O_2	350–500	24	Fe_2O_3/Fe_3O_4 + spinel/Cr_2O_3/Ni-enrichment/316 from the outer to inner layer [10]
	316	3.0% H_2O_2	250–420	39	The maximum corrosion rate was 12.5 $mmyr^{-1}$ at 375 °C; H_2O_2 improved corrosion significantly [11]
	316	5 wt.% NaCl	300–500	30	Chromizing treatment decreased the corrosion rate, and corrosion mode was pitting [12]
	316	100% H_2O_2/2,4-dichlorophenol	400	25	Surface was mainly covered by carbon-contaminated and iron oxide layers [13]
	316	3% $Ce[NO_3]_3 \cdot 6H_2O$	400	30	Corrosion protection was mainly due to the formation of a Ce oxide layer [14]

(continued)

Table 6.1 (continued)

Material	Material	Oxidant/acid/alkali/salt	*T* (°C)	*P* (MPa)	Corrosion results
	316	100% H_2O_2/200% NaOH/ 2-Chlorophenol	320/380	22/25	The elevation of destruction efficiency was greater at subcritical than supercritical; the various granule metal oxides covered the surface, mainly included Fe oxides [15]
	316	HCl	350–500	25	Weight losses were significantly increased with the increasing hydrogen partial pressure [16]
	316	1–5 wt.% NaCl + 0.01–1% H_2O_2	350–500	40	Mass loss was increasing with NaCl content; layer enriched with Cr and Mo, depleted in Ni and Fe [17]
	316	0.025 mol kg^{-1} H_2SO_4 + 0.025 mol kg^{-1} O_2	400	25–60	Good corrosion resistance at 400 °C both at 30 MPa and 40 MPa; intergranular corrosion occurred and the corrosion rate increased by more than two orders of magnitude when pressure at 60 MPa [18]
	316/625/C276	0.001 mol L^{-1} HCl/ 0.01 mol L^{-1} NaOH	400	25	Cracking occurred under both feed solutions [19]
	316 + TiO_2	5vol.% air + 0.1/0.2/ 0.5 mmol L^{-1}HCl	–	–	TiO_2-coated suggested the applied coatings were permeable to subcritical and supercritical water [20]
Nickel-based	Ni-based	0.01 mol/kg H_2SO_4	400–650	30	The reduction in corrosion resistance was more significant for alloys with higher Cr content [21]
	Ni-based	0.1 mol L^{-1} HCl	400	25	Cracking susceptibility of Ni-based alloys decreased with increasing Cr content of alloys [22]

(continued)

Table 6.1 (continued)

Material	Material	Oxidant/acid/alkali/salt	*T* (°C)	*P* (MPa)	Corrosion results
	Ni-based	1.5 g L^{-1} NaCl/10 wt% H_2O_2	400–500	22–25	Chloride leaded to crack initiation and propagation during the first few hours of constant load tests [23]
	Ni-based	pH = 1, 2, 7	300/360/425	24.1	pH = 1: Chromium was selectively dissolved and only molybdenum forms stable oxide; pH = 2: Dealloying of Ni and Fe, and oxidation of Cr and Mo were observed; pH = 7: Dealloying was not observed at any temperature; SCC in acidic feed streams was observed at all temperatures [24]
	Ni-based	0.01 mol/kg H_2SO_4+3 ppm to 800 ppm O_2	400	30/60	Fe became major contributor to corrosion resistance when oxygen concentration of the environment was increased up to 800 ppm [25]
	Ni-based	0 01 mol kg^{-1} H_2SO_4 + 0.025 mol kg^{-1} C_2	400	25–60	Corrosion rates of the Ni-based alloys at 400 °C significantly increased with increasing pressure [18]
	Ni-based	HCl	300–600	–	G-30 exhibited the highest corrosion resistance; most pronounced degradation being at high subcritical temperatures [26]
	Ni-based	CH_2Cl_2/ H_2O_2 + 2000 ppm chloride	420	40	Weight loss of all the samples varied between 1.3 mm/y for G-30 and 31.6 mm/y for alloy 686; G-30 showed good corrosion behavior due to formation of a protective oxide layer during preoxidized [27]

(continued)

Table 6.1 (continued)

Material	Material	Oxidant/acid/alkali/salt	*T* (°C)	*P* (MPa)	Corrosion results
	625	2.0% H_2O_2	450–500	24	$Ni(OH)_2/NiO/NiCr_2O_4/Cr_2O_3$ from outer to inner layer [28]
	625	Deionized water	500	24	Own the thinnest oxide layer and exhibited a low density of micrometer-sized pits A two-layer structure with a chromium-rich inner and an outer layer with higher iron content [5]
	625	100%H_2O_2/200%NaOH/ 2,4-dichlorophenol	340/380	25	The mass reduction at supercritical (0.074%) was higher than subcritical (0.026%) [29]
	625	4.8 wt.% H_2O_2/2 wt.% H_2SO_4/2.4 wt% NH_3	380–390	23	Corrosion rate was 18 $mmyr^{-1}$ at this condition [9]
	625	HCl (pH = 2)	355/350/ 299	24.1	Maximum thickness of the dealloyed oxide layer was 7.5/5/4.3 μm; the maximum thickness occurred at high-temperature subcritical [30]
	625	HCl (pH = 2.0)	300–425	24.1	SCC at subcritical temperature; Ni was dissolved, and Cr forms stable oxides [31]
	625	0.1 mol L^{-1} $C_6H_4N_2O_5$, 25 mol L^{-1} $[NH_4]_2SO_4$, and 1.3 mol L^{-1} NH_3	≤ 380	22.1	Ammonia and sulfate leaded to solution acidification and NiO instability The alloy ultimately failed by transpassive dissolution [32]
	625	100% H_2O_2/200% NaOH/ 2-chlorophenol	320/380	22/25	Subcritical contributed to the elevation of destruction efficiency was greater than supercritical [29]

(continued)

Table 6.1 (continued)

Material	Material	Oxidant/acid/alkali/salt	*T* (°C)	*P* (MPa)	Corrosion results
	625	Deionized water	400–550	24	The oxide on 625 was too thin to measure [7]
	625	H_2SO_4/$NaHSO_4$/Na_2SO_4	150–500	38	Severe material loss occurred between 300 and 390°C; corrosion phenomena in oxygenated [33]
	625	$Cl^- + O_2$	130–500	38	In neutral or alkaline solutions and in deaerated hydrochloric acid (HCl), corrosion rates of transpassive dissolution decreased drastically [34]
	625	0.2 mol kg^{-1} H_2SO_4 + 1.44 mol kg^{-1} O_2	250–500	38	The upper temperature limit of severe corrosion at 24 MPa and 390 °C; as temperature increased further and the density of the solution dropped to low values, only slight corrosion was detected [35]
	625 chromium	$HCl + H_2O_2 + NaCl + CH_3OH$	500	46.5	On 625, a strong transition of Ni into the fluid phase; on pure chromium, Cr_2O_3 scales composed of several layers were formed and spallation was observed [36]
	C-276	HCl	350–550	25	Containing both Cr and Mo were more resistant to corrosion than Ni–Cr binary alloys [16]
	C-276	Oxygen concentration<10 ppb	500–600	25	An outer NiO layer and an inner Cr_2O_3/$NiCr_2O_4$-mixed layer [37]
	C-276	100% H_2O_2/200% NaOH/2-chlorophenol	340/400	22.29/25.33	The depth of oxygen penetration in metal at supercritical was deeper than at subcritical and a greater loss of alloys at the subcritical than at the supercritical [38]

(continued)

Table 6.1 (continued)

Material	Material	Oxidant/acid/alkali/salt	*T* (°C)	*P* (MPa)	Corrosion results
	C-276	700% H_2O_2/ 2,4-dichlorophenol	400	25	Carbon-contaminated and iron oxide on layers; a significant depletion of chromium and nickel [13]
	C-276	60 wt.% 1,2-$C_2H_4Cl_2$/ 40 wt% C_2H_6SO/ 35 wt.% H_2O_2	538	31	Dealloying of Ni and Fe and cracking of the burst-disk [39]
	C-276	PCBs/670wppm chloride	400	30	Iron was detected at 4 mg/g of effluent. Nickel (100), molybdenum (17), chromium (27), and cobalt (1.1 mg/g); the temperature range most violently corroded in the SCWO system was 372–271 °C [40]
	800	25 ppb dissolved oxygen	500	25	Grain boundary engineering (GBE) enhanced spallation resistance and reducing oxidation rate [41]
	Alloy690	1500 mg L^{-1} H_3BO_3/ 2.3 mg L^{-1} LiOH H_2O	325	15.6	Triple-layer oxide films; the outmost layer was big oxide particles, the interlayer with loose oxides, and the inner layer with incompact cellular oxides [42]
	Alloy690	Deionized water	400–550	–	Outer layer was iron-rich, and the inner layer was composed of Cr_2O_3, NiO and $Ni(Cr,Fe)_2O_4$ [7]

(continued)

Table 6.1 (continued)

Material	Material	Oxidant/acid/alkali/salt	T (°C)	P (MPa)	Corrosion results
	G-30	pH = 2–6, 2000wppm of chloride/H_2O_2	100–400	25–48	At pH 6, the corrosion attack in subcritical region increased with increasing pressure between 150 and 250 °C. A reduced attack at high pressures, whereas for 250 bar, a strong increase was obvious At pH = 2, corrosion attack increased with decreasing pressure between 100 and 380 °C. Above the critical point, the corrosion attack increased with increasing pressure or density of the fluid [43]
	G-30/Cr–Fe	CH_2Cl_2/ H_2O_2 + 2000wppm chloride	420	40	Preoxidized would improve the corrosion resistance of Cr–Fe; 20 μm oxide film on Cr–Fe and 300 μm oxide film on G-30; G-30 lost nickel down to a depth of 200 μm [44]
	Ni–Cr alloys	CH_2Cl_2 + H_2O_2	100–415	40	Kinetics for binary Ni–Cr alloys were compared with the predictions of a mathematical model [45]
	Ni–Cr–Mo	Extraction from the fly ash of incinerated municipal solid waste	400	25	Exhibiting preferable corrosion resistance, the Cr content was 30 mass% at least under this environment [46]

(continued)

Table 6.1 (continued)

Material	Material	Oxidant/acid/alkali/salt	*T* (°C)	*P* (MPa)	Corrosion results
	Fe–Cr (10–40% Cr)	1 mmol L^{-1} HCl^{-1} + 0.6 mmol L^{-1} H_2O_2	350–500	29.4	Increasing the Cr content improved oxidation resistance but accelerated transpassive dissolution at 500 °C; the addition of Mo (1–2%) to Fe–Cr alloys (>25%Cr) was found to be effective for improvements of the corrosion resistance; transpassive dissolution rate of Cr was high at temperatures around 350 °C and above 500 °C, but was low at temperatures of 400–450 °C [47]
	Ni–Cr (10–30% Cr)				
Titanium-based	Gr2	100% H_2O_2/200% NaOH/ 2-chlorophenol	340/400	24	Titanium Gr2 has the more corrosion resistance than Hastelloy C-276 and Zirconium 702 [38]
	Gr2	700% H_2O_2/ 2,4-dichlorophenol	400	25	Titanium Gr2 was more corrosion-resistant than 316 SS, titanium G2, and Zirconium 702 [13]
	Gr2/Gr5/Gr7/Gr12/P-C	O_2/HCl/H_2SO_4/H_3PO_4	600	27	Corrosion in chloride containing solution was low; In the presence of sulfate or phosphate, corrosion of Gr2 became severe and Gr12 showed a low corrosion rate [48]
	Ti alloy	C_4H_5Cl/ion-exchange resin + 2% HCl/2% H_2SO_4 + 2% H_2O_2	400	28.5	Ti alloys would be applied as reactor materials for organic waste decomposition [49]

(continued)

Table 6.1 (continued)

Material	Material	Oxidant/acid/alkali/salt	T (°C)	P (MPa)	Corrosion results
	Ta, Ti, and Ti alloys	Condition 1: 2% H_2SO_4 + 2% H_2O_2 Condition 2: 1.2% H_2SO_4 + 2% H_2O_2	350	30	Ti alloys were sufficient to be applied as reactor materials [50]
	TAI/ZTA/ALC	0.1 mol kg^{-1} H_2SO_4 + 0.1 mol kg^{-1} H_3PO_4 + 0.1 mol kg^{-1} HCl	240–500	27	In H_2SO_4, the corrosion rate was still high at higher temperatures and formed a non-protecting sale;
					In H_3PO_4, less corrosion due to the formation of $AlPO_4$ on the surface of the specimens;
					In HCl, dissolution of the alumina grained corrosion phenomenon at temperatures of 240 °C -290 °C, intergranular corrosion at higher temperatures;
					The order of resistant against intergranular corrosion as follows: TAI > ZTA > ALC [51]
	Four types	Deionized water	450	45	Corrosion resistance as follows: Si_3N_4 < SiC < mullite < alumina < sapphire. High-purity alumina ceramics showed the highest corrosion resistance in the ceramics [52]
Ceramics	Sialon ceramics	Deionized water	450	45	α-sialon exhibited better corrosion resistance than β-sialon and α/β-sialon, strength of β-sialon decreased to 65% of its original strength and α-sialon decreased to 90% [53]

(continued)

Table 6.1 (continued)

Material	Material	Oxidant/acid/alkali/salt	*T* (°C)	*P* (MPa)	Corrosion results
	Many types	0.44 mol kg^{-1} O_2 and 0.05 mol kg^{-1} HCl	465	25	HIP-BN, B4C, TiB_2, Y_2O_3, and Y-TZP disintegrated; SiC- and Si_3N_4-based materials showed a large weight loss, up to above 90%; Al_2O_3- and ZrO_2-based materials did not corrode severely [54]
	Multilayered ceramics/nickel alloys	O_2/chlorinated cutting oil	300–650	–	The experimental nickel alloys performed better than the baseline nickel alloys; a titania-multilayered ceramic system had withstood such an environment without significant loss [55]
	Cr_2O_3-based ceramics	Deionized water	650	25	Adding proper amounts (65 wt.%) of MgO and CeO_2 can enhance the stability of Cr_2O_3-based ceramics, because of formation of $MgCr_2O_4$ or reactive element Ce. When the addition content increased to 8 wt.%, the degradation of Cr_2O_3-based ceramics increased [56]
Pure metal and metal oxide	Niobium	0.05 mol kg^{-1} HCl + 0.48 mol kg^{-1} O_2 0.05 mol kg^{-1} H_2SO_4 + 2 mol kg^{-1} O_2	350/500	24	Niobium could be an applicable material for reactors in subcritical water and not be used as reactor material for supercritical water under oxidizing conditions [57]
	Niobium	4.8 wt.% H_2O_2/2 wt.% H_2SO_4/2.4 wt% NH_3	380–390	23	The rate of corrosion of the niobium coupons was low [9]

(continued)

Table 6.1 (continued)

Material	Material	Oxidant/acid/alkali/salt	*T* (°C)	*P* (MPa)	Corrosion results
	Tantalum	0.05–0.2 mol kg^{-1} HCl/ H_2SO_4/H_3PO_4 + O_2	360–500	24	The corrosion rates in HCl and H_2SO_4 were low at 360 °C and unacceptably high at supercritical temperatures, corrosion rates in H_3PO_4 were low at all temperatures [58]
	Fe/Ni/Cr	HCl + O_2	400	25–40	The corrosion rate of chromium was the smallest; iron oxide solubility was the highest, and chromium oxide solubility was the lowest among these metals [59]
	Fe/Mo/Cr	5 wt.% NaCl	300–450	30	Pure Fe was covered with a thick oxide film; the mass changes of pure Cr were very small; the mass loss of pure Fe is higher than that of pure Mo at 300–350 °C, but is lower at 400–450 °C [60]
	Fe/Ni/Cr/ Mo/Ti	1 mmol L^{-1} HCl	500	29.4	Oxidation rates: Fe > Ni > Ti > Mo ≫ Cr; the addition of H_2O_2 to HCl solutions accelerated the corrosion of Fe, Ni, Cr, and Mo; Ni showed the largest weight gain among the five metals examined [61]
	Cr_2O_3/Fe_2O_3/NiO	Deionized water	450	26	Chromium oxide solubility in high-temperature region was the lowest among the three oxides; metal oxide solubility increased with increasing temperature [62]

(continued)

Table 6.1 (continued)

Material	Material	Oxidant/acid/alkali/salt	*T* (°C)	*P* (MPa)	Corrosion results
	Ti/Zr/Nb/Ta/Au/SS/ Hastelloy C	0–0.02 mol kg^{-1} Acid 0.1–1 mol kg^{-1} O_2 0–0.2 mol kg^{-1} NaCl	200–400	26	Nb and Zr rapidly converted to oxides; Ti proved resistant to dissolution, but was subject to pitting corrosion; Au was subject to galvanic effects; Ni and Mn of alloys preferentially dissolving out of the metal and Cr and Mo remaining begin; Ta proved to be the most resistant metal tested [63]
	Ni/Mo/Cr	HCl	350	24	Ni and Mo showed strong material loss; corrosion products formed at the surface consisted of oxygen and chromium [64]

iron-based alloys would be rich in Fe_2O_3/Fe_3O_4 and Cr/Ni/Mo oxides, respectively, in either SCWO or SCW without chloride [5, 7, 60, 65, 66]. The depletion of Fe was also observed on outer layer when solution contains the chloride. Iron-based alloys showed a weaker corrosion resistance than nickel-based alloys in previous SCWO corrosion studies, especially in the chlorinated solution [67]. The 316L alloys showed the highest corrosion rate about 50 mmy/r among all alloys, indicating that austenitic stainless steel was not suitable for chlorinated feed in SCWO [26]. When the test was with decomposed organic matter, it was always found that the carbon-contaminated was covered the surface of alloys due to the incomplete degradation of organic matter [13, 68]. It is known that the high-temperature subcritical water has higher corrosivity and solubility than supercritical water. Alloys always get a higher corrosion rate at subcritical condition than supercritical condition. The higher destruction efficiency of organic matter at subcritical condition can be explained by the more dissolved catalytic alloying elements by severe corrosion [24, 26, 29, 30, 34, 38, 39, 43, 69–72]. The high pressure 60 MPa could make intergranular corrosion and corrosion rate by more than two orders of magnitude than 25 MPa, indicating the importance of density of water on corrosion behavior of alloys [18]. With the improving oxidant concentration and adding H_2SO_4, the corrosion rate of 316L SS could reach 39 mmyr^{-1} and the high rate of 316L tubing corrosion due to the fluid mechanical effects associated in the dense supercritical fluid region [9]. The Ce^{3+} could form a Ce oxide layer to prevent the alloy from further corrosion [14]. Depending on the above results, iron-based alloys could endure the corrosion conditions of supercritical water with or without oxidant [18]. However, iron-based alloys displayed a high corrosion rate under both subcritical water and supercritical water when the feed contained aggressive feeds (Cl^-, SO_4^-, and PO_4^{3-}).

(2) **Nickel-based alloy**

The nickel-based alloys are generally considered important for severe service application; however, they cannot withstand some aggressive feeds in SCW. The studies for corrosion behaviors of nickel-based alloys were conducted among Inconel 625, Hastelloy C-276, Hastelloy G-30, Inconel 690, and others under supercritical or subcritical conditions. Increased chromium concentrations resulted in increased corrosion resistance of nickel-based alloys under SCWO condition, indicating the protective oxide Cr_2O_3 formed on the alloys [21, 30, 36, 46, 47]. The increasing chromium content also accelerated transpassive dissolution at 350 and 500 °C, but was low at temperatures of 400–450 °C [47]. Stress cracking corrosion (SCC) was observed at a low pH value; however, it was not observed in the neutral solution [24]. In the corrosion test of the nickel-based alloys exposed to an aqueous solution containing chloride and oxygen, pitting corrosion was observed at temperatures above 130–215 °C and the transpassive dissolution dominated at the critical temperature; however, transpassive dissolution corrosion rate of Inconel 625 decreased drastically in neutral, alkaline, and deaerated hydrochloric acid (HCl) solution [34]. Corrosion rates of the Ni-based alloys significantly increased with increased pressure in an aqueous solution containing H_2SO_4 and O_2 at 400 °C [18]. In supercritical oxidizing water, the oxides of nickel-based alloys were generally composed of duplex-layer structure, and the inner layer was a chromium-rich layer containing Cr_2O_3 or $NiCr_2O_4$, and the outer layer was a nickel-rich layer containing NiO or Ni(OH), indicating that the diffusivity of Ni is fast [5, 7, 25, 28, 31, 37, 42, 73–75]. However, the acidic solution would lead to NiO instability in high temperature [31, 32]. The most mass reduction and the most severe morphological change were both observed under subcritical condition [26, 33, 38, 64, 76]. In contrast to 304L SS, 316L SS, and Inconel 690, the oxide on Inconel 625 was too thin to be measured in deaerated supercritical water over the temperature range 400–550 °C [7]. However, the Inconel 625 also presented unacceptably high corrosion rate 18 $mmyr^{-1}$ when exposed to heavily oxygenated ammoniacal sulfate solutions in the high-density supercritical region [9]. The G30 exhibited the highest corrosion resistance and lowest corrosion rate 5 $mmyr^{-1}$ among various alloys exposed to an aggressive chlorinated SCWO feed [26]. The Cr-based alloy Ducrolloy displayed a reduction of weight loss by air preoxidation before SCWO corrosion test; however, the alloy G30 showed no significant effect of the preoxidation on corrosion rate [44].

(3) **Titanium-based alloy**

Titanium and titanium alloys (Gr2, Gr5, Gr7, Gr9, and G12) were candidate alloys for conducting corrosion tests at SCWO conditions. However, the titanium could not reach the required mechanical strength for high-temperature and high-pressure applications, and it was suited for using as liners in SCWO systems with high chloride concentration [48–50, 63, 72, 77–81]. The formation of TiO_2 oxide layer played a major role in titanium dissolution in high chloride [82, 83]. Corrosion of Gr2 became severe, and Gr12 showed a low corrosion rate among candidate titanium-based alloys in the presence of sulfate or phosphate [48]. The titanium Gr2 showed the more resistant to corrosion than other alloys (Inconel 600, Hastelloy C-276, Monel 400, 316 SS, and Zirconium 702) in the SCWO process for decomposing halogenated compounds [13, 38, 77]. The titanium displayed more resistant to corrosion than 304 SS in high-temperature subcritical solution with HCl [84]. Titanium oxyphosphates and titanium oxide composed the corrosion products generated on pure titanium in treatment of chlorpyrifos by SCWO [72]. When HCl, H_3PO_4, and oxidant were presented in SCW, titanium-based alloys were appropriate corrosion-resistant materials; however, titanium-based alloys showed creep and suffered severe corrosion by H_2SO_4 [77, 79].

(4) **Ceramic material**

Zirconia toughened alumina (ZTA), TAI, ALC, sialon ceramic, HIP–BN, B_4C, Si_3N_4, TiB_2, Y_2O_3, Y–TZP, SiC, mullite, alumina, sapphire, Cr_2O_3, and titania-multilayered ceramics were applied for corrosion test at condition of SCWO. Corrosion resistance for the ceramics is as follows in this order: Si_3N_4 < SiC < mullite < alumina < sapphire in deionized water at 450 °C and

45 MPa [52]. The SiC, alumina, Y-stabilized zirconia, Si_3N_4, BN, aluminosilicate, and cordierite–mullite showed poor corrosion resistance in supercritical water gasification of sugar beet slurry [85]. In oxidizing supercritical water, materials containing HCl, SiC, Si_3N_4, HIP–BN, B_4C, TiB_2, Y_2O_3, and Y–TZP-based showed severe corrosion; however, Al_2O_3- and ZrO_2-based materials did not corrode severely [54]. Due to the formation of non-protecting corrosion products, the ZTA ceramics suffered severe corrosion in high-temperature aqueous solutions containing H_2SO_4 or HCl; however, the formation of berlinite ($AlPO_4$) prevented corrosion of the specimens in H_3PO_4 solution [51]. The Cr_2O_3-based ceramics with yttrium oxide or yttrium oxide-stabilized zirconium oxide added suffered obvious degradation with oxygen concentrations (4 vol.% hydrogen peroxide) at 650 °C [86]. The titania-multilayered ceramic system remained intact after 120–180 h when exposed to a chlorinated cutting oil TrimSol in a different SCWO environment, ranging from 300 to 650 °C [55]. The α-sialon displayed better corrosion resistance and strength than β-sialon on exposure to supercritical water at 450 °C under 45 MPa for 2–50 h [53]. The proper adding amounts (≤ 5 wt%) of MgO and CeO_2 could enhance the stability of Cr_2O_3-based ceramics in SCW at 650 °C and 25 MPa [56].

(5) **Pure metal and metal oxide**

Niobium, zirconium, tantalum, gold, iron, nickel, chromium, titanium, molybdenum, and chromium oxide were also selected for conducting corrosion tests in SCWO process, and these tests are important to investigate the corrosion mechanism. Niobium was resistant to corrosion in oxidizing subcritical water with acid and oxidizing supercritical water without acid; however, niobium or zirconium would rapidly convert to oxides which eroded into the fluid in oxidizing supercritical water with acid [9, 57, 63]. Tantalum displayed the corrosion rates increased strongly above the critical temperature in HCl and H_2SO_4, and corrosion rates in H_3PO_4 were low at all temperatures [58]. The tantalum was also proved to be the most resistant metal among Hastelloy C, titanium, zirconium, niobium, tantalum, and gold in high-temperature oxidizing solution containing acid and NaCl [63]. The gold would not be suitable for applying in a SCWO system, due to the galvanic effects of Au [63]. In supercritical pure water, chromium showed the lowest oxidation rate among all tested metals, and their oxidation rates were in the following order: Fe > Ni > Ti > Mo $\gg$ Cr [61]. In addition, the addition of H_2O_2 to HCl solutions accelerated the corrosion of Fe, Ni, Cr, and Mo, and Ni showed the severest corrosion among five metals [61]. The soluble metal cations could precipitate to fine particles to form a protecting layer at supercritical conditions, indicating that the addition of ionic light metal salt could reduce corrosion [87]. The bare uncoated tube showed severer breakaway corrosion than TiN and $Ti_{0.35}Al_{0.65}N$ coated tube after exposure to supercritical water [88].

The research on stability of metal oxides was important for investigating the process of oxidation corrosion in high-temperature water. The solubilities of three tested metal oxides were in the following order: iron oxide > nickel oxide >

chromium oxide, and metal oxide solubility increased with increasing temperature at a higher pH region and at neutral conditions [59, 62].

There is no material can endure all SCWO conditions, and different materials show different corrosion properties under the SCWO conditions. Accordingly, the most suitable material will be selected for constructing the SCWO system by taking corrosion resistance and cost into account. As Table 6.1 shows, the corrosion tests of iron-based alloys, nickel-based alloys, titanium alloys, ceramic, pure metals, and metal oxides were conducted, and the test results are useful for selecting proper material in SCWO plant. The nickel-based alloy presented more resistance to corrosion than iron-based alloy; however, it also showed the severe corrosion under conditions with aggressive SCWO streams. Titanium-based alloy and partial ceramic material presented outstanding resistance to corrosion; however, they were always used as liners due to the weak mechanical strength. The corrosion tests of pure metal and metal oxide were conducted for investigating corrosion mechanism, and niobium was recommended to apply at subcritical condition.

6.1.2 Effect of Alloying Elements on Corrosion Behavior

Under the same condition, different materials always display different corrosion properties, which mainly depend on different alloy elements. The recent studies have focused on behaviors of different element in candidate materials on the condition of SCWO. The content variation of alloy element, the stability of metal oxide, and the element migration were the main contents of the research. The important elements included iron, chromium, nickel, molybdenum, and carbon.

(1) **Iron**

The iron is usually used to improve economy of iron-based alloys, nickel-based alloys, and titanium-based alloys. In supercritical water, Fe shows the highest oxidation rates to form stable oxide Fe_3O_4 among the other metal elements (Fe, Ni, Ti, Mo, and Cr), and it also shows the major contribution to corrosion resistance at the existence of oxygen in supercritical water [5, 7, 61, 89]. However, the corrosion rate of Fe in the oxidizing HCl solution is three times higher than that in supercritical HCl solution, indicating that the acidic chlorinated solution can also promote dealloying Fe in high-temperature aqueous solution [17, 39, 61]. At low temperature 400 °C, Cr can form a protective oxide, indicating Cr oxide is more stable than Fe oxide and Fe oxide solubilities are the highest among Fe, Ni, and Cr [14, 59]. The Fe always has a faster diffusion velocity than other elements so that the Fe is rich in the outer oxidation layer and Fe oxides make up the majority of the outer layer at temperature higher than 400 °C [7, 8, 90, 91]. The mass loss of Fe is higher than that of Mo at 300–350 °C and lower than that of Mo at 400–450 °C in supercritical water with NaCl [60]. The Fe oxides could hardly exist as the temperature increased stably, especially over 500 °C. The Fe oxides could dissolve at high temperature, so the outer layer becomes loose and desultory. Fe would change to Fe_2O_3, Fe_3O_4, and $\alpha(\gamma)$-FeOOH under the SCWO condition.

(2) **Chromium**

The Cr is considered as the most important alloying element to improve corrosion resistance in SCW [21]. It was found that 316 steel stainless would decrease the corrosion rate greatly after the chromizing treatment under SCWO condition using 5% NaCl solution. The test result indicated alloy suffered the most serious corrosion in high-temperature subcritical water and chromium had excellent corrosion resistance to chloride ion from subcritical to supercritical condition. The higher Cr concentration would make lower penetration rate of alloys. The chromium oxide presented the lowest solubility among the oxides of chromium, iron, and nickel [17, 21, 31, 59, 62, 64]. The Cr showed the most stable state among Fe, Cr, and Mo in 5 mass% NaCl at temperature of 300–500 °C [60]. The dissolved chromium ions could form compact chromium oxide layer to provide a protective function against corrosion at 400 °C and 30 MPa in H_2SO_4 aqueous solution [25, 92]. The cracking susceptibility of nickel-based alloys also decreased as the Cr content increased in supercritical water with either HCl or NaOH [19]. The higher chromium-containing alloys presented the lower oxide thickness than the higher chromium-containing alloys [74]. The oxygen affinity of chromium was stronger than iron and nickel, and Cr would be selectively oxidized to $Cr(OH)_3$ and Cr_2O_3. The Cr(III) of Cr_2O_3 would be transformed into Cr (VI) by further oxidation in the high-temperature oxidizing water, and the Cr(VI) would loss by the dissolution. A more continuous and stable Cr_2O_3 layer was formed at higher temperature (450 and 500 °C) than critical temperature (400 °C) in oxidizing supercritical water [28]. As a result of the more slow diffusion velocity than Fe, the Cr was usually rich in the inner layer [5, 7, 91]. The Cr oxides were also unstable and dissolved in supercritical strongly acidic solution (pH = 1). The Ni–Cr–Mo alloys contained Cr at least 30 mass% showed better resistant to corrosion in a supercritical water including chloride [46]. The increasing of Cr content improved corrosion of the alloy, and alloy also displayed good corrosion resistance even at the Cr contents as high as 40% [30, 47]. The NiCr25 alloy performed best resistant to corrosion among binary Ni–Cr alloys (0–45 mass% Cr) in an aqueous solution resulting from the oxidation of CH_2Cl_2 at 40 MPa and temperatures between 100 and 415 °C [45].

(3) **Nickel**

The toughness, strength, and corrosion resistance of the steel will be enhanced when it contains Ni element. The oxygen affinity order for metal element is Cr > Fe > Ni, indicating the Cr and Fe can easily form the oxides under the SCWO conditions. The nickel-based alloys often show a severe depletion of nickel. In this simplified model, nickel ions migrated to solution and ultimately changed to dissolved form. Oxygen ions migrated to the matrix and combined with the chromium to form chromium oxide. In contrast to other metal elements like Fe, Cr, Mo, the Ni usually presents a worst serious depletion and the highest concentration of nickel is detected in the corrosion effluent, especially in acidic chlorinated solution [13, 17, 30, 31, 36, 39, 40, 44, 93]. The pure chromium presents greater corrosion rate than Inconel 625 in supercritical water containing HCl, NaCl, and H_2O_2 [36].

(4) **Molybdenum**

The Mo can improve high-temperature strength resistance to pitting corrosion and resistance to crevice corrosion in high-temperature environment. The Mo can form more stable Mo oxides (MoO_2, MoO_3) than other metal in strong acidity and chlorinated solution at subcritical or supercritical temperature [17, 24]. The Mo improves the resistance to pitting corrosion in subcritical salt solution obviously [94]. In nickel-based alloys, the synergistic effect of Mo and Cr shows more resistant to corrosion than Ni–Cr binary alloys [16, 95].

(5) **Carbon**

The hardness of steel stainless increases and the malleability decreases as the carbon content increases. Carbon contamination is significant in treatment of organic wastes by SCWO, implying the destruction reaction occurs also on the material surface. Therefore, the high content of carbon can be observed in the outer layer oxides [13]. Other elements like niobium, titanium, zirconium, and tantalum are also investigated under the SCWO conditions. Titanium, niobium, and tantalum are stable as oxides in subcritical condition, and these elements can enhance the corrosion resistance in subcritical condition [9]. Ta_2O_5 and Nb_2O_5 are both the stable states in subcritical condition. However, the Zr oxides are more easily dissolved at the supercritical condition than at the subcritical condition.

6.2 Early Oxidation of Heat-Resistant Steels in Near-Pure SCW

6.2.1 Early Oxidation Behavior of Austenitic Stainless Steel TP347H

(1) **Oxidation kinetics**

The weight gains (ΔW) of the austenitic steel TP347H samples exposed to SCW at 25 MPa, and three temperatures of 465, 540, and 580 °C after the corrosion tests were gravimetrically measured, as shown in Fig. 6.2. The weight gains increased as the exposure time and temperature increased. Equation (6.1) was used to fit the weight-gain data,

$$\Delta W = kt^n \tag{6.1}$$

where k is a rate constant, and n is a time exponent. Those data produced good fitting results with an R-square higher than 0.96 at the three temperatures. For the

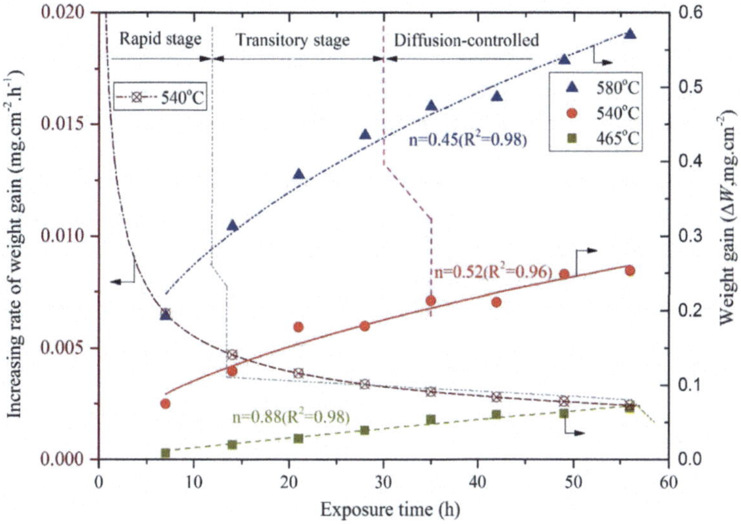

Fig. 6.2 Weight gain of alloy TP347H versus exposure time in SCW at 25 MPa and different temperatures [101]

465 °C, a time exponent of 0.88 was obtained, which indicates that the oxidation kinetics may follow a near-linear rate law during the early oxidation for exposure time less than 56 h. In contrast, the data obtained at 540 and 580 °C exhibited reasonable fitting results with n = 0.52 and 0.45, respectively. It suggests that the oxidation rates can be characterized by near-parabolic laws, which are consistent with the extensive results of a parabolic oxidation rate of TP347H at 538–565 °C from both tests in coal-fired power plants and long-term laboratory tests for up to tens of thousands of hours [1, 96]. Under the 540 °C condition, the weight gain rapidly increased with the exposure times before 14 h, and then its increase rate decreased gradually in the second period from 14 to 35 h. These two processes can be referred to as the rapid oxidation and transitory stages, respectively. After 35 h, the weight gain exhibited an increase at a slower rate than that in the transitory stage; this stage is referred to as the diffusion-controlled stage because in this period, the oxidation process may be controlled by a diffusion process [97–100]. When the exposure temperatures increased from 540 to 580 °C, the transitory stage occurred earlier. A possible explanation for this is that proper higher temperatures may promote the formation of a continuous and protective chromium-rich oxide layer due to higher oxygen affinity of chromium. Meanwhile, lower temperatures are not beneficial for the production of Cr-rich oxides by the selective oxidation of chromium, and such temperatures would hinder the transformation from the rapid oxidation stage to the transitory stage, which can reasonably explain why the oxidation process at 465 °C was still at the rapid stage for exposure times up to 56 h.

(2) Early oxidation evolution

At 540 °C (Fig. 6.2), the oxidation process of alloy TP347H achieved conversion from the rapid oxidation stage to the diffusion-controlled stage within 35 h. To explore this transformation process in detail, the early oxidation evolution of alloy TP347H was analyzed by SEM, AFM, Raman microscopy, and XPS. Plan-view SEM images of the SCW-exposed TP347H samples at 540 °C for 7, 14, and 35, are presented in Fig. 6.3. For each representative condition, two images with a low magnification (left) and high magnification (right) are depicted. Cuboidal Fe-rich oxides were distributed on top of an inner Cr-rich oxide scale. The cubic oxides grew with an increase in exposure time. Even after the exposure to SCW for 35 h, at which point the diffusion-controlled oxidation process had been activated, the scattered Fe-rich oxides were still visible rather than impinging and combining with each other to form a continuous and relatively complete outer oxide layer. This may suggest the preferential formation of a relatively uniform inner Cr-rich subscale as a diffusion barrier to metal ions moving outward [96]. In addition, pits with diameters of ~1 μm were also detected on the sample surface after an exposure of approximately 14 h, probably due to the presence of inclusions such as niobium carbide [102].

Detailed three-dimensional AFM topographies and characteristic parameters, which reveal the evolution of the geometric characteristics of the oxide scales obtained from alloy TP347H exposed to SCW at 540 °C for 7, 14, and 35 h, are displayed in Fig. 6.4. The 3-D AFM topographies (Fig. 6.4a–c) clearly depict a columnar oxide growth. As the exposure time increased from 7 to 35 h, an intuitive reduction in the peak density was observed. Figure 6.4 also presents the area peak height count distribution and the peak height parameters of the TP347H samples for three exposure times, which were obtained from a statistical analysis of the AFM topographies with a scanning area of 10 × 10 μm. Before 14 h, the peak height mode of the area, peak height count distribution, and the peak height parameters, such as the average roughness (*Sa*), area peak-to-valley height (*St*), maximum area peak height (*Sp*), and maximum area valley depth (*Sv*) all increased with increasing

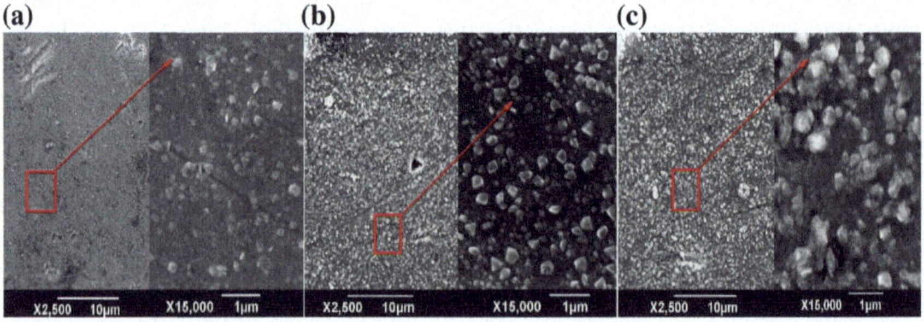

Fig. 6.3 Surface morphologies of the TP347H samples after exposure to SCW at 540 °C and 25 MPa for the three exposure times **a**: 7 h, **b**: 14 h, and **c**: 35 h. Images with different magnifications are presented

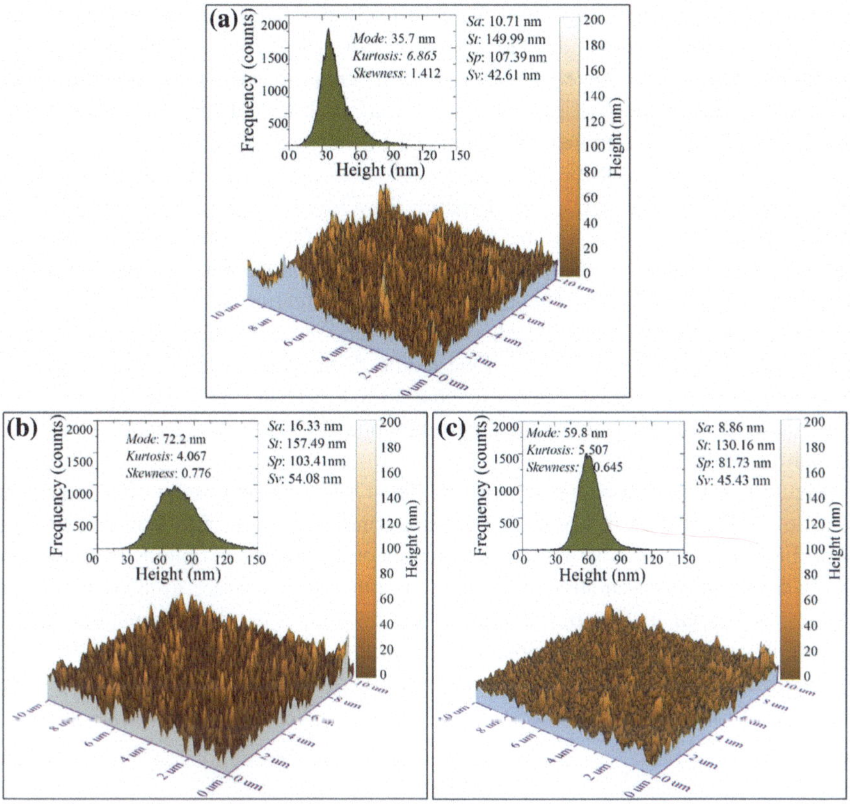

Fig. 6.4 Typical 3-D morphologies, area peak height count distribution, and peak height parameters (*Sa, Sp, Sv, and St*) of alloy TP347H after exposure to SCW at 540 °C and 25 MPa for **a** 7 h, **b** 14 h, and **c** 35 h, respectively. *Note Sa, Sp, Sv, and St* are all 3-D parameters expanded from the roughness (2-D) parameters, which express the average of the absolute values of deviations from the mean surface, the maximum value of peak height, the maximum value of valley depth, and the sum of *Sp* and *Sv*, respectively, on the surface in the measured area

exposure times. These phenomena can be attributed to the rapid formation and the predominantly longitudinal growth of the metal oxide nucleus, which agrees well with the observation in the rapid oxidation stage before 14 h shown in Fig. 6.2. From 14 to 35 h, the area peak height mode decreased, and the variation of peak height parameters (*Sa, St, Sp,* and *Sv*) exhibited the same tendency, which was likely due to the gradual formation of a uniform and compact Cr-rich protective inner layer scale. This was also reflected by the oxidation kinetics and surface evolution of the alloy TP347H exposed to SCW at 540 °C (see Figs. 6.2 and 6.3). The kurtosis and skewness were typically used to characterize the variability and location of a data set. Their calculated data are also presented in Fig. 6.4. The positive kurtosis values (representing a steeper distribution compared with the normal distribution, as the kurtosis is zero for the normal distribution) decreased from 7 to 14 h and then increased from 14 to 35 h. The positive skewness values for three exposure times indicated a lack of symmetry, or more precisely, a predominant number of lower peaks, which decreased with the exposure time. Presumably, the initial fine-grained oxides may collide and fuse with each other, thereby weakening the numerical advantage of the lower peaks.

The Raman spectra of the three oxides selected as comparative samples such as magnetite (Fe_3O_4), $FeCr_2O_4$, and hematite (Fe_2O_3) are displayed in Fig. 6.5. The primary Raman features for magnetite and $FeCr_2O_4$ are observed at 669 and 685 cm^{-1}, respectively [103], while hematite has characteristic features at 1320, 410, and 293 cm^{-1} [104]. The prominent Raman feature for chromia occurs at 553 cm^{-1} (not shown here) [103], which is very close to the secondary Raman feature of magnetite at 550 cm^{-1}. Due to the disturbance resulting from the predominant presence of the magnetite phase on the tested TP347H surface, it is difficult to determine whether the chromia phase exists only by observing the Raman patterns.

Figure 6.5 is also associated with the TP347H surfaces formed after exposure times of 7, 14, and 35 h. For each exposure condition, the surface components are described by three or four typical Raman spectra, which are obtained by successively selecting representative areas from the inner layer surface to the outer larger oxides, as shown in Fig. 6.5. These results provide a good confirmation on the surface heterogeneity. After exposure to SCW at 540 °C for 7 h (Fig. 6.5(10)–(7)), magnetite (Fe_3O_4), $FeCr_2O_4$, and hematite (Fe_2O_3) were detected on the inner layer

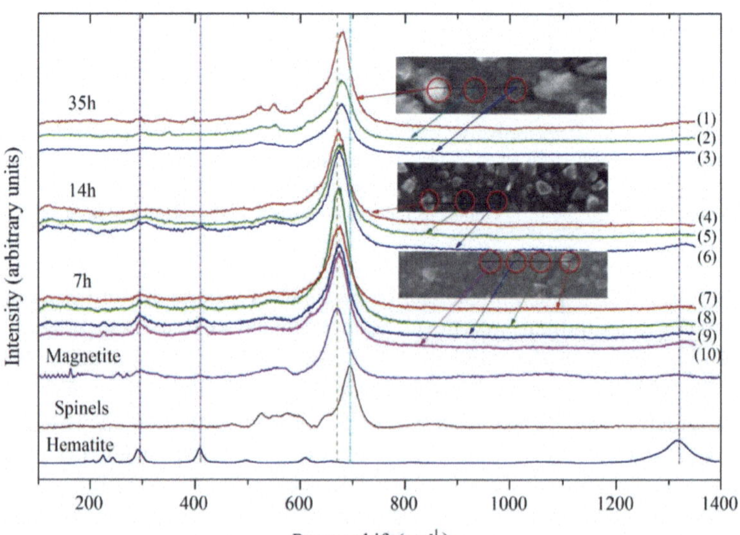

Fig. 6.5 Raman spectra for standard samples and TP347H samples exposed to SCW at 540 °C and 25 MPa for 7, 14, and 35 h, respectively. The numbered spectrum is obtained from representative areas from the inner layer surface to the outer larger oxides, as depicted in the figure

surface, while only magnetite (Fe_3O_4) and minute amounts of the $FeCr_2O_4$ phase were detected when the detector was focused on the outer oxides. Presumably, the isolated Fe-rich oxides may primarily consist of the $(Fe, Cr)_3O_4$ phase. The formation of magnetite and spinel on typical austenitic steels after exposure to SCW agrees with previous reports [10, 103, 105]. The Raman spectrum group at 14 h (Fig. 6.5(6)–(4)) indicates a similar variation tendency to that of the 7-h spectra, but the characteristic peaks of hematite on the inner layer surface (Fig. 6.5(6)) were relatively weak. The Raman features of hematite completely disappeared when the exposure time was further prolonged to 35 h (Fig. 6.5(3)). A comparative analysis of the three inner layer spectra of Fig. 6.5(10), (6), and (3), as shown in Fig. 6.6, indicates that the characteristic peak position of the spectrum in the range of 600–750 cm^{-1} shifted to higher Raman as the exposure time prolonged, and the $FeCr_2O_4$ percentage among magnetite and $FeCr_2O_4$ increased gradually. These results can provide a possible explanation for the preferential formation of $FeCr_2O_3$ compared to Fe_3O_4, which may be derived from the solid solution of Fe_2O_3 and Cr_2O_3, as well as the favorable reaction between Fe and Cr_2O_3 due to the decreasing oxygen partial pressure at the scale/environment interface as the exposure time increased.

Figure 6.7 shows XPS survey scans of the surface of the TP347H samples exposed to SCW for 7, 14, and 35 h. The results show the presence of chromium in

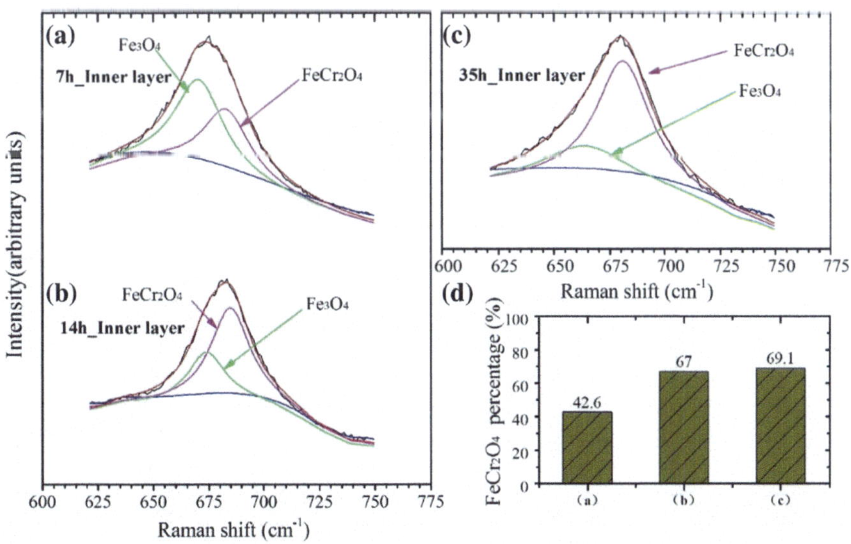

Fig. 6.6 Raman spectra in range of 600–750 cm^{-1} for the inner layer oxides of alloy TP347H with three exposure times, obtained from **a** 7-h spectrum (10), **b** 14-h spectrum (6), and **c** 35-h spectrum (3), respectively, in Fig. 6.5, and **d** relative content variation of $FeCr_2O_4$

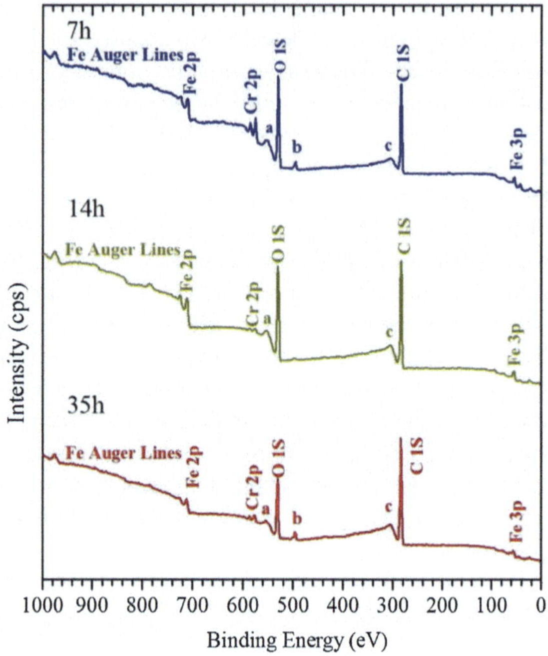

Fig. 6.7 XPS analysis showing the surface chemistry of TP347H samples exposed to SCW at 540 °C and 25 MPa for 7, 14, and 35 h, respectively

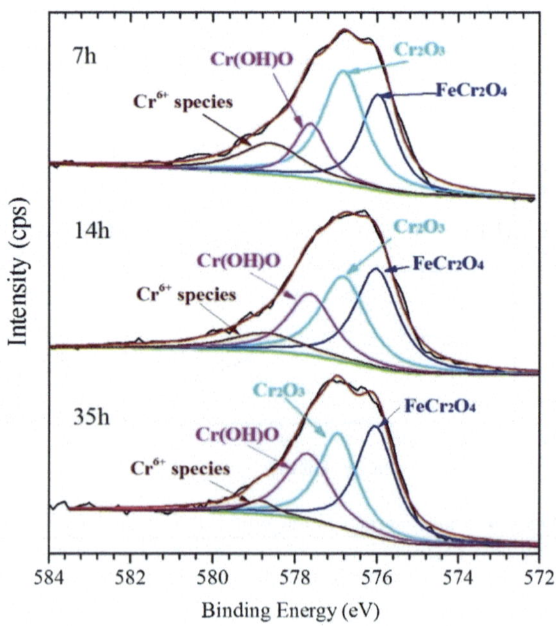

Fig. 6.8 XPS Cr $2p_{3/2}$ core-level spectra (and their decomposition) of the scales formed on alloy TP347H at 540 °C and 25 MPa

addition to iron, oxygen, and carbon on the sample surfaces ('C' is likely derived from contamination or fingerprints), further confirming the incomplete nature of the outer iron-rich layer. Small features corresponding to the iron Auger emission were also clearly observed. Notably, the signals for *a* and *b* are likely derived from the energy loss peaks or ghost peaks of O and C, respectively. However, Ni, as the third element in the substrate, was not detected regardless of the exposure time, due to its lower outward diffusion rates within the substrate and lower oxygen affinity compared to Cr and Fe [97, 106]. To identify the chemical states of the major elements of Cr, Fe, and O, their high-resolution spectra (obtained at binding energy intervals of 0.1 eV) are presented in Figs. 6.8, 6.9, and 6.10, respectively.

The results of the high-resolution Cr $2p_{3/2}$ core-level spectra in Fig. 6.8 reveal the presence of four components, most likely $FeCr_2O_4$ at approximately 576.0 ± 0.2 eV [107], Cr_2O_3 located at BE of 576.6 ± 0.4 eV [108, 109], Cr(OH)O at 577.2 ± 0.2 eV [110], and Cr^{6+} species at approximately 578.9 eV [111], although the latter three components were not detected using the above-mentioned characterization methods. With the increasing exposure time, the relative content of

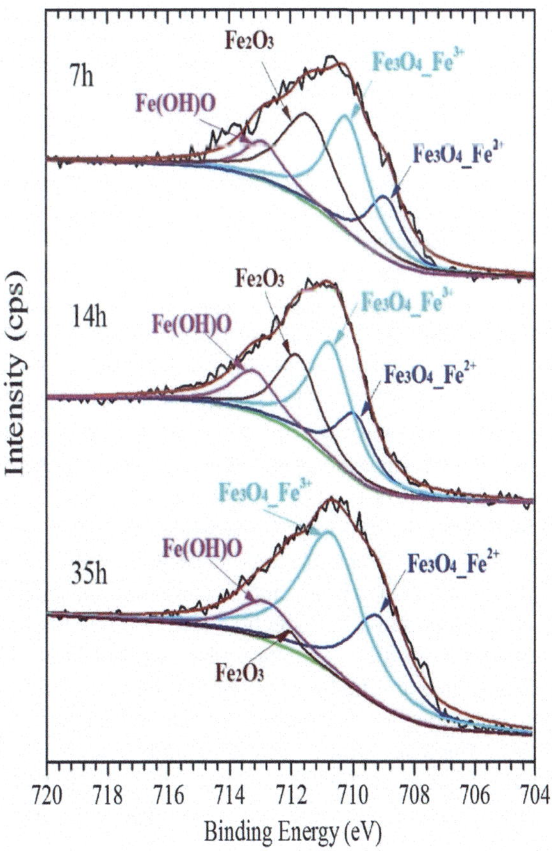

Fig. 6.9 XPS Fe $2p_{3/2}$ core-level spectra (and their decomposition) of the scales formed on alloy TP347H at 540 °C and 25 MP

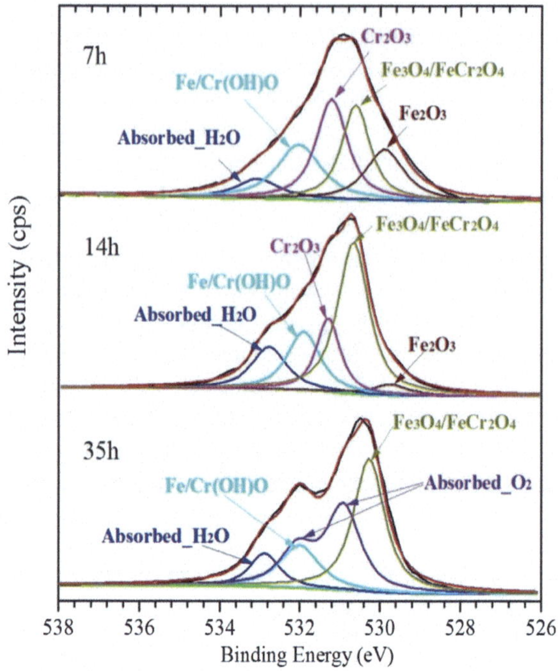

Fig. 6.10 XPS O 1s spectra and representative components of the oxide films obtained at 540 °C and 25 MPa after various exposure times

$FeCr_2O_4$ increased, while that of Cr_2O_3 decreased, which was indicative of the gradual conversion of Cr_2O_3 to $FeCr_2O_4$ by solid reactions [98, 112]. The presence of the Cr^{6+} species was unexpected, and the related signal intensity decreased as the exposure time increased.

Figure 6.9 displays narrow scans of Fe $2p_{3/2}$, which also indicate the peak fitting for all the samples [97, 113, 114]. Due to the multivalent nature of the Fe in Fe_3O_4, the used peak fitting methodology fitted peaks for Fe^{2+} and Fe^{3+}, and fixed the intensity ratio of Fe^{2+} to Fe^{3+} at approximately 1:2 [103]. The binding energy of the Fe in $FeCr_2O_4$ was typically between that of Fe^{2+} and Fe^{3+} in Fe_3O_4 phases, and thus, the characteristic peaks of Fe in $FeCr_2O_4$ were not presented in Fig. 6.9 for simplification [113, 114]. There was no notable variance in the iron hydroxide content with the exposure time, while the Fe_3O_4 (actually representative of Fe_3O_4 and $FeCr_2O_4$) concentrations increased. The Fe_2O_3 content decreased as the exposure time increased, and it tended to disappear at 35 h, which, to a certain extent, supported the deduction from the results shown in Fig. 6.6 that a transformation from Fe_2O_3 to other phases (most likely $FeCr_2O_4$) [100, 115].

From the identified primary hydroxides and oxides of Cr and Fe in Figs. 6.8 and 6.9, based on the observed O 1s binding energy in Fe(OH)O at 531.5 ± 0.3 eV [108], which is almost consistent with that in Cr(OH)O, and because the O 1s binding energy in Fe_3O_4 and $FeCr_2O_4$ are all located at ~530.2 eV [107], the representative components at three exposure times were selected separately as fitting objects, as shown in Fig. 6.10. The relative contents of the absorbed water and hydroxides were almost unchanged over prolonged exposures [109], while the $FeCr_2O_4$ and Fe_3O_4 became primary components as the concentrations of Fe_2O_3 and Cr_2O_3 decreased. At 35 h, the emergence of O 1s spectrum as a multi-peak photoelectron spectrum was unexpected, and it may be due to the ionization of non-valence electrons, such as (σ1s)2 or (σ*1s)2 within the O_2 molecules [116]. The molecular configuration of O_2 is (σ1s)2 (σ*1s)2 (σ2s)2 (σ*2s)2 (σ2px)2 (π2py)2 (π2pz)2 (π*2py)1 (π*2pz)1, with two unpaired valence shell electrons. Once any electron belonging to (σ1s)2 or (σ*1s)2 within an oxygen molecule was excited, the isolated electron remaining in the corresponding 1s orbital would trigger spin–spin coupling with the unpaired valence electrons ((π*2py)1 or (π*2pz)1) to form multiplets. There was no rational means for an increase in the oxygen partial pressure at the scale–SCW interface due to the sustained oxygen consumption as the exposure time increased, but the absorbed oxygen molecules were only detected on the TP347H samples exposed to SCW for 35 h but not on those exposed for a much shorter time such as 7 or 14 h. This phenomenon may be attributed to the limited inward transport of oxygen molecules. As noted in extensive investigations on corrosion behavior of austenitic and F-M stainless steels in SCW, the corrosion process was mainly controlled by the outward diffusion of iron ions through the inner Cr-rich oxide layer as a result of the more rapid inward transport of oxygen carriers to the substrate/scale interface [99, 100, 115, 117].

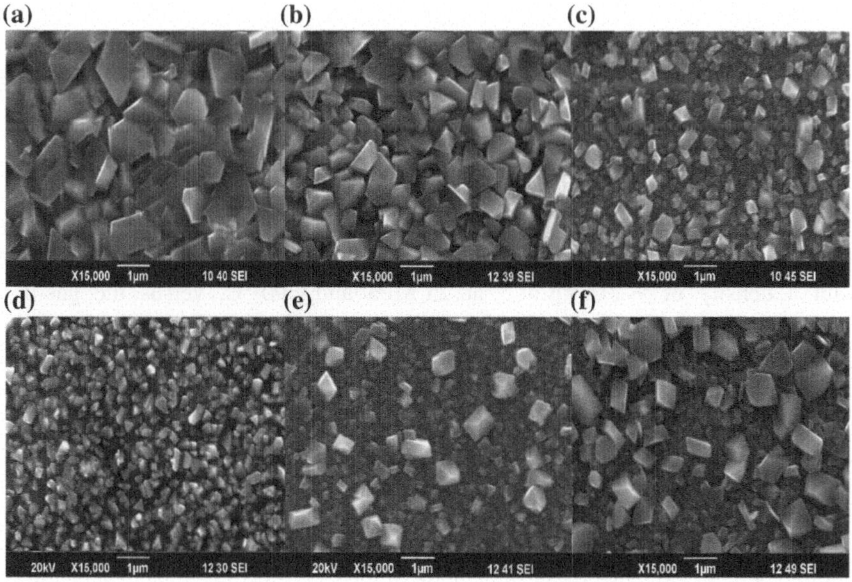

Fig. 6.11 Surface morphologies of TP304H samples after a 40-h exposure at **a** 25 MPa, 580 °C, **b** 25 MPa, 540 °C, **c** 25 MPa, 465 °C, **d** 25 MPa, 390 °C, **e** 17 MPa, 390 °C, and **f** 17 MPa, 540 °C

Thus, it appears that oxygen molecules may not be the predominant oxygen supplier for the inward growth of the inner layer in the diffusion-controlled stage.

(3) **Influences of temperature and pressure**

When the exposure time was extended to 40 h, the oxidation process of steel TP347H at 540 and 580 °C was in the diffusion-controlled stage for 5 and 10 h, respectively (see Fig. 6.2), while the oxidation process of the TP347H specimens exposed at 465 °C or lower temperatures was still in the rapid oxidation stage. Therefore, it is important to explore the effects of temperature and pressure on corrosion behavior of TP347H steel for a typical exposure time of 40 h. SEM images of TP347H specimens subjected to SCW or steam at 390, 465, 540, and 580 °C and different pressures for 40 h are depicted in Fig. 6.11. Figure 6.11a–c indicates that the scales formed on the TP347H samples exposed to SCW at 25 MPa from 580 to 465 °C are clearly a duplex-layer structure. The inner layer was closely packed with relatively fine-grain Cr-rich oxides, while the loose surface layer consisted of bulky, cubic Fe-rich particles that are approximately perpendicular to the sample surface. The distribution density of the cubic Fe-rich oxides decreased as the temperature increased, while their grain sizes increased from 0.5 to 1.5 μm. Figure 6.11d and e depict plan-view images of the oxide scale produced on the TP347H samples at 390 °C and pressures of 25 and 17 MPa, respectively. At 25 MPa, there was a negligible difference in the magnitude of the oxide particles, which suggests the absence of preferential growth of some oxide granules in the high-density SCW. For Fig. 6.11e, the obtained oxide scales were very similar to those generated in SCW at higher temperatures (Fig. 6.11a–c), except for a smaller density distribution of the larger Fe-rich oxides, which implies an analogous oxidation mechanism, i.e., the solid-state growth mechanism generally occurs in high-temperature steam and SCW with lower densities (typically less than 200 kg/m^3) [1, 118]. This explanation was also verified by the similar surface morphologies of the 540 °C specimens at 25 and 17 MPa, as shown in Fig. 6.11b, f. The clear difference between Fig. 6.11d and e appears to indicate that when the pressure increased from 17 to 25 MPa at 390 °C, a variation of the oxidation process of TP347H occurred. This may be attributed to the liquid-like behavior of the SCW with a density of $\sim$215 kg m^{-3} at 25 MPa and 390 °C, versus the gas-like behavior of the medium used in Fig. 6.11a–c and e–f [115]. The former may be more inclined to trigger another corrosion process of metal dissociation/oxide precipitation [91, 104].

Figure 6.12 illustrates the general relative atomic contents of Fe, Cr, Ni, and O on the surfaces of the tested samples, where the Fe proportion among the three main metals of Ni, Cr, and Fe gradually increased from 465 to 580 °C at 25 MPa (shown in Fig. 6.12c and e–f), which is in good agreement with the enrichment phenomenon of the Fe-rich oxides with the temperature, as shown in Fig. 6.11a–c. Figure 6.11c and e–f also show an increase of the oxygen content and a decrease of Cr and Ni on the sample surface as the temperature increases. A possible explanation for this is that higher temperatures can accelerate outward diffusion of the Fe ions to the surface, promoting the formation of cubic Fe-rich oxides on scale/gas

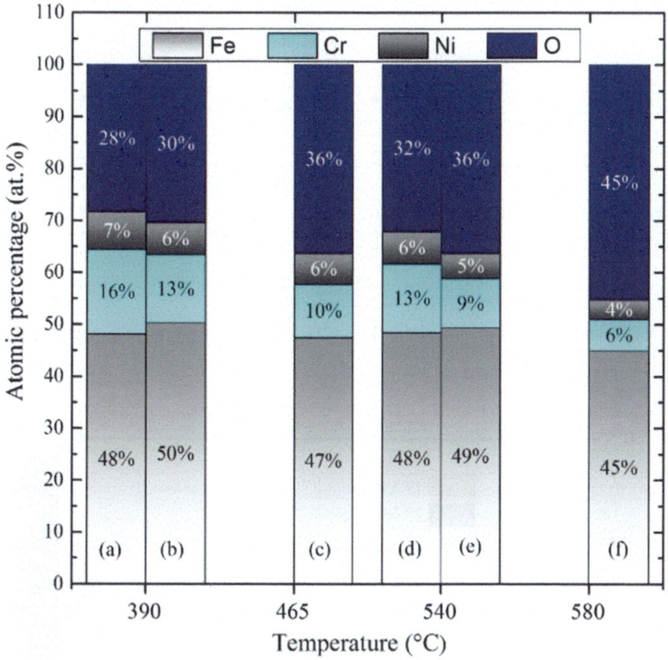

Fig. 6.12 Relative atomic percentages of Fe, Cr, Ni, and O on TP347H surfaces after a 40-h exposure at four temperatures and two pressures: (**a**) (**d**) at 17 MPa, others at 25 MPa

interface, and thus deteriorate the oxidation resistance of the tested alloys [104, 119, 120]. When the TP347H samples were tested at either 390 or 540 °C for 40 h (see Fig. 6.12a–b and d–e), the relative oxygen contents of the oxide scales formed on the sample surfaces increased when the environmental pressure increased from 17 to 25 MPa.

Figure 6.13 shows the surface X-ray diffraction patterns of the tested TP347H samples after exposure to SCW at 25 MPa for 40 h at three temperatures. Two major characteristic peaks were identified: (a) the Fe–Cr–Ni peak represents the substrate and (b) the magnetite (Fe_3O_4) and spinel-phase peaks corresponding to Fe $(Cr, Fe)_2O_4$ oxides. Figure 6.13 indicates that the majority of peaks can be associated with magnetite/spinel and the base alloy. When the temperature increased from 465 to 580 °C, the peak intensity of magnetite/spinel phases increased, while the characteristic peaks of the base alloys weakened, which implies the thickening of the scales.

Figure 6.14 macroscopically shows more severe oxidation at higher temperatures. The weight gains of the TP347H samples after a 40-h exposure increased with an increase in temperature from 390 to 580 °C at 17 and 25 MPa. It also shows that the weight gain of the TP347H samples after a 40-h exposure increased

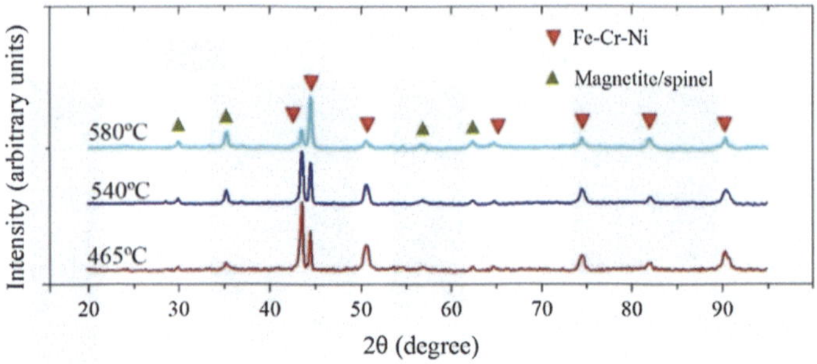

Fig. 6.13 XRD patterns of the corroded TP347H surfaces at three temperatures for 40 h at 25 MPa

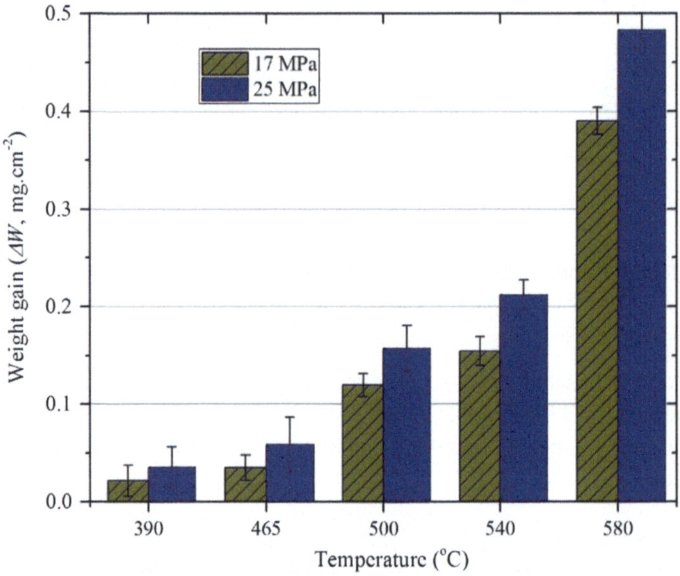

Fig. 6.14 Weight gain of TP347H samples exposed to SCW for 40 h versus temperatures at 17 MPa and 25 MPa. The "error bars" superimposed onto the chart mark the highest and lowest values recorded for each set of three samples

with an increase in pressure from 17 to 25 MPa regardless of the experimental temperature. The variations of the weight gain, and the oxygen partial pressure, and their increasing rates with the pressure at 540 °C are presented in Fig. 6.15. The weight gain of the TP347H samples after exposure for 40 h and its increasing rates slowly increased with the pressure from 17 to 23 MPa and then sharply increased from 23 to 25 MPa. The various experimental oxygen partial pressures (p_{O2}) were calculated from the decomposition equilibrium of H_2O using the HSC software and database [117, 121]. With an increase in pressure, p_{O2} increased gradually, while its

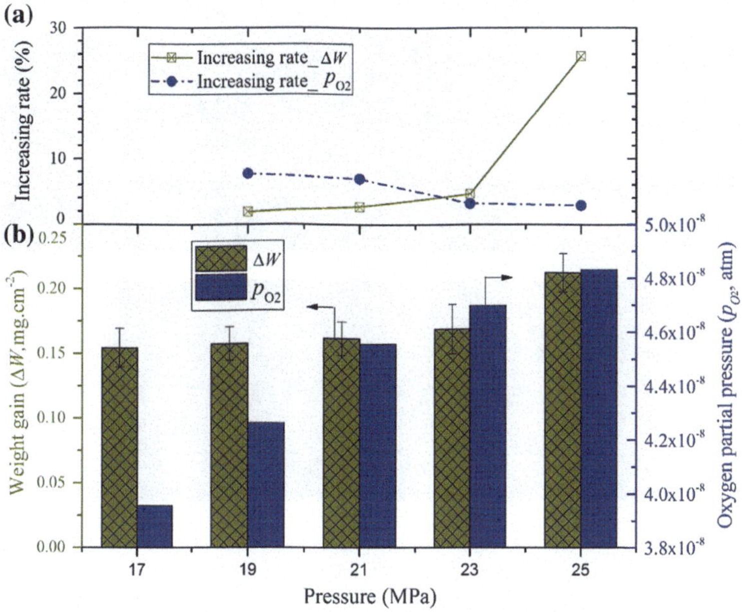

Fig. 6.15 **a** Weight gains (ΔW), oxygen partial pressures (p_{O2}), and **b** their increasing rates as a function of experimental pressure at 540 °C. The "error bars" superimposed onto the chart mark the highest and lowest values recorded for each set of three samples

increasing rate decreased rapidly from 17 to 23 MPa and then decreased gradually. The clear variation divergence between the increasing rates of the weight gain and oxygen partial pressure implies that the effect of the experimental pressure on the oxidation behavior of TP347H may not be embodied solely on the oxygen partial pressures.

(4) **Cross-sectional structure**

To explore the inward growth behavior of the inner subscales and compare the results with those reported in the literatures, the cross-sectional characteristics of the TP347H samples oxidized in SCW at 540 °C and 25 MPa for 120 h were analyzed by SEM/EDS (as shown in Fig. 6.16). An adherent, uniform duplex-layer scale was identified (Fig. 6.16a). The EDS maps of the major components show the O enrichment within the entire scale, and Fe, Cr, and Ni were clearly depleted in the inner layer, outer layer, and entire scales, respectively, relative to the metal substrate, as observed in Fig. 6.16b–e. This perceptive description is proven by the EDS line scans (Fig. 6.17) of the cross sections at the special region, which is schematically marked in Fig. 6.16a. Figure 6.17 also clearly shows the presence of a Ni-rich layer near the interface between the scale and matrix, which is consistent with the results obtained by Otoguro [122], Was [97], and others [1, 112, 123]. They reported that the Ni-enriched layer may increase the oxidation resistance of

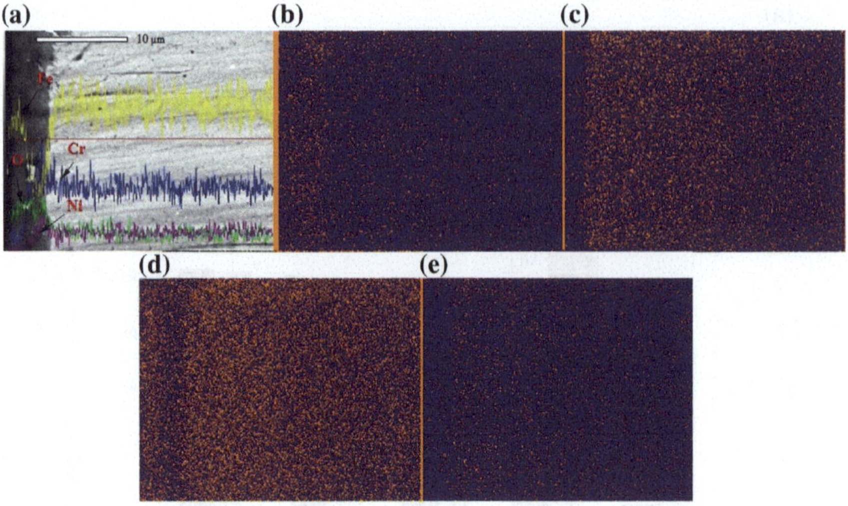

Fig. 6.16 A SEM image and corresponding EDS maps of the cross-sectional TP347H exposed to the SCW at 540 °C and 25 MPa for 120 h. **b** O, **c** Cr, **d** Fe, and **e** Ni

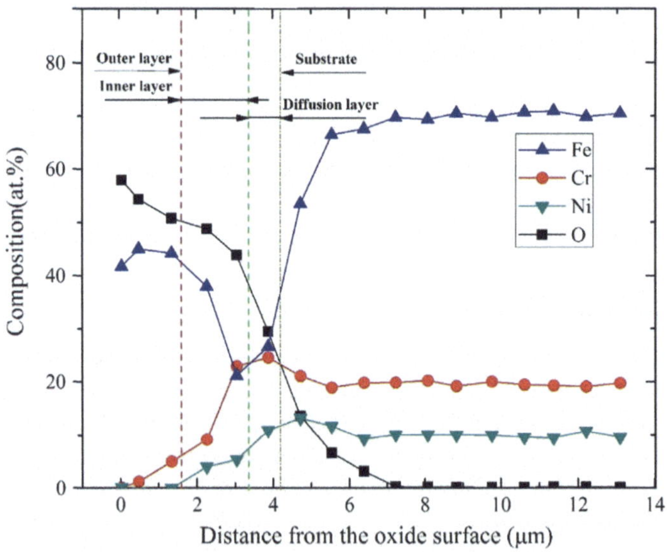

Fig. 6.17 Oxide composition profiles for the alloy TP347H tested in SCW at 540 °C and 25 MPa for 120 h

the alloys by suppressing the outward diffusion of metal ions, especially Cr, due to the lower diffusion coefficient of Cr than that of Fe in the "pure" Ni layer [1, 112]. In addition, because the thermal expansion coefficient of the Ni-enriched zones is generally between those of the alloys and Fe-rich oxides, they can relax the thermal induce stress generated at the substrate/scales interface [122]. However, if an Ni segregation occurs in the form of Ni-enrichment cluster at the inner terminus of alloy grain boundaries due to the selective Cr-rich oxide formation, it may trigger or intensify the grain boundaries' cracking due to further embrittling of the grain boundaries [124]. By associating the EDS profiles with above identified oxide phases, it can be concluded that the crystal structure of the scale formed on the TP304H surfaces is predominant magnetite for the outer layer and Cr-rich spinels for the inner layer with a small amount of dissolved Ni. This duplex-layer scale is typical of that formed on austenitic steels after long-term SCW exposure [97, 99, 103, 104, 115]. Combined with the analysis in part 3.2, it can be speculated that as the exposure time extended further (beyond 35 h), the initial isolated Fe_3O_4 particles distributed on the inner layer may predominantly grow laterally, and they may impinge and combine with each other to eventually form a relatively complete outer Fe-rich oxide layer [125].

In fact, the interface between the inner layer and the substrate was very irregular for the alloy TP347 exposed to SCW [96, 123, 125]. Another cross-sectional image of the TP347 after the same exposure time of 120 h at 540 °C, and its oxygen map is shown in Fig. 6.18. Figure 6.18a exhibits a duplex-layer scale. A local oxidation front of the inner layer has stretched into the substrate approximately 5 μm along the alloy grain boundary, which may be a "local internal oxidation zone" due to higher inward permeation ability of oxygen along the alloy grain boundaries than in the alloy grains. The oxygen map corresponding to Fig. 6.18a further identified the local inner layer inserting matrix to a certain extent, as shown in Fig. 6.18b. Table 6.2 gives the relative atomic percentage of Cr, Ni, Fe, and O for several representative points on the cross section of alloys TP347 (as observed in Fig. 6.18a). Point 1 depicts the relative atomic content of the three main alloying elements, Cr, Ni, and Fe, in the substrate: 19.8Cr-9.6Ni-70.6Fe (at. %). The Cr content increased from Point 5 to Point 2, i.e., from the outer layer to the internal

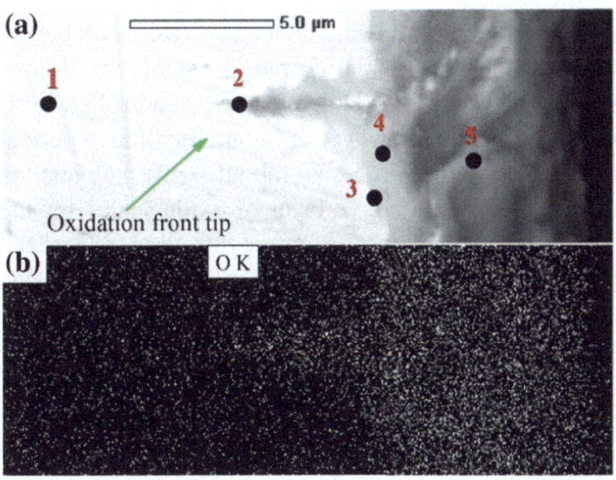

Fig. 6.18 **a** Another cross-sectional image and **b** corresponding oxygen map of the alloy TP347H sample obtained at 540 °C and 25 MPa

Table 6.2 Relative atomic percentage of Cr, Ni, Fe, and O for several representative points on the cross section of alloy TP347 (as shown in Fig. 6.18a)

Locations	Relative atomic content			
	Ni	Cr	Fe	O
Point 1	9.6	19.8	70.5	0.1
Point 2	1.3	36.6	3.7	58.5
Point 3	6.0	22.2	19.0	52.8
Point 4	5.4	19.9	23.1	51.6
Point 5	0.1	0.1	41.8	58.0

Note Trace elements such as Si, Mn, and C are not taken into consideration

oxidation front tip, while the Fe content decreased, which suggests Cr-rich oxide was the main component of the inner layer. It is worth mentioning that the total content of iron and nickel was approximately less than 5% at the internal oxidation front tip, namely Point 2, which is indicative of the presence of Cr_2O_3 phase at the oxidation front tip. This result is consistent with the existing research that unlike in fine-grained austenitic stainless steels, Cr_2O_3 can only be identified around the intersections between the inner layer and alloy grain boundaries in the coarse-grained alloys, such as TP347H [1].

6.2.2 Investigation on Early Formation and Evolution of Oxide Scales on Ferritic–Martensitic Steels

(1) **Oxidation kinetics**

Weight gains (Δw, in mg cm^{-2}) of the T91 samples exposed to SCW and steam at 540 °C for 1, 10, 20, 40, and 120 h, and the fitted plots of Δw versus exposure time (t, in hours), are shown in Fig. 6.19. The weight gains of the T91 samples in SCW were obviously higher than those of the specimens exposed to the high-temperature steam, while the weight gains all increased with an increase in exposure time for the two environments [100]. Oxidation rates can be characterized by near-parabolic laws with similar oxidation rate constants, implying that the oxidation process of steel T91 in two environments may be controlled by a similar predominant diffusion process [98, 99]. However, in steam the time exponent was 0.35, which was smaller than that obtained in SCW (about 0.42), suggesting that relatively fewer defects existed within protective layer of oxide scales for the former [100], assuming that the diffusion of defects is rate-controlling. In order to comparatively analyze the temperature effects on early weight gains of F-M steels, some Δw data and kinetic curves of typical F-M steels at several temperatures obtained from available literatures [98, 100, 117, 126] are also shown in Fig. 6.19.

In this paper, the experimental results are consistent with the extrapolated curves. In SCW, the weight gains of common F-M steels at 540–550 °C are around two times higher than that obtained at 500 °C, while the 600 °C data generally are

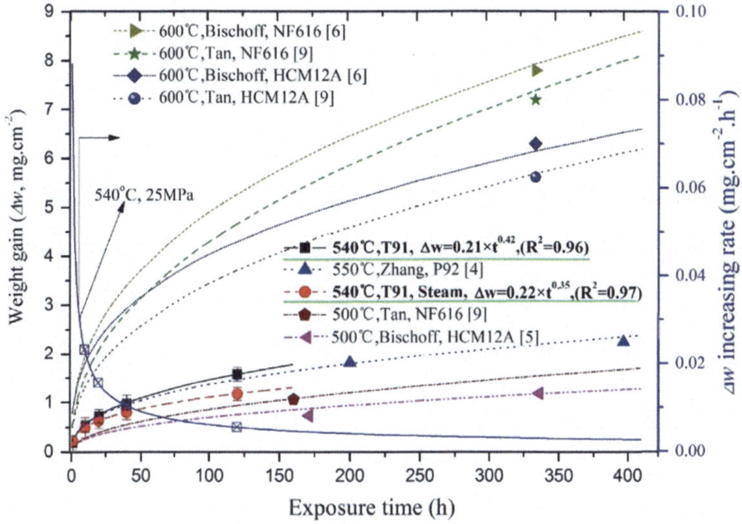

Fig. 6.19 Weight gain (*Δw*) as a function of exposure time (*t*) for F-M steels exposed to steam (9.9 MPa and 540 °C) and SCW (25 MPa and 500 °C, 540 °C, 550 °C, and 600 °C) [127]

about 5–6 times larger than those at 500 °C. The increasing rate of Δw ($\mathrm{d}\Delta w/\mathrm{d}t$), for T91 steel in SCW at 540 °C, is also presented in Fig. 6.19. The increasing rate first decreases significantly in the initial stage and then falls slowly with exposure time, eventually the curve of increasing rate tends to a constant. Depending on the magnitude of second derivative of Δw in exposure time ($\mathrm{d}^2\Delta w/\mathrm{d}t^2$), the early oxidation of F-M steels in SCW at 540 °C can be categorized into three stages: a rapid oxidation stage ($\mathrm{d}^2\Delta w/\mathrm{d}t^2 > 10^{-3}$ mg cm^{-2} h^{-2}), a steady-state growth stage ($\mathrm{d}^2\Delta w/\mathrm{d}t^2 < 10^{-5}$ mg cm^{-2} h^{-2}), and a transitory stage between these two. Thus, the rapid oxidation within 20–40 h occurred as the consequence of fast oxygen transport from the SCW to the steel surface and rapid interfacial reaction, while the diffusion-controlled stage, i.e., the steady-state growth stage, did not appear until at least up to 120 h [117, 126]. Relative to the austenitic steels, which exhibit starting times of the transitory stage and the steady-state stage of approximate 14 and 35 h, respectively, the occurrence of each oxidation stage for T91 in SCW was delayed [128].

(2) **Influence of exposure times**

Surface morphologies of the corroded T91 samples for several exposure times are depicted in Fig. 6.20. For each condition, two images at low magnification and one at high magnification are shown. The surface oxide layers of each sample presented in Fig. 6.3 were all detected as the chromium-depleted by EDS analysis. In SCW at 540 °C, when the exposure time was 1 h, cubic Fe-rich oxide particles with a relatively uniform size approximate 1 μm essentially covered the whole sample surface, as shown in Fig. 6.3a. The accumulation density of the Fe-rich oxides

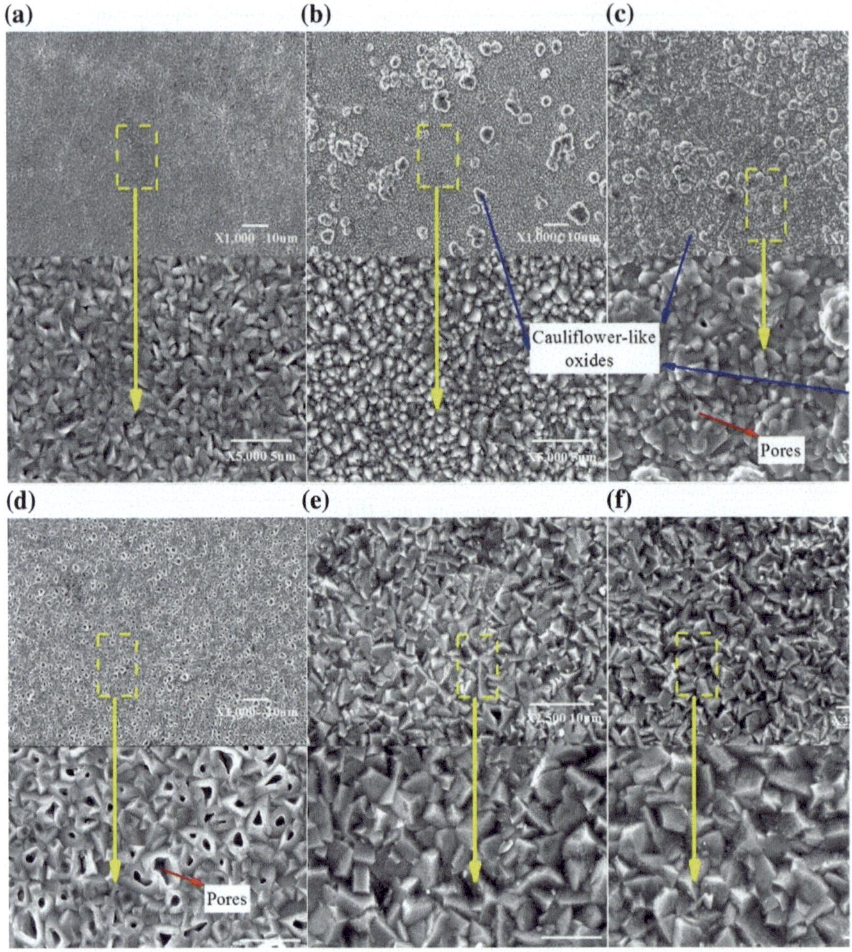

Fig. 6.20 SEM images of T91 samples exposed to SCW at 540 °C for **a** 1 h, **b** 10 h, **c** 20 h, **d** 40 h, and **e** 120 h; and **f** steam at 540 °C for 120 h. Surface morphologies with different magnification are presented

increased as the exposure time increased from 1 to 10 h; however, there was no obvious size variation for the Fe-rich particles. For 10 and 20 h, cauliflower-like Fe-rich oxide particles with a size of ~4 μm occurred on the sample surfaces, while distribution density of the cauliflower-like oxides was larger for the 20-h exposure, as Fig. 6.3b, c shows. The cauliflower-like particulates may be derived from the preferentially rapid formation of new oxides at the defects such as oxide grain boundaries and/or pores [99]. What worth noting is the presence of some small pores in several oxide grains for 20 h. The cauliflower-like oxides disappeared as the exposure time increased to 40 h, while the pore size increased. However, when the exposure time further extended to 120 h, the disappearance of the previous pores was observed (Fig. 6.3e). Surface pores occurred at shorter exposure times but disappeared as the exposure time increased, which is consistent with the previous research on F-M steels [129]. In addition, Fig. 6.3f presents a plan-view image of the oxide scale produced on the T91 samples that were exposed to 9.9 MPa steam. From Fig. 6.3e, f, we conclude that there may be no obvious difference in the surface morphologies between the samples obtained in SCW and steam.

Figure 6.21 presents typical two- and three-dimensional AFM morphologies and features of the T91 samples after exposure to SCW at 540 °C for 10, 20, and 40 h. These two-dimensional images of the T91 samples for three exposure times were in good agreement with their corresponding SEM images, as shown in Fig. 6.20b–d. From the 3-D topographies, it can be seen that the average roughness (*Sa*) and area peak density (*Sds*) of the sample surfaces decreased with an increase in exposure time, which can be generally attributed to the continuous coalescing and fusion of the Fe-rich oxide particles. When the exposure time was 10 h, the sample surfaces were almost completely covered by the oxide particles, and some had merged completely. At 20 h, the bulging cauliflower-like oxides formed, which may result in the occurrence of some wide but shallow valleys to a certain extent. As the exposure time further extended to 40 h, further fusion of the oxide grain brought about the formation of more compact, outer magnetite layer, and the appearance of some pores due to the vacancy condensation [130].

The standard Raman spectra of magnetite (Fe_3O_4) and hematite (Fe_2O_3) selected to compare with the Raman spectra of the T91 sample surfaces subjected to three

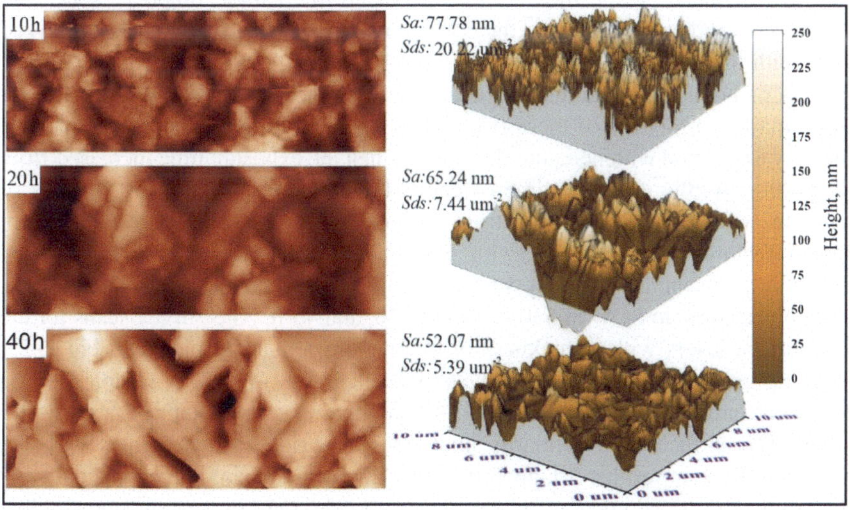

Fig. 6.21 2-D (left) and 3-D (right) AFM topographies and characteristic parameters (*Sa* and *Sds*) of the T91 samples exposed to SCW at 540 °C for 10, 20, and 40 h. *Note Sa, the average roughness*, expresses the average of the absolute values of deviations from the mean surface; *Sds, the area peak density*, is the number of summits per unit area making up the surface

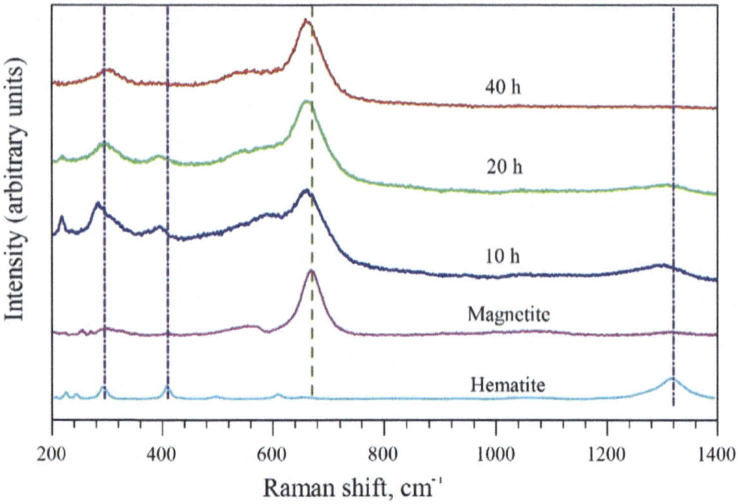

Fig. 6.22 Raman spectra for standard Fe_2O_3 and Fe_3O_4, and the T91 surfaces exposed to SCW at 540 °C for three exposure times

conditions are displayed in Fig. 6.22. The primary Raman feature of magnetite occurs at $\sim$669 cm^{-1} [103], while hematite features appear at 1320, 410, and 293 cm^{-1}. For each exposure condition, Raman spectra were recorded for 20 unique spots located within the selected surface area of approximate 30 μm × 30 μm. The obtained 20 spectra were then averaged to obtain a composite spectrum, which was used to characterize the surface components throughout the analyzed area. With the exposure of 10 and 20 h, the Raman spectra contained the primary features of both magnetite and hematite, which indicates that the thinner surface scales may be a mixture of these two oxides. It is worth mentioning that the peak intensity associated with hematite decreased when the exposure time increased from 10 to 20 h. As the exposure time extended to 40 h, the characteristic features of hematite within the Raman spectrum disappeared completely, and the present Raman features are consistent with that of standard magnetite, implying that magnetite is essentially the only surface component for the 40-h exposure.

After exposure to SCW at 540 °C for 10, 20, and 40 h, XPS survey scans for the T91 surface layers with a thickness of few nanometers are depicted in Fig. 6.23. From this graph, it can be seen that the surface chemistry of the T91 samples for the three exposure times are similar, regardless of the relative content of several elements. The surface layer primarily comprises Fe oxides, which agrees with the foregoing Raman results: Magnetite with/without small amount of hematite is predominantly the component of the outer subscale. One interesting point is that traces of Mo were detected in the surface layer due to the higher elemental resolution of XPS analysis. The formation of Mo oxides on Mo-containing alloys exposed to SCW was also observed in previous research [102, 118], but we noted

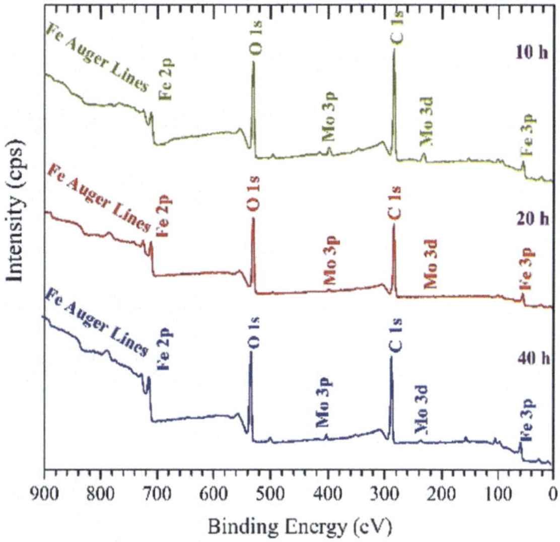

Fig. 6.23 Surface chemistry of T91 samples exposed to SCW at 540 °C for a variety of exposure times

that in our experiments the material of construction of the pressure vessel (Inconel 625) also contained molybdenum. Additionally, carbon was observed on the surfaces, which can be attributed to deposited organics on the samples during transfer to the spectrometer. The high-resolution spectra (obtained at binding energy intervals of 0.1 eV) of elements of principal interest, such as Ni and Cr, Fe, O, and Mo, are given in Figs. 6.24, 6.25, 6.26, and 6.27, respectively, in order to identify their detailed chemical state.

Figure 6.24 presents the binding energy regions of Ni 2p and Cr 2p. Ni as the third element contained in the substrate was not detected regardless of the exposure time, which may be due to its lower outward diffusion rates within the substrate and smaller oxygen affinity relative to Cr and Fe [131]. However, typical Cr oxide features were observed in these high-energy resolution spectra of Cr 2p, which indicates that the outer layer of the scale also contains slight amounts of chromium [132] despite being predominant magnetite [98, 100, 117, 126].

The high-energy resolution of the Fe $2p_{3/2}$ core-level spectra and their deconvolutions are shown in Fig. 6.25. For simplification, three components, Fe_3O_4, Fe_2O_3, and γ-FeOOH, where the Fe binding energy increased in turn [113, 114], were required to fit the Fe $2p_{3/2}$ spectra. Due to the multivalence nature of Fe in Fe_3O_4, the peak fitting methodology employed in this work fitted peaks of Fe^{2+} and Fe^{3+} and meanwhile fixed the intensity ratio of Fe^{2+} to Fe^{3+} at approximately 1:2 [103]. Effect of the exposure time on the iron hydroxides content was not obvious, while the Fe_3O_4 concentrations increased as the exposure time increased. The Fe_2O_3 peak weakened with an increase in exposure time, and essentially disappeared at 40 h, which suggests a transformation from Fe_2O_3 to other phases (such as Fe_3O_4 and $Fe_xCr_{3-x}O_4$) [100, 115].

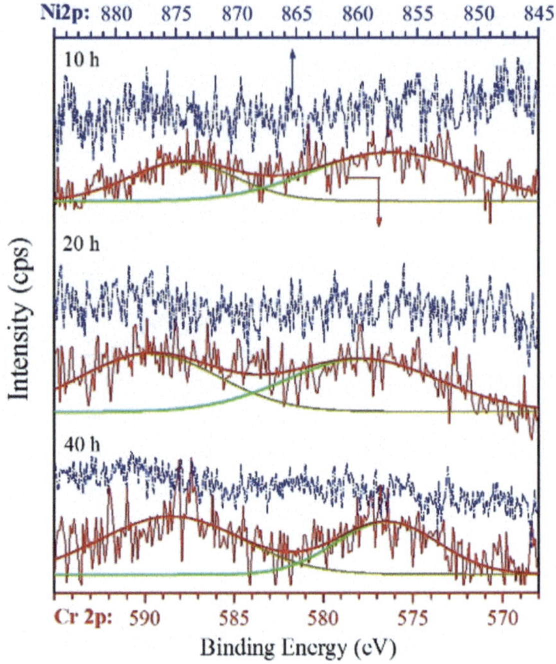

Fig. 6.24 High-resolution spectra of Ni 2p (blue, dotted line) and Cr 2p (red, solid line) of the surface scales on steel T91 exposed to SCW at 540 °C

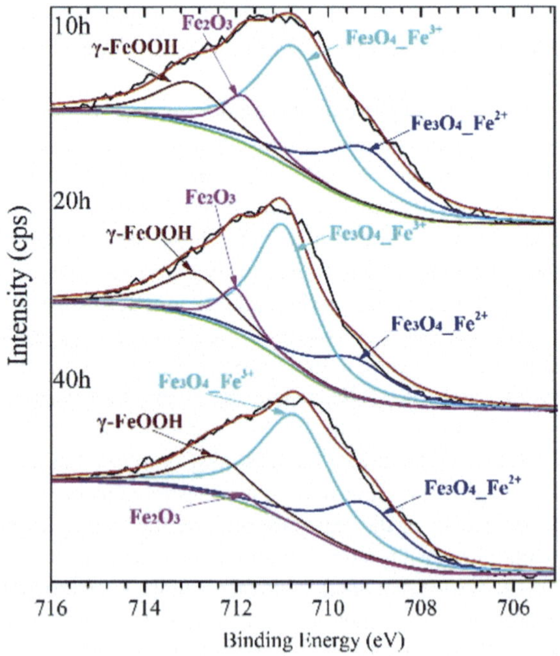

Fig. 6.25 Deconvolution of Fe $2p_{3/2}$ core-level spectra for the T91 samples exposed to SCW at 540 °C

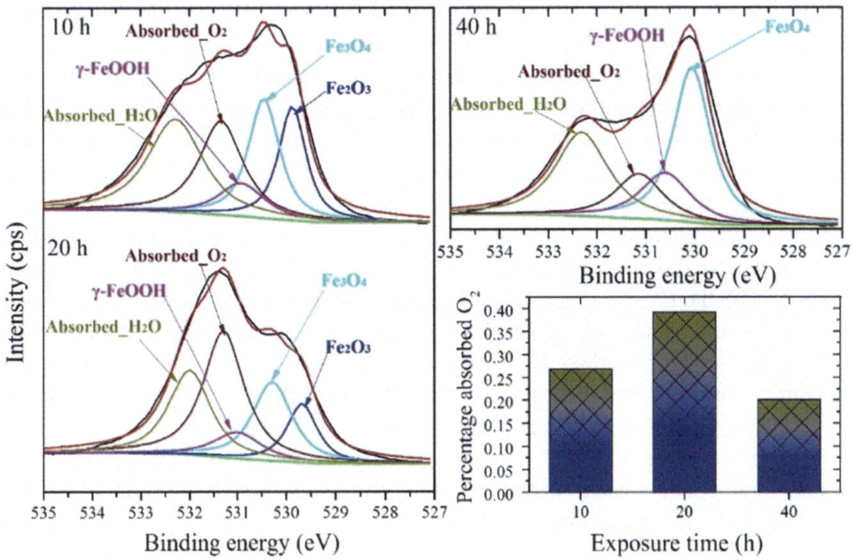

Fig. 6.26 Deconvolution of O 1s spectra on the T91 surfaces in SCW at 540 °C for 10, 20, and 40 h; and relative content of the absorbed O_2

Fig. 6.27 XPS narrow scans of Mo 3d spectra on the T91 samples exposed to SCW at 540 °C for three exposure times, and their deconvolution parameters

Sample	Label	BE(eV)	FWHM	Sample	Label	BE(eV)	FWHM
10 h	I	235.85	1.75	40 h	I	235.72	1.01
	II	232.65	1.13		II	232.52	1.02
20 h	I	235.73	1.77	MoO_3 [29]	$3d_{3/2}$	235.85	-
	II	232.53	1.40		$3d_{5/2}$	232.65	-

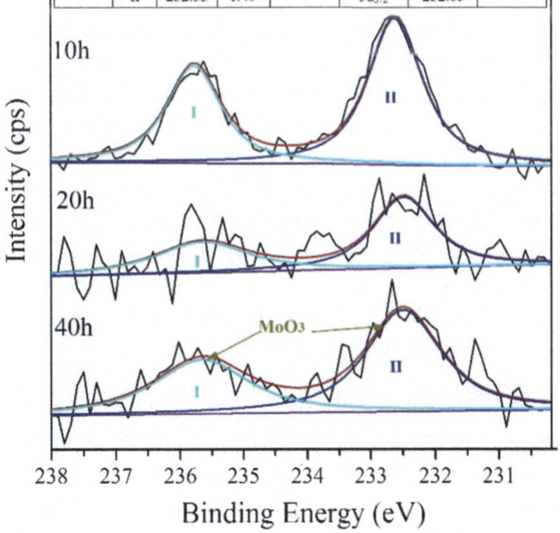

The O 1s spectra of the T91 samples after exposure to SCW for 10, 20, and 40 h are presented in Fig. 6.26. These O 1s peaks are systematically decomposed into four or five components, including absorbed H_2O, γ-FeOOH, O_2, Fe_3O_4, and Fe_2O_3 [133]. During the decomposition and fitting process, it is worth mentioning that the content ratio of oxygen in γ-FeOOH and Fe_3O_4 depends on that of iron in the two oxides as calculated by Fig. 6.25, and the former is always 0.5 times larger than the latter. There were no obvious content variations of the absorbed water and Fe-hydroxides regardless of the exposure time [109], where the latter are consistent with the results shown in Fig. 6.25. The fraction of the total oxygen in the system (H_2O, γ-FeOOH, O_2, Fe_3O_4, and Fe_2O_3) that is in Fe_3O_4 decreases with increasing exposure from 10 h to 20 h, and then increases as the exposure time further extends to 40 h, while the absorbed O_2 within surface layer of the scales varied in the opposite manner. Figure 6.26 also presents the percentage of absorbed O_2 versus the exposure time. The O_2 content observed at 20 h of exposure was 0.39.

For exposure time of 10, 20, and 40 h, the high-energy resolution spectra of Mo 3d and their corresponding deconvolution parameters are indicated in Fig. 6.27. The core-level spectra of Mo $3d_{3/2}$ and Mo $3d_{5/2}$ in MoO_3 were located at the binding energies of 235.85 and 232.65 eV, respectively [134]. No other molybdenum oxides, besides a very small amount of MoO_3, were observed on the detected surface, and there was no significant variation of MoO_3 content as the exposure time increased. The slight differences of MoO_3 concentrations among the three conditions may be due to randomly selected detection areas and uneven Mo distribution within the oxide layers. This seems to indicate that the existence of MoO_3 on the F-M steels exposed to SCW during early oxidation is attributable to its presence in the samples although the autoclave cannot be discounted as the source. However, we could find no literature reporting the presence of Mo oxides on F-M steels after a long enough exposure to SCW [100, 135–137]. In the case of high molybdenum alloys, Mo-rich oxides appeared as isolate particles and/or single layer [102, 118]. Generally, Mo as an alloying element, which has the effect of increasing Cr activity, can promote the formation of protective Cr-rich oxide layer on the alloy surface [138]. However, this beneficial effect of molybdenum may be limited because of the relatively little Mo in the T91 substrate and probably disappeared completely with thickening. This may be an explanation for the above-mentioned absence of MOx within the oxide films formed on F-M steels after a long exposure to SCW. In addition, Chen et al. [139] and Yin et al. [130] reported that Mo was generally found to be depleted in the outer layer while being enriched in the inner layer.

(3) **Temperature Effects**

Surface morphologies obtained from SCW-exposed T91 samples for 40 h at 390, 465, 540, and 580 °C are presented in Fig. 6.28. At 390 °C, cuboidal Fe-rich oxide particles intricately piled up on the sample surface, which generally constituted the outer oxide layer of duplex-structure scales formed on typical F-M steels in SCW [100, 132, 136, 137], was observed. The distribution density of the Fe-rich oxides

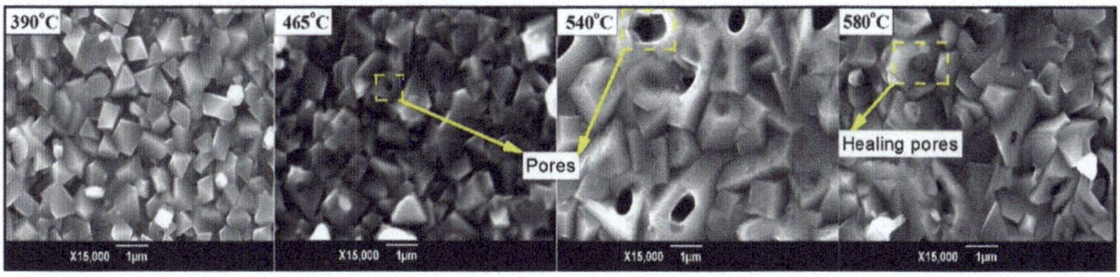

Fig. 6.28 Plan-view SEM images of the T91 samples exposed to SCW for 40 h

increased with an increase in temperature from 390 to 465 °C, while there was no obvious variation of the particle size. At 540 °C, the cubic Fe-rich particles coalesced and fused with each other into a more complete, compact outer Fe-rich layer; meanwhile, some pores having a diameter of approximately 0.5–1 μm were also observed [99, 135]. When the experimental temperature increased to 580 °C, the larger pores seemed to disappear, but some interesting pore-healing was evident, which was characterized by some smaller oxide particle existing at the bottom of the original larger pores.

Standard X-ray diffraction patterns of magnetite and iron, and XRD results of the T91 samples after exposure to SCW for 40 h at the four typical temperatures above are presented in Fig. 6.29. These data indicate that all components existing on the sample surface were as magnetite and a spinel (the latter has almost the same diffraction patterns as magnetite) [130, 132], and/or the alloy substrate roughly represented by "Iron" in Fig. 6.29. The peak intensity of the substrate decreased with an increase in temperature, while the characteristic peaks of magnetite strengthened, which implies the formation of thicker scales on the samples at higher temperatures. When the experimental temperature increased to 580 °C, the features of the substrate's pattern almost disappeared completely, which suggests that the incident X-rays may fail to reach the substrate because of the thicker scale.

The relative content of the predominant surface elements, such as Fe, Cr, O, and Ni, and weight gains of the T91 samples exposed in SCW for 40 h, as functions of the experimental temperature are given in Fig. 6.30. There was a clear positive correlation between the weight gains of the T91 samples and temperatures. The effect of temperature on the Ni atomic content did not appear to be significant, while the Cr concentration seemed to slightly decrease as the temperature increased. These results can be attributed to the slower outward diffusion of Cr than that of Ni and continuous oxygen accumulation on the surface oxide layer. With an increase in temperature from 390 to 580 °C, the O/Fe atomic ratio ascended from 1.6 to 2.3, which is larger than the O/Fe ratio of any iron oxides (magnetite was the predominant component on the oxidized samples as shown in Fig. 6.29). This finding probably implies that the incident X-rays may only slightly penetrate into the scales, but not reach the substrate, and that there may be abundant iron vacancies present within the outer Fe-rich layer. Defect factor of Fe (DF_{Fe}) in the outer layer, compared to normal Fe_3O_4, was defined (Table 6.3) and calculated by taking into account the relative atomic contents of main elements such as O and Cr, as shown

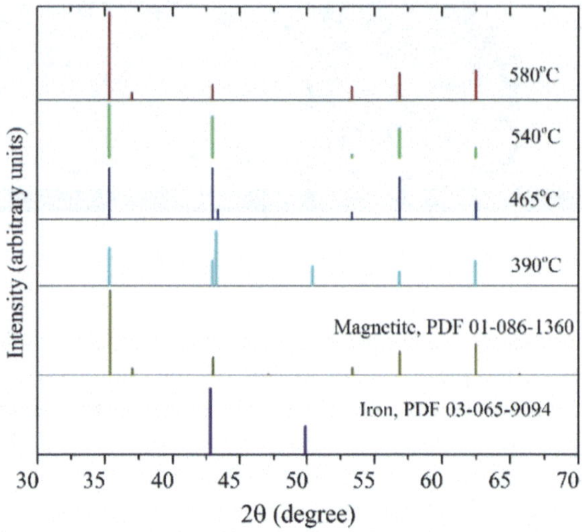

Fig. 6.29 Standard PDF cards of magnetite and iron, and XRD patterns of the T91 samples exposed to SCW for 40 h

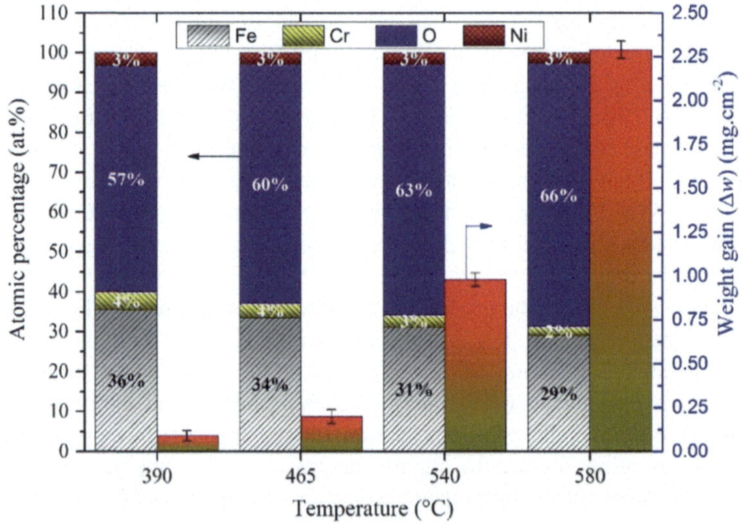

Fig. 6.30 EDS analysis and weight gain of the T91 specimens after a 40-h exposure in SCW at four typical temperatures

Table 6.3 Defect factor of Fe in the surface subscales compared to normal magnetite versus experimental temperature

Temperatures (°C)	390	465	540	580
DF_{Fe}	0.09	0.32	0.58	0.87

Notes DF_{Fe} = ([O] − 2*[Cr])/([Fe] − [Cr]/2) − 4/3, 4/3 is the O/Fe ratio of normal Fe_3O_4

in Fig. 6.30. The positive DF_{Fe} increased gradually with an increase in the experimental temperature. These results confirm a larger iron vacancy density within the outer layer of oxide scales at higher temperatures, which may enhance the outward diffusion of metal ions, leading to higher oxidation rates of the T91 samples.

(4) **Scale Structures**

The SEM images and the corresponding elemental depth profiles for the cross sections of the scale on the T91 samples oxidized in SCW at 540 °C for 1, 40, and 120 h are shown in Fig. 6.31. At 1 and 40 h, the samples exhibited a duplex-layer oxide structure. According to the cross-sectional morphologies and EDS analysis, and the above-mentioned results, the outer layer primarily comprises magnetite, while the inner layer was the spinels, while the Cr content was very close to that of the substrate. However, for the 120-h results relative to that of the 1- and 40-h samples, the content variation of the targeted elements, such as Cr, Fe, and O, across the substrate/scale interface was more gradual, which indicates the generation of an additional diffusion layer between the substrate and the inner layer. Besides, the assembled-pore layers were periodically shown in the SEM image of the inner layer [98, 100, 126], implying the growth periodicity of the inner Cr-rich layer during the oxidation process [140]. The assembled-pore line near the inner/diffusion layer interface was characterized by a larger pore size. As the exposure

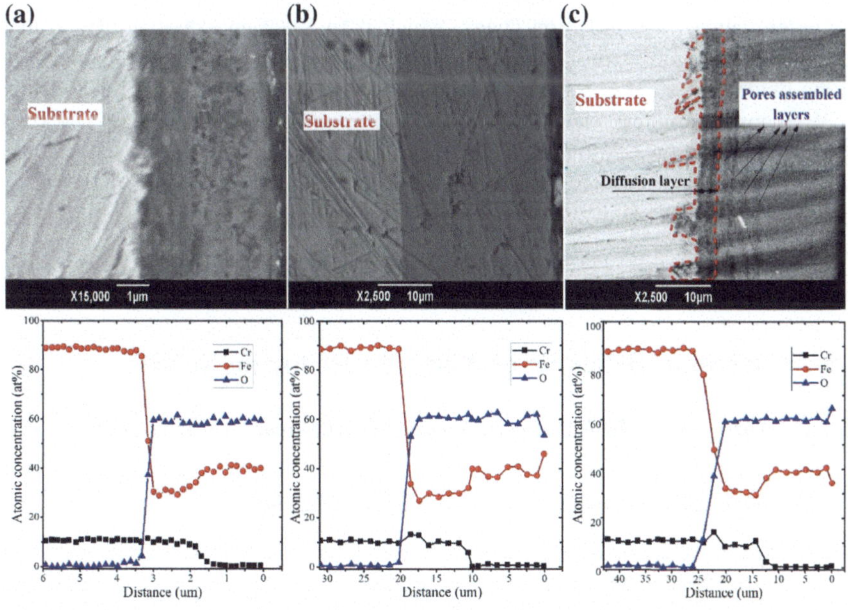

Fig. 6.31 Cross-sectional images (left: substrate, right: scales) and their corresponding elemental depth profiles of the T91 samples exposed to SCW at 540 °C for **a** 1 h, **b** 40 h, and **c** 120 h

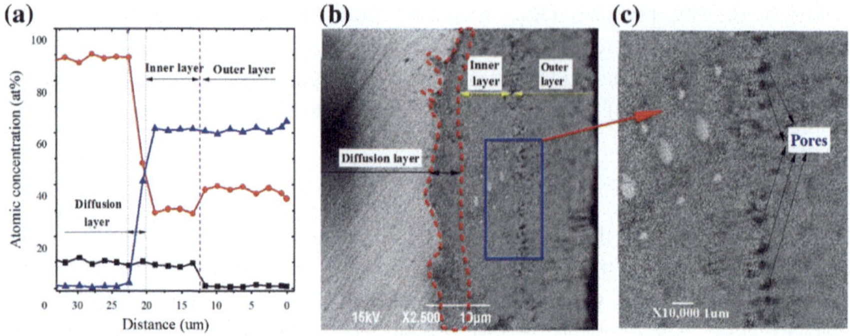

Fig. 6.32 Cross section of the T91 sample exposed to steam at 540 °C for 120 h. **a** elemental depth profile, **b** cross-sectional image, and **c** higher magnification image of the pores assembled zones

time increased from 1 to 40 h and 120 h, the scale thicknesses increased from approximately 3–20 and 25 um, respectively. The scale thickness seemed to exhibit an increase at a slower rate during the transitory stage, i.e., approximately from 40 to 120 h.

To compare the oxide scales of typical F-M steels exposed to SCW and high-temperature steam, Fig. 6.32 presents the cross-sectional characteristics of the T91 samples exposed to steam at 9.9 MPa and 540 °C for 120 h. The SEM image (Fig. 6.32b) and corresponding EDS analysis (Fig. 6.32a) indicated that the samples exposed to steam at 9.9 MPa and 540 °C exhibited a three-layer oxide scale: the outer-, inner-, and diffusion layers, which is to a certain extent similar to that found on the samples exposed to SCW at 540 °C for 120 h. However, the thickness of the oxide scales ($\sim$22.5 um) was slightly thinner to approximate 25 um for the SCW-exposed samples, which suggests the better oxidation resistance of T91 steel in steam than in SCW. Additionally, in contrast to the samples exposed to SCW, no periodic assembled-pore layers were found in the inner oxide layer of the samples in steam at 9.9 MPa. However, the accumulation of pores along the inner and outer oxide layer interface was obviously observed in Fig. 6.32c [1, 100].

6.3 Corrosion Behavior of Alloys in Reducing SCWs

6.3.1 Corrosion Behavior of Inconel 600 and Incoloy 825

(1) **The effect of exposure times on corrosion behavior**

The corrosion behavior of the Inconel 600 and Incoloy 825 was investigated in SCW at a sulfide content of 12,800 ppm at 520 °C and a pressure of 25 MPa. The change of weight gain with duration of exposure time from 1 to 3 h and parabolic plots of the corrosion kinetics with an exposure time of 3 to 60 h are shown in

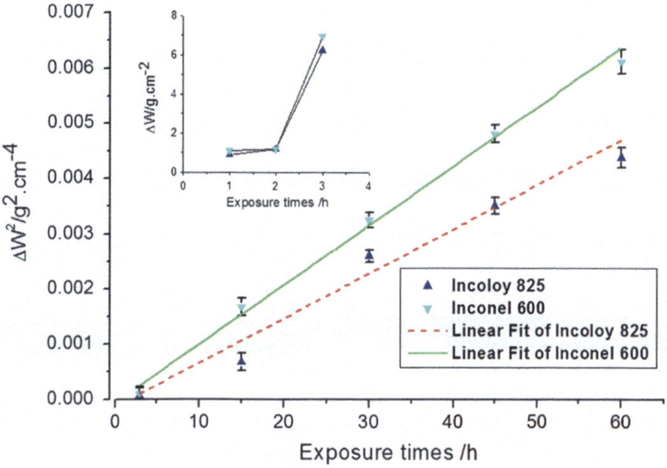

Fig. 6.33 Corrosion kinetics and fitting curves (from 3 to 60 h) for Inconel 600 and Incoloy 825 tested in SCW with sulfides at 520 °C [143]

Fig. 6.33. As shown in the figure, the weight of Incoloy 825 was less than Inconel 600. Clearly, the corrosion resistance of Incoloy 825 was slightly better than that of Inconel 600. For two alloys, over a certain period between 2 and 3 h, the change of weight gain changed from a low rate to a high rate, which might suggest the occurrence of breakaway corrosion, i.e., accelerating weight-gain kinetics. The fitting curves with an exposure time from 3 to 60 h, as shown in Fig. 6.33, showed a linear relationship between the squared values of weight gain per unit area, denoted as ΔW^2 and exposure times, with the correlation coefficients of up to 0.99 and 0.96 for Inconel 600 and Incoloy 825, respectively. Thus, the functional relationship from 3 to 60 h can be noted as $\Delta W^2 = k \cdot t + b$, where t represents variable exposure times in h, while k is the corrosion constant that approximately equates to 1.07×10^{-4} and 8.07×10^{-5} g^2 cm^{-4} h^{-1} for Inconel 600 and Incoloy 825, respectively. For two alloys, the investigation results that the weight gain of per unit area approximately followed a parabolic law, indicating that the corrosion rate from 3 to 60 h was controlled by a diffusion process [141, 142].

(2) **Surface morphology**

Figure 6.34 shows the surface morphology of scales on Inconel 600 and Incoloy 825 specimens after 3- and 60-h exposure. After 3-h test, thin scales might have been produced on sample surfaces of Inconel 600 and Incoloy 825. The former probably owned a thicker scale than that of Inconel 825 with minor scratches left by abrasive paper still being visible. All corrosion products present on sample surface layer of two alloys approximately showed a dual-phase type with dark and bright contrast (Fig. 6.34a, b). EDS analysis indicated that the dark and bright phases (marked in Fig. 6.34a, b) were likely to be sulfides and oxides, respectively. Further, it may be noted that the dark phases occurred on the Inconel 600 specimen,

Fig. 6.34 SEM images of the surface morphology of Inconel 600 and Incoloy 825 specimens after exposure to SCW with sulfide of 12,800 ppm at 520 °C for 3 h and 60 h **a** Inconel 600 for 3 h, **b** Incoloy 825 for 3 h, **c** Inconel 600 for 60 h, and **d** Incoloy 825 for 60 h

of which particle size was relatively less than bright phases, evenly covered almost the whole surface, while a portion of the Incoloy 825 surface was not occupied by the dark sulfide phases. The fast outward diffusion of cations coupled with a high sulfur potential in present atmosphere led to formation of sulfides with fast growth rates, so they occupied a large portion of the scales surface. After exposing for 60 h, it was found that the surface of the two alloys was covered by a thick layer of corrosion products with a color of metallic light bronze. Comparing to Inconel 600, the surface color of Incoloy 825 was more somber. In addition, the corrosion products on Inconel 600 were brittle and easy to be peeled off from the substrate under slight tapping, while the scales formed on Incoloy 825 showed good adhesion to the matrix. The surface morphology after 60 h is seen in Fig. 6.34b, c. The surface scales of Inconel 600 and Incoloy 825 had a composition of 71.1Ni-O1.7-28.9S and 4.8Cr-69.6Ni-1.3Fe-6.5O-17.8S in wt%, respectively, as shown in Fig. 6.35.

(3) **Cross-sectional analysis**

Figure 6.36 shows the cross-sectional BSE image with EDS mapping of Inconel 600 after testing for 60 h. It can be seen from Fig. 6.36a that the scales on alloy

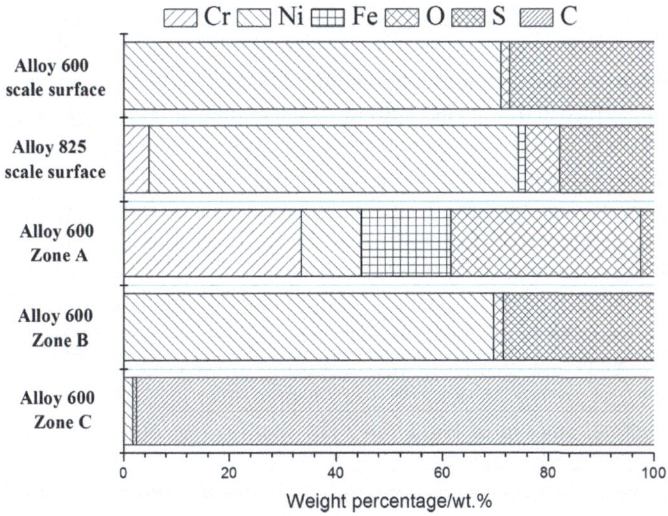

Fig. 6.35 Elements' distribution on several specified areas of Inconel 600 and Incoloy 825 exposed to SCW containing sulfides

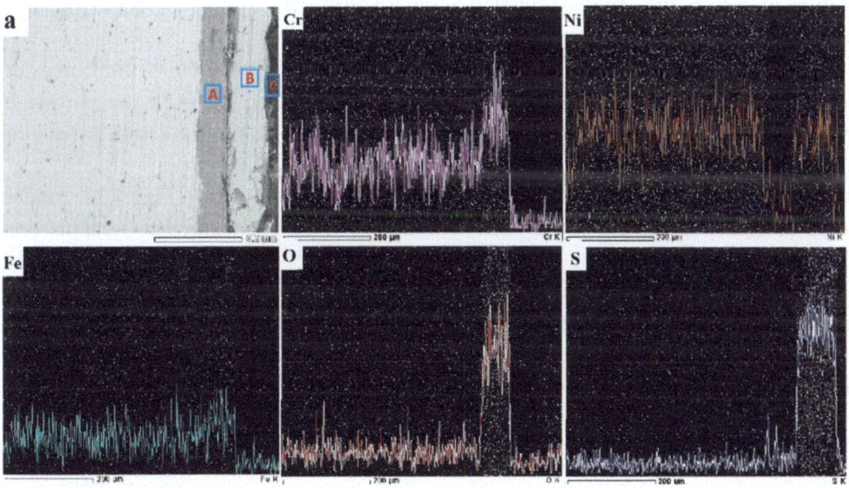

Fig. 6.36 Backscatter electron image along with EDS element intensity maps for Cr, Ni, Fe, O, and S from the cross section of Inconel 600 sample exposed to SCW (520 °C, 25 MPa) containing 12,800 ppm sulfide for 60 h

Inconel 600 consisted of an inner and an outer layer indicated as A, B in the image. The interface between two zones was approximately straight, which may be the initial alloy surface. The inner and outer layer thicknesses of the Inconel 600 scale were about 80.2 and 121.9 μm, respectively. The composition of layers A, B, and C can be found in Fig. 6.35. As shown in the figure, the inner A had a composition of 33.6Cr-11.2Ni-16.8Fe-35.9O-2.5S (in wt%) suggesting that it mainly consisted of chromium-rich oxides and a small amount of sulfides. The outer layer B was predominantly consisted of nickel sulfide with a composition of 69.8Ni-1.8O-28.4S (in wt%). Layer C was epoxy resin layer with a carbon content up to 97.5 wt%. The results of EDS element mapping correspond to the composition of each layer.

XRD analysis was conducted on the surface scales of Inconel 600 at different scale depths, as shown in Fig. 6.37. The Ni_3S_2 layer was detached upon slight hitting, and the remaining inner layer was analyzed again. The inner layer consisted primarily of spinel-type phase, with a small amount of Ni_3S_2. Since XRD patterns for spinels containing Fe, Cr, and Ni were very similar, the spinel type could not be identified by XRD analysis. However, according to the composition of inner layer A 33.6Cr-11.2Ni-16.8Fe-35.9O-2.5S (wt%), the spinel-phase $FeCr_2O_4$ might exist with a very small amount of Ni sulfide. The main corrosion products in the inner layer were significantly different from those of the inner subscale appearing on the Inconel 600 exposed to pure supercritical water. For the latter, the inner layer was generally composed of $NiCr_2O_4$ phase. In the present work, the absence of the $NiCr_2O_4$ phase in the inner layer might resulted from lower oxygen partial pressure in this work than that in pure supercritical water. The compact Ni_3S_2 layer was thick but not very adherent, an obvious gap occurred at inner–outer layer interface, as shown in Fig. 6.36a, suggesting that outward diffusion was involved in the corrosion process [144].

The cross-sectional BSE image and EPMA line-scan profiles of the corrosion products formed on alloy Incoloy 825 after the exposure for 60 h are shown in Fig. 6.38. The scales formed on the Incoloy 825 were also multilayered, consisting of an inner Cr-rich oxides and a Ni-rich outer scale. XRD analysis was also conducted at different scale depths of alloy Incoloy 825. As shown in Fig. 6.39, the top scale consisted of Ni_3S_2 and a small amount of $(Ni,Fe)_3S_4$ and Cr_2O_3. This corresponded to alloy surface composition of 4.8Cr-69.4Ni-13Fe-6.5O-17.8S. The presence of Ni_3S_2, $(Ni,Fe)_3S_4$, Cr_2O_3, and spinel-type phase was all detected in the intermediate and inner layer, except that CrS was only detected in the inner layer. However, the Cr_2O_3 was more dominant in the inner layer compared to that in the intermediate. Figure 6.40 shows the high-magnification images of alloy Inconel

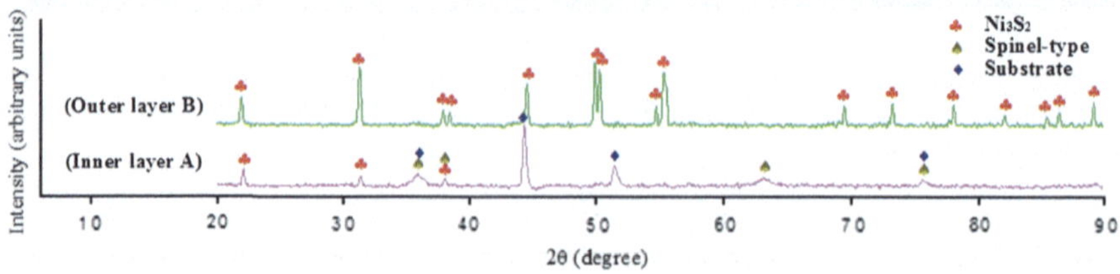

Fig. 6.37 XRD patterns of the scales formed on the Inconel 600 exposure to SCW with sulfides for 60 h

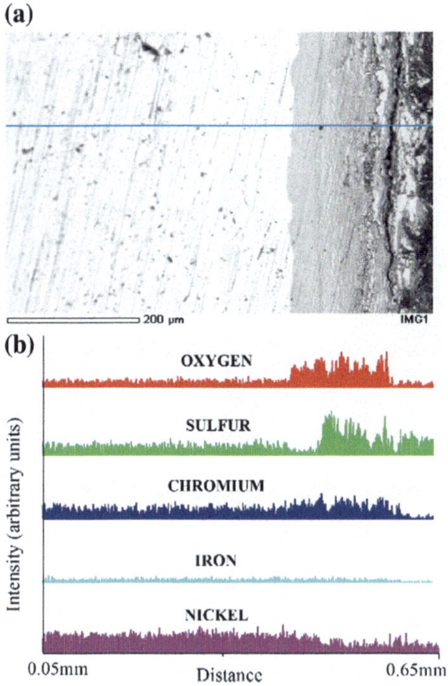

Fig. 6.38 Cross-sectional **a** back-scattering electron image and **b** EPMA line-scan profiles of the corrosion layer on the Incoloy 825 after the exposure for 60 h in SCW (520 °C, 25 MPa) with 12,800 ppm sulfide

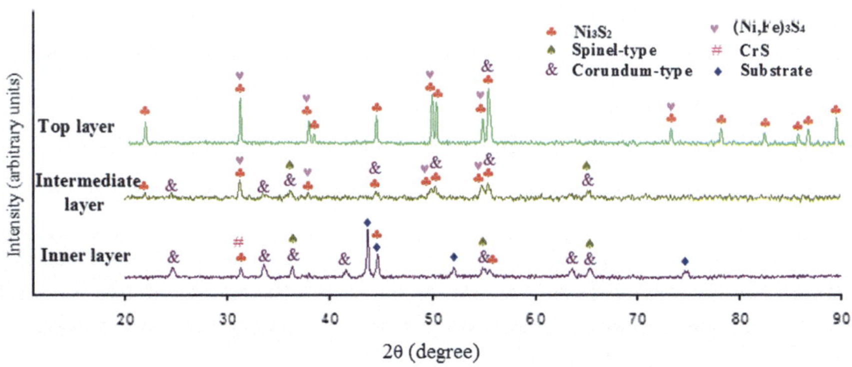

Fig. 6.39 XRD patterns of the scales formed on the Incoloy 825 exposed to SCW with sulfides at different scale depths for 60 h

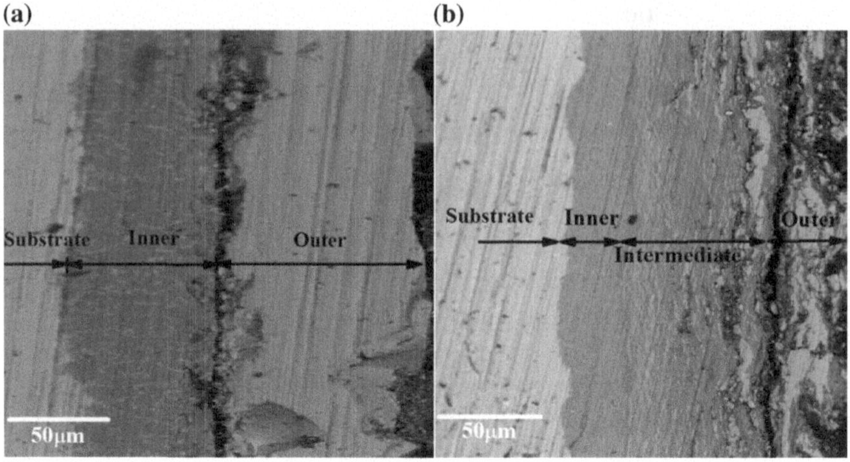

Fig. 6.40 Corrosion layer structures representing well maintained scales formed on **a** Inconel 600 and **b** Incoloy 825 under higher magnification

Table 6.4 Mass change date of samples after exposure tests in SCW containing NH_4Cl at 400 °C for 70 h

Alloy	625	600	800	316
Mass change (mg/cm^2)	−3.13	−2.69	−1.28	−3.70

600 and Incoloy 825 after testing for 60 h. As shown in the figure, for alloy Incoloy 825, the thickness of each layer toward the substrate was measured to be 25.1, 70.2, and 28.4 um, respectively.

6.3.2 *Effect of Ammonium Chloride on Corrosion Behavior*

(1) **Mass change**

The mass change data before and after the testing is shown in Table 6.4. All the samples showed obvious mass loss. Weight loss may be caused by two reasons. Firstly, ammonium chloride would produce acids in supercritical water environment [145], as shown in Eq. (6.2).

$$NH_4Cl + O_2 \longrightarrow HNO_3 + HCl + H_2O \tag{6.2}$$

Obviously, the concentration of oxygen is a significant consideration when looking at the acid concentration in ammoniated solutions. For the reason that most of the oxygen in the reactor space had been eliminated by nitro gas bubble, dissolved oxygen (DO) content in solutions was little. Hence, the concentration of acid would be limited. The measured pH of the effluent was 6.7, which confirmed the inference mentioned above.

The appearance of acid may significantly induce transpassive dissolution of Cr oxide formed on both Ni-based alloy and stainless steel [146], which may cause grain boundaries attack. Moreover, any type of nickel oxide is unstable in acidic, high-temperature aqueous solutions. Nickel may be dissolved as Ni(II) ions under the present experimental conditions.

The other reason for the weight loss may be due to the formation of ammine $Ni(NH_3)_n^{2+}$ by the following equation [147, 148]:

$$Ni + \frac{1}{2}O_2 + nNH_3 + H_2O \longrightarrow Ni(NH_3)_n^{2+} + 2OH^- \quad (6.3)$$

However, due to the lack of oxygen in the reaction space and the low ion product constant in supercritical water, the degree of the reaction mentioned above was limited. Moreover, formation of ammines would just occur at a temperature up to 200 °C. Nickel ammines were not likely to form beyond this temperature. As a result, Ni element in samples may just form nickel ammines during the heating and cooling-down process.

(2) **Surface morphology analysis**

After 70-h exposure, there were lots of solid particles deposited on the top of autoclave (shown in Fig. 6.41). Morphologies of the deposited matter were observed by SEM, and chemical elements were characterized by EDS. The results are shown in Fig. 6.42. It can be confirmed that the solid particles are absolutely Fe oxide. And it can be inferred that the Fe oxides are mainly from the leaching of Fe in the autoclave. As a result, the samples may also be covered with Fe oxide deposits.

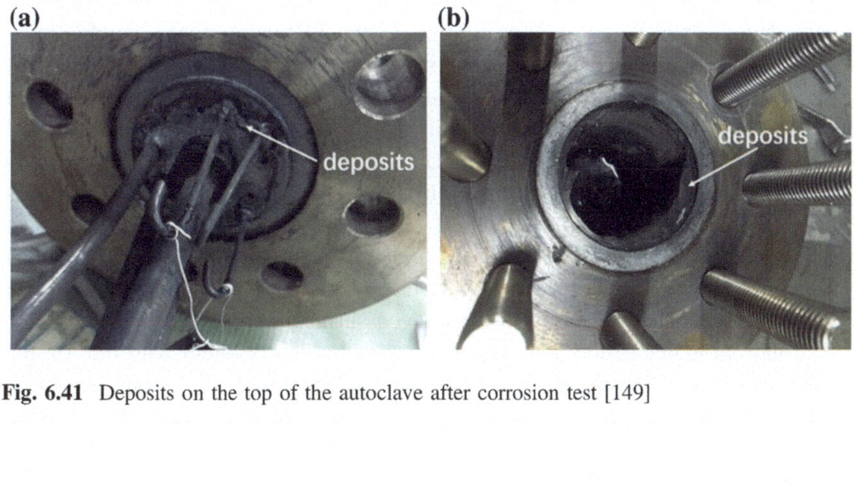

Fig. 6.41 Deposits on the top of the autoclave after corrosion test [149]

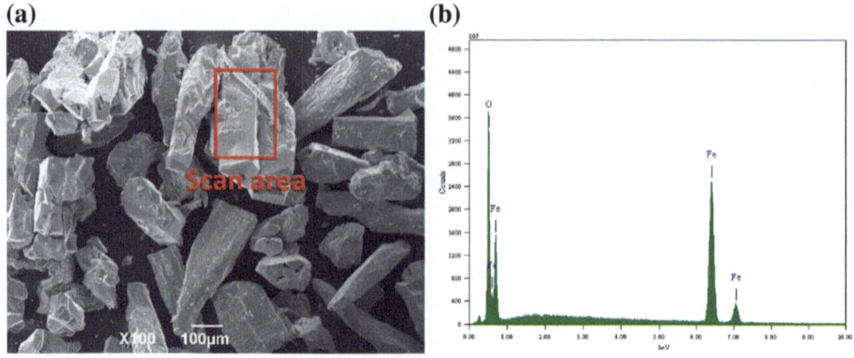

Fig. 6.42 SEM observation and EDS analysis of the deposits on the top of the autoclave

Surface morphologies of the oxide films on samples after exposing in SCW at 400 °C are shown in Fig. 6.43. Pitting corrosion was found on 316 SS, and other parts of 316 SS were relatively smooth. EDS analysis revealed that both the smooth and pitting areas mainly contained O, Fe, and Cr, which can be concluded that the oxide films were mostly Fe oxide and Cr oxide. Although the Ni content in 316 SS is about 12 wt%, nearly no Ni elements were found by EDS analysis. Besides, the concentration of Ni in the pits area was larger, and Cr was less than that in the smooth area, which may be caused by the adsorption of chloride ions in the pits or other uneven regions area and the ensuing selective dissolution of Cr oxide. The surface roughness would remarkably affect the pit growth rate in the early stage of pitting corrosion. Chloride may firstly accumulate in the defects on the surface and cause the pitting corrosion. Note that the samples were mechanical polished by SiC paper down to 2500 grit, the surface of samples was relatively smooth. As a result, at the beginning of corrosion, the number of the pits was limited. And the anodic current was distributed among the limited number of growing pits, leading to relatively larger growth rate of the individual pit [146]. Hence, the depth of the pits on 316 SS was overwhelming. The surfaces of Ni-based alloys were covered with plenty of deposit-like particles. According to the SEM figure, the roughness of surface of Ni-based alloy samples had significant correlation with the concentration of Fe in alloys. Inconel 800, which had the most Fe concentration, showed the most even surface. And Inconel 625, Fe concentration of which was only 5 wt%, had lots of deposits-like particles on the surface. This may be due to the stability of alloy elements at 400 °C. According to the reports by Asselin et al. [150], Fe oxide was the most stable oxide in ammoniacal sulfate solutions at high-density supercritical zone (between 375 °C and 400 °C) based on the EDS profiles of 316 SS. Hence, Ni oxides and Cr oxides on Ni-based alloy surface were dissolved, and Fe oxides were formed or participated, resulting in the uneven surface morphology.

The morphologies of the deposits on samples looked different with that on the top of autoclave. EDS results of the surface of Ni-based alloys are shown in Fig. 6.44. It can be inferred that the black scales covering on Ni-based alloy

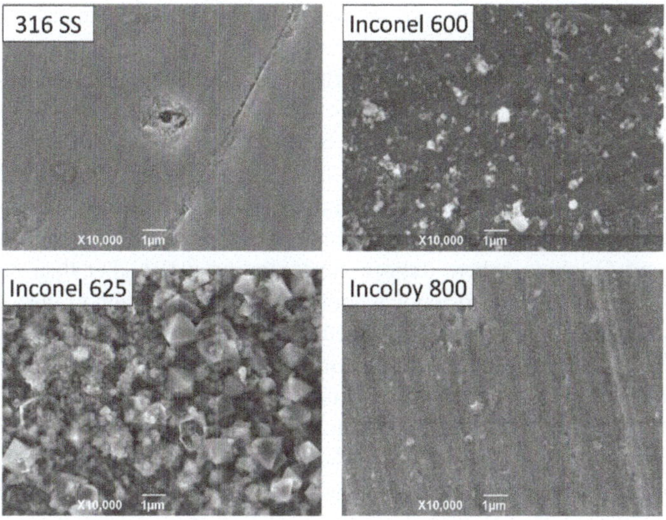

Fig. 6.43 Surface morphology of samples after 70-h exposure to 400 °C supercritical water environment containing 6000 ppm NH_4Cl

Fig. 6.44 EDS results of the deposits on the surface of Ni-based alloys after 70-h exposure to 400 °C supercritical water environment containing 6000 ppm NH_4Cl

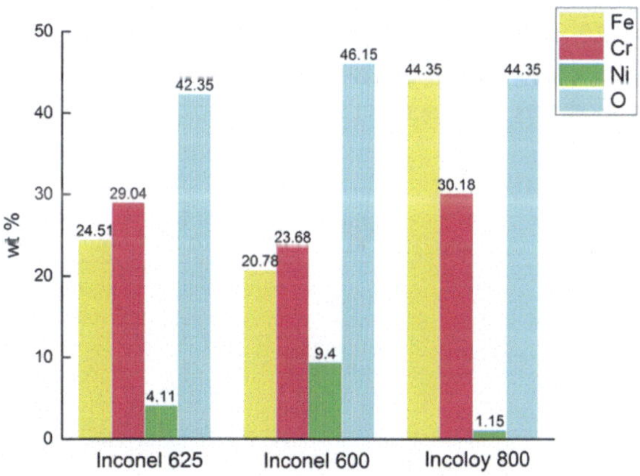

coupons were mostly Fe oxide and Cr oxide. Based on the measurement of particle sizes, the deposits on samples were obviously smaller than that on the top of autoclave. The deposits on the top of autoclave can be the result of an aggregation of particles from the samples and the structural materials of the autoclave. Moreover, the deposits on the top of autoclave contained only Fe and O, and the deposits on the samples were both Fe oxide and Cr oxide. For the reason that the superficial area of the inner space of the autoclave was significantly larger than that of the samples, more deposits should be the corrosion product of the autoclave, rather than the coupons. Further study should be done on an autoclave made of nickel-based alloys or titanium alloys rather than stainless steels, in order to exclude the interference of Fe oxide.

The composition and structure of oxide films are important information of the corrosion path. SEM and EDS analysis of cross section of samples were done in order to have an overall understanding of corrosion. The analysis of cross section of 316 SS is shown in Fig. 6.45. The oxide film of 316 SS could be divided into two parts. The outer layer was rich in Fe and Cr, which was nearly the same alloy composition of the base alloy. And the inner part of the oxide film was filled with Fe oxide and Cr oxide, where oxygen concentration was higher than other parts. Moreover, intergranular corrosion was found on the cross section of the sample (the discontinuous dark zone), which indicated that the corrosive ions can readily pass through the oxide films and diffuse along grain boundaries in the inner parts, namely the oxide films formed on the samples were failed to separate the corrosive ions. Therefore, it can be concluded that 316 SS underwent serious corrosion in supercritical water containing NH_4Cl. Though a thick oxide film (about 150 μm) was formed after 70-h exposure, the corrosive ions were able to diffuse into the base alloy through the oxide film and attack the brain boundary, leading to the failure of structural materials.

The oxide films formed at Inconel 625 was very thin. Nearly, no obvious films were formed except for some discontinuous pitting-like holes. SEM observation of a distinct pit and EDS map of typical elements is shown in Fig. 6.46. The depth and width of the pit were about 21 and 23 μm, respectively. There was clearly Fe and O enrichment in the pit; however, nearly, no Ni was found in the oxide. The observation was in coincident with the report by Asselin et al. [151]. It is known that NiO

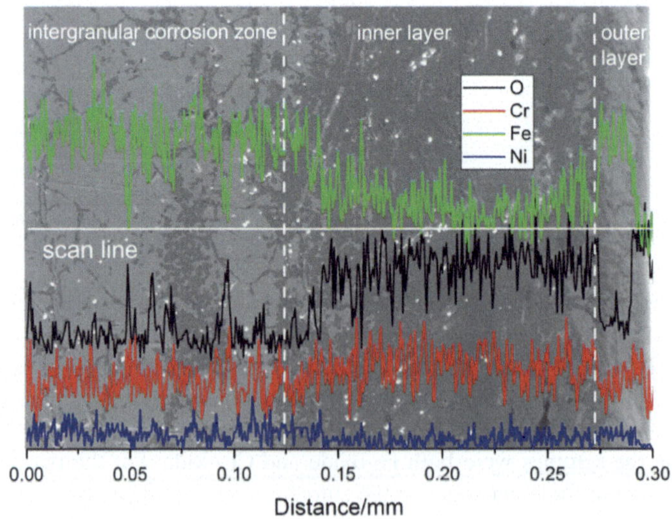

Fig. 6.45 Cross-sectional SEM images and EDS line scan of the oxide film on 316 SS exposed to SCW for 70 h at 400 °C

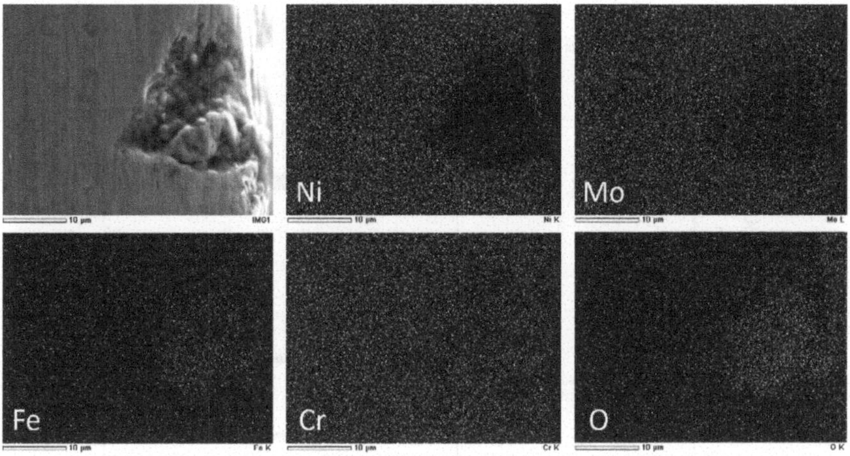

Fig. 6.46 SEM observation of a distinct pit and EDS map of elements in Inconel 625 after 70-h exposure to 400 °C supercritical water environment containing 6000 ppm NH_4Cl

is stable enough to protect the alloy 625 in high-temperature alkaline solution, even after transpassive dissolution has occurred to chromium which dissolved as hexavalent chromate [152]. However, in acidic environments, Ni oxides lost the protection ability, dissolving readily into Ni^{2+}. Moreover, as mentioned in Sect. 3.1, Ni would form complex compounds with NH_3 in NH_4Cl solutions, leading the exhaustion of Ni in oxides. It was interesting that Mo was depleted in the pits formed at the Inconel 625, which was inconsistent with the researches by Kritzer et al. [153] that Mo and Cr concentrations in pits were both up to 20 at.%. However, the experiment conducted by Kritzer et al. was at oxidizing subcritical water containing hydrogen chloride. The relatively high temperature and the presence of ammonia may cause the dissolution of Mo oxide. Mo is an alloy element which can significantly improve the pitting potential and inhibit intergranular corrosion in non-oxidizing acid environment. Nevertheless, Mo has a synergistic effect on corrosion resistance with Cr, which means Mo cannot function without the aid of Cr at both subcritical and supercritical water [154, 155].

As the oxide film of Inconel 600 was thin, the active scan area (namely the oxide film) covered little proportion of the whole cross-sectional view. Thus, the accuracy of line scan was poor if being used in EDS analyzing the oxide film of Inconel 600. Therefore, point scan of EDS was used, and some points among the oxide layer and base alloy were selected to evaluate the alloy composition (shown in Fig. 6.47). A sizable gap which had a length of about 1.5 μm was found between the oxide film and the base alloy, which was a transition zone between base alloy and oxide layer based on the point scan date. In this region, Ni was selectively dissolved, and Fe and Cr oxides were formed. Fe, Ni, and Cr can diffuse through the grain boundaries, and the speeds of the atoms are Fe > Ni > Cr [156]. Therefore, Fe oxide was preferentially formed and deposited at the outer space of the transition zone.

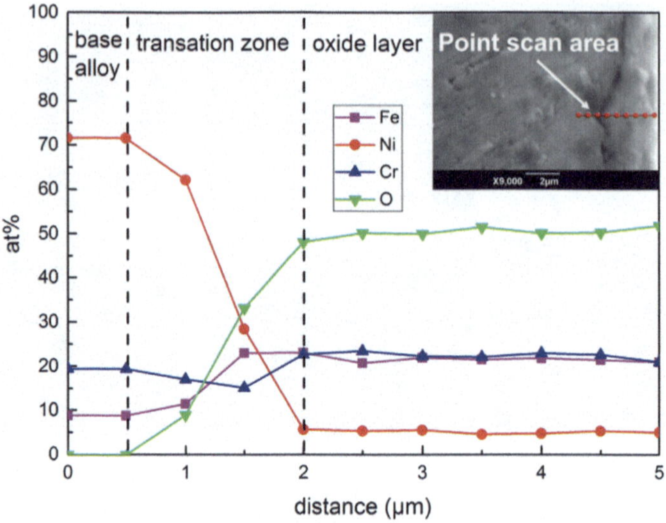

Fig. 6.47 Elemental distribution and cross-sectional SEM images of the oxide film of Inconel 600 after 70-h exposure to 400 °C supercritical water environment containing 6000 ppm NH_4Cl

(3) **Surface chemical analysis**

XRD patterns obtained for the samples after 70-h exposure in supercritical water containing NH_4Cl are shown in Fig. 6.48. Fe oxides, Ni oxides, and spinel were found on 316 SS, Incoloy 800, and Inconel 600. However, the intensities of character peaks on them were different. On 316 SS, the peaks corresponding to Fe oxide and spinel were stronger than that corresponding to base alloy, which can be concluded that the oxides formed on 316 SS were relatively thick. Few peaks were

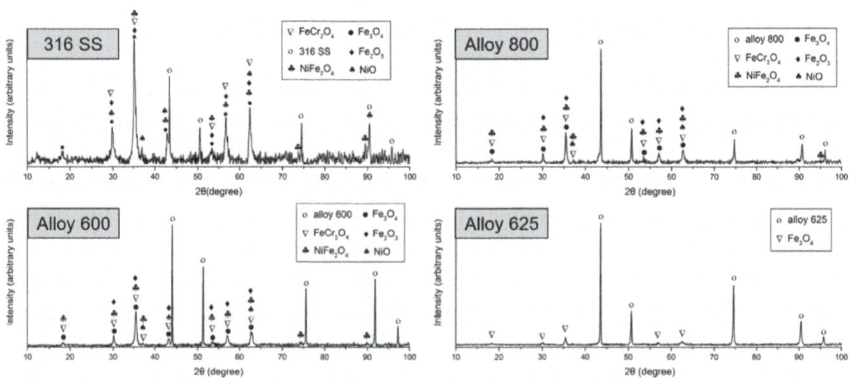

Fig. 6.48 XRD profiles of samples after exposing in supercritical water for 70 h

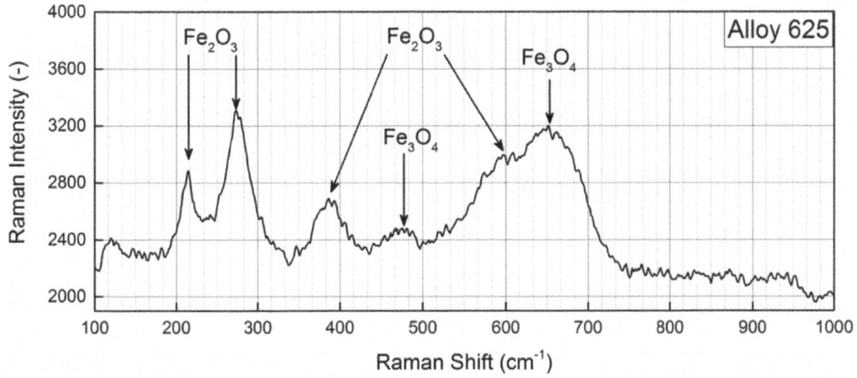

Fig. 6.49 LRS result of Inconel 625 after exposing in supercritical water for 70 h

formed on Inconel 625 except for the peaks corresponding to base alloy. However, the morphology of Inconel 625 obtained by SEM showed that the surface of the sample was covered by deposits. And the cross-sectional analysis showed that the oxide films of alloy 625 were very thin, except for pitting areas. Therefore, the oxide observed by SEM and detected by XRD is mainly deposited on the surface of the sample, rather than formed by the sample itself.

For the reason that the characteristic peak of alloy 625 in XRD analysis was limited, laser Raman spectroscopy was used to supplement the detection of samples. The LRS result of Inconel 625 is shown in Fig. 6.49. Apparently, Raman peaks corresponded to Fe_2O_3 and Fe_3O_4 were detected, which were in accord with that of XRD analysis. Raman spectroscopy was widely used in composition recognition for the surface materials. In consideration of the fact that all the samples were covered by Fe oxides, the Raman result of other samples was also mainly Fe oxide, which cannot be convinced. Hence, chemical cleaning should be done before LRS detection, and it would be discussed in the further articles.

6.4 Corrosion Behavior of Alloys in Oxidizing SCWs

6.4.1 Roles of Dissolved Oxygen on Oxidation Behavior of Ni-Based Alloys 625 and 825

(1) **Morphology analysis**

Figure 6.50 shows the surface morphologies of the oxide films on Ni-based alloy 625 after exposure times of 100, 300, and 500 h. The low-magnification observation revealed the uniform morphology of the corroded alloy 625 sample. During the initial exposure time of 100 h, acicular oxide was observed on the alloy 625 surface. The latter two images captured after 300- and 500-h exposure revealed similar

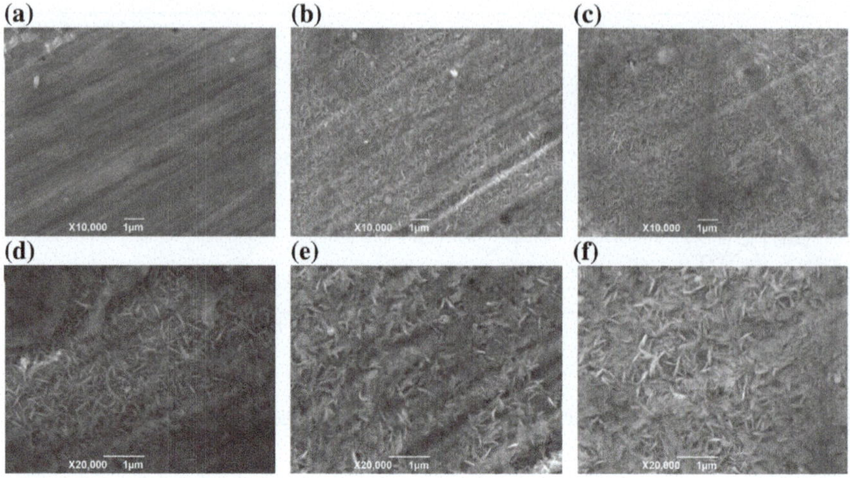

Fig. 6.50 Surface morphologies of the oxide films on Ni-based alloy 625 after exposure time of **a** 100 h, **b** 300 h, **c** 500 h in oxidizing SCW at 500 °C and 24 MPa. **e**, **d**, and **f** correspond to high-magnification figures of **a**, **b**, and **c** [157]

Fig. 6.51 EDS analysis for surface composition of alloy 625 after 100-, 300-, and 500-h exposure in oxidizing SCW at 500 °C and 24 MPa

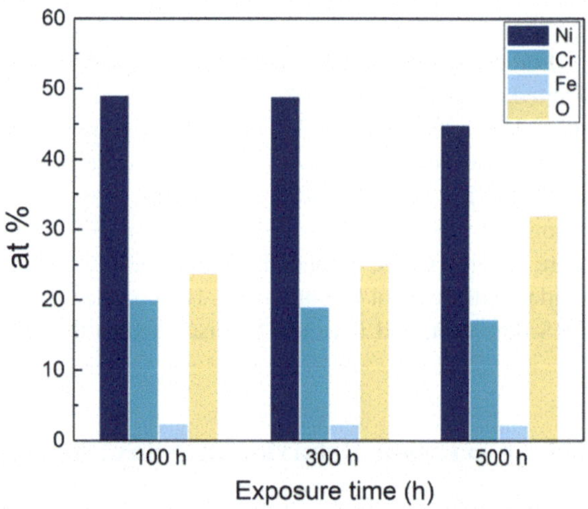

surface morphologies. The distinct scratches on the sample surface indicated that the presence of an extremely thin oxide films and all of the exposed samples were covered with acicular fine oxide particles. Hence, all 625 exhibited some signs of general corrosion after immersion in SCW for up to 500 h. Figure 6.51 shows the EDS scan of surface films on samples at exposure time of 100, 300, and 500 h. The concentration of oxygen detected on the corroded alloy surface increased with the exposure time.

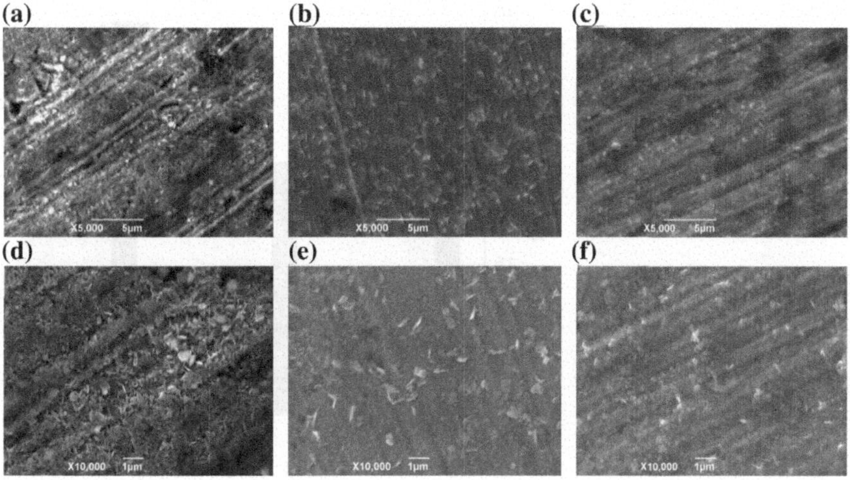

Fig. 6.52 Surface morphologies of the oxide films on Ni-based alloy 825 after exposure times of **a** 100 h, **b** 300 h, **c** 500 h in oxidizing SCW at 500 °C and 24 MPa. **e**, **d**, and **f** correspond to high-magnification figures of **a**, **b**, and **c**

Figure 6.52 shows the SEM images of Ni-based alloy 825. As opposed to alloy 625, alloy 825 revealed a clear rough surface with severe corrosion. After 100 h of exposure, discontinuous films were formed, and some needle bar-shaped granules were observed on the surface of alloy 825. As the diffusion rates of typical metal ions in Ni-based alloys follow the order of Fe > Ni > Cr [156], high concentration of iron was first detected on the sample surface, which dissolved and precipitated as oxides or hydroxides. Figure 6.53 shows the EDS scan of the surface film on alloy 825 after 100, 300, and 500 h of exposure. Compared with the chemical composition of the as-received alloy 825, the amount of Fe was almost constant. Hence, the films of alloy 825 surface were thought to contain iron oxide. This deduction was verified by XRD and LRS results which will be discussed in a later section. Rod-like oxide particles also accumulated on the surface of alloy 825 as a function of time. In addition, the outer layer was filled with low amount of scale-like oxide particles, predominantly comprising iron and oxygen as confirmed by EDS analysis. Pitting-like phenomena were observed on both alloy 625 and 825 samples. The result is in agreement with the observation in previous reports [158, 159]. The numbers of pits observed on alloy 625 were lower than that on alloy 825 when the samples were examined by SEM in a same magnification. An obvious pit, with a depth and width of approximately 1.755 and 3.451 μm, respectively, was observed in the cross-sectional morphology of alloy 825 (Fig. 6.54).

(2) **Cross-sectional analysis**

It is difficult to characterize the oxide films formed on Ni-based alloys in SCW because of their remarkably small film thickness. Thus, alloys after an exposure

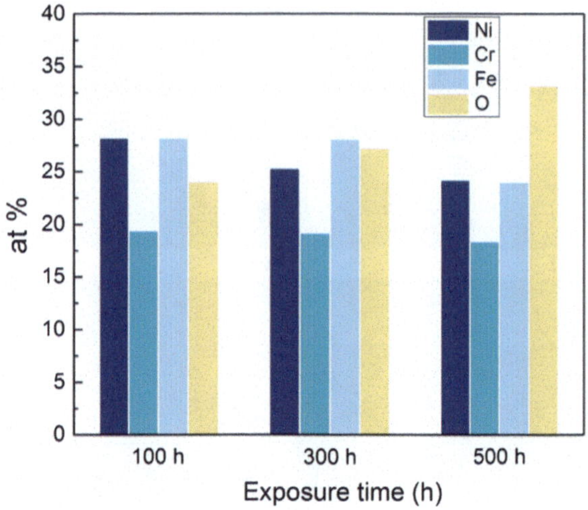

Fig. 6.53 EDS analysis for surface composition of alloy 825 after 100, 300, and 500 h of exposure in oxidizing SCW at 500 °C and 24 MPa

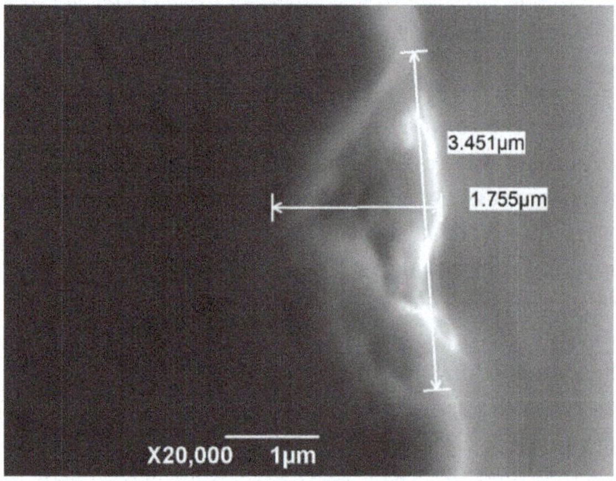

Fig. 6.54 SEM cross-sectional view of a pitting morphology on alloy 825. The measurement of its size shows that the depth and width of the observed pit are about 1.755 and 3.451 μm, respectively

time of 500 h were selected as the test samples for cross-sectional analysis. EDS was used to characterize the composition of the oxide films, and EDS line-scan analysis was typically used for cross-sectional analysis. However, because of the limited resolution of the SEM instrument, the active scan area (namely the oxide film zone) only covered a marginal proportion of the entire SEM image. Thus, the accuracy of the EDS line scan was poor for measuring extremely thin films in this study. Hence, EDS point scan was carried out, and several points along a straight line were selected for analyzing the elemental composition.

According to the previous study by other researchers, a Cr-rich layer was observed at the interface between the oxide film and base alloy in [37, 160, 161]. In addition, a Cr-deficient layer underneath the oxide layer was reported [162]. However, Cr-rich or deficient layers were not observed in our study, although the

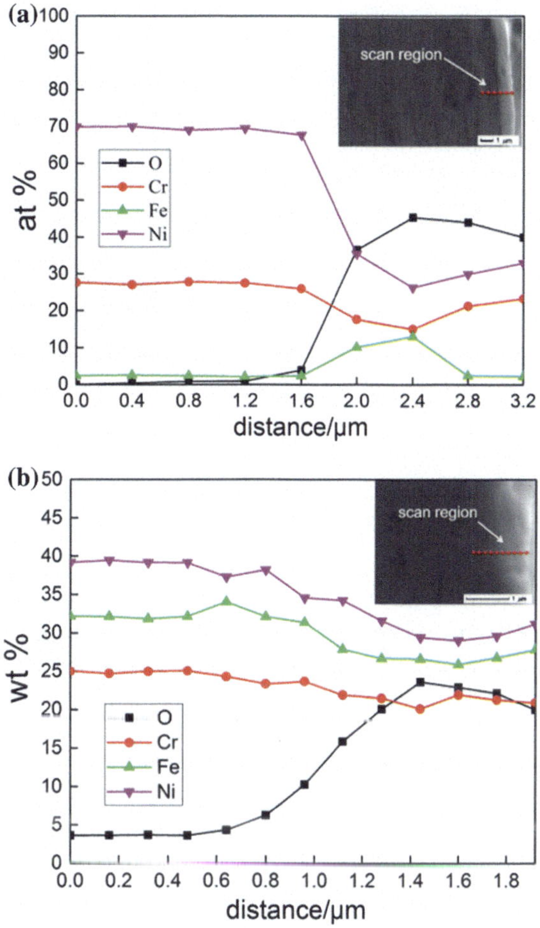

Fig. 6.55 EDS point scan profiles of oxide films grown on samples after exposure times of 500 h in oxidizing SCW at 500 °C and 24 MPa: **a** alloy 625, **b** alloy 825. Inset images in **a** and **b** show the point scanning line

Cr ion concentration slightly decreased between 1.6 and 2.8 μm. Figure 6.55 shows the EDS profiles of samples at the location corresponding to the area shown by the dotted lines in the inset image. For alloy 625, the middle zones of the oxide layer were enriched with Fe. A negligible amount of Cr was detected under the entire layer. However, the outer space of the oxide film showed higher oxidation of Cr, based on calculation for the proportion of Fe, Ni, and Cr along the oxide film of alloy 625 (Fig. 6.56). The different elemental distributions were possibly related to their different outward diffusion rates in the oxide layer, different solubilities of their corresponding oxides in SCW, and different affinities for oxygen. Similar results were obtained for alloy 825. However, Ni-rich layer was not observed for the oxide film of alloy 825, and oxygen was concentrated between the outside layer and metal matrix.

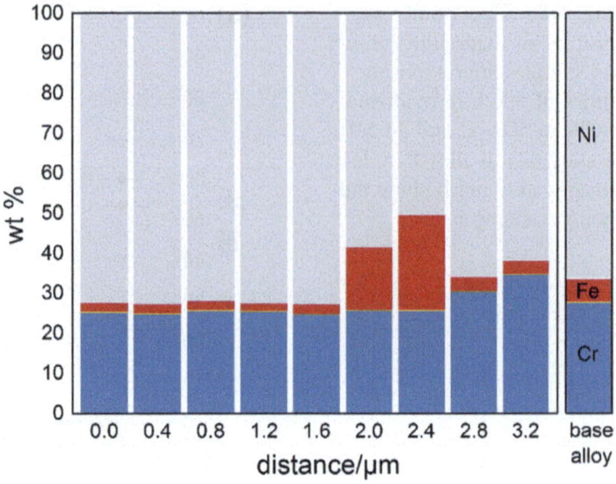

Fig. 6.56 Percentage of Ni, Fe, and Cr in the oxide film of alloy 625 after exposure in oxidizing SCW for 500 h. Percentage on Ni, Fe, and Cr of the as-received sample 625 is also shown on the right side of the figure

(3) **Oxide phases**

Figure 6.57 shows the XRD patterns of the samples after 500 h of exposure. NiO (according to PDF card 44-1159), Fe_2O_3 (according to PDF card 33-0664), and $NiFe_2O_4$ (according to PDF card 10-0325) were detected on the alloy 625 sample surface based on the characteristic peaks. Significant peaks corresponding to the Ni–Cr–Fe alloys were observed, possibly caused by the oxide films with an extremely low thickness. Strong peaks corresponding to Fe_2O_3 and Cr_2O_3 were observed at similar positions. However, the weak peaks distinguishing Fe_2O_3 and Cr_2O_3 were not observed in XRD patterns of alloy 625 because of the presence of multiple substances on the surface and intrinsic noise of the instrument. Thus, it was difficult to distinguish Fe_2O_3 and Cr_2O_3 by XRD in this study, and the results need to be verified by LRS.

Alloy 825 exhibited an XRD pattern similar to that of alloy 625. In addition, peaks corresponding to NiO and Fe_2O_3, except for $NiFe_2O_4$, were detected. Distinct XRD intensities for NiO peaks were observed. In general, the intensities of the characteristic peaks were not proportional to the concentrations of corresponding materials. Various factors such as scanning speed, exposure intensities, and dryness level of the samples could also affect peak intensities. Nevertheless, the different XRD peaks intensities of alloy 625 and 825 were possibly related to the fact that Ni content in alloy 625 was greater than that in alloy 825, assuming that the experiment was carried out in a similar environment.

LRS was also employed to detect the oxide phases on the surfaces of alloy 625 and 825 after exposure to SCW containing 5000 ppm of DO at 500 °C for 500 h. Inhomogeneous sample surface morphologies were observed, with several discontinuous dim spots, which were possibly formed by the aggregation of alloy elements. Four spots at different surface regions were chosen for Raman scattering, and the results were considered accurate only if similar data were obtained in four experimental runs. Figure 6.58 shows the modified Raman spectra of the oxide

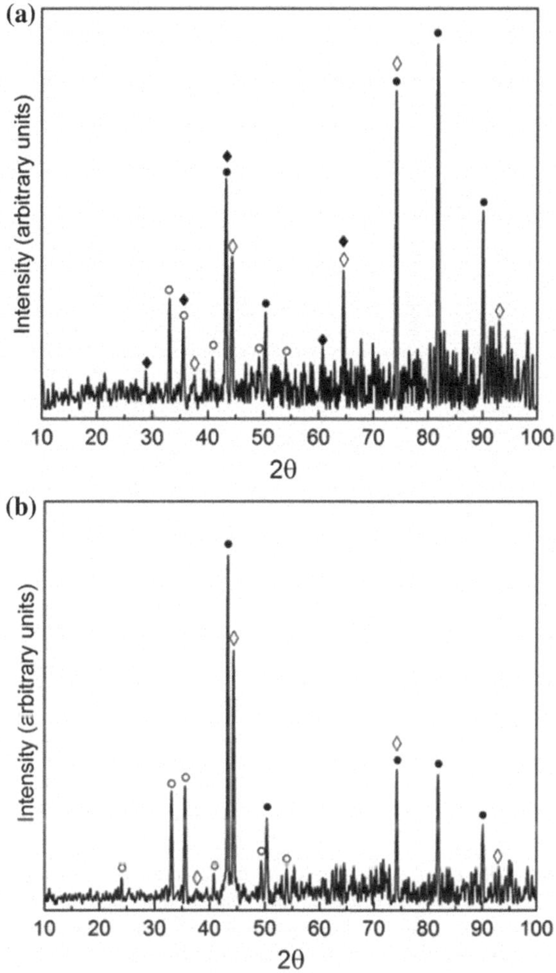

Fig. 6.57 XRD pattern of the samples after 500 h of exposure in oxidizing SCW at 500 °C and 24 MPa: **a** alloy 625 and **b** alloy 825. (•), (◦), (O), and (♦) denote main diffraction lines of matrix, NiO, Fe_2O_3, and $NiFe_2O_4$, respectively

films. Raman peaks corresponding to Fe_2O_3, Fe_3O_4, NiO, and $NiFe_2O_4$ were detected on the alloy 625 surface, consistent with the XRD results.

Figure 6.58 shows the Raman spectra of alloy 825. The spectra revealed a predominant intense sharp peak centered at 660 cm^{-1}, corresponding to Fe_3O_4. Moreover, the Raman peaks observed at 221, 290, 405, and 620 cm^{-1} corresponded to Fe_2O_3. NiO was not detected on the surface of alloy 825. The detection depth of Raman spectroscopy was dozens of nanometers, only characteristic peaks

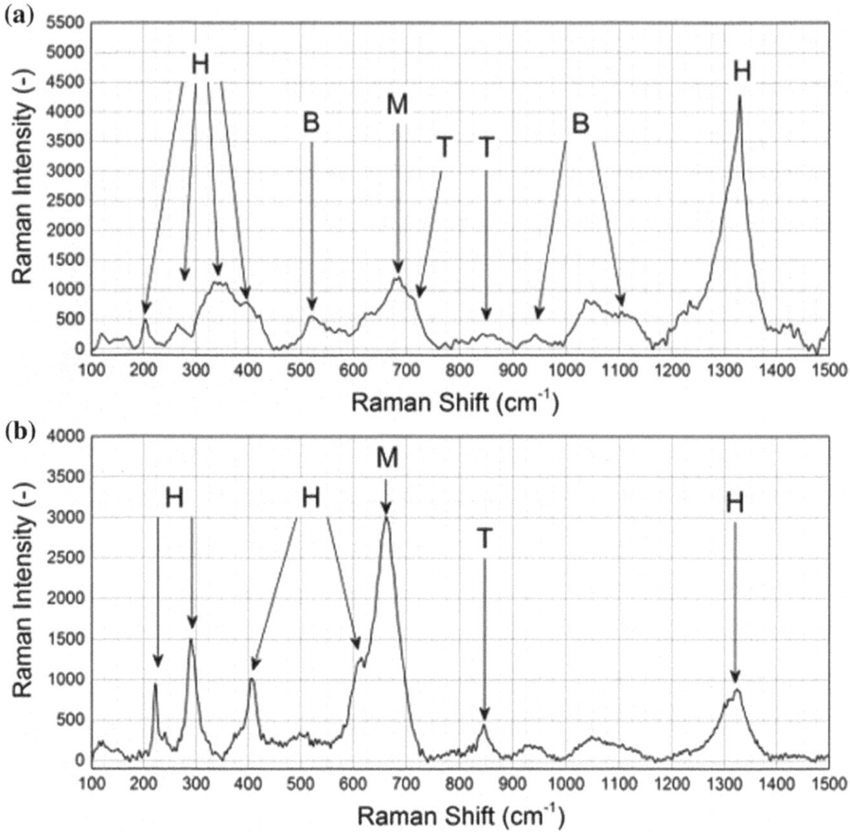

Fig. 6.58 Laser Raman spectra of the samples after exposure time of 500 h in oxidizing SCW at 500 °C and 24 MPa: **a** alloy 625 and **b** alloy 825. The spectra were recorded on the smooth surface rather than pits. Laser power was 17 mW, and the excitation wavelength was 633 nm. (H), (M), (B), and (T) denote main Raman line of Fe_2O_3, Fe_3O_4, NiO, and $NiFe_2O_4$

corresponding to the outmost layer were detected. Hence, the outer layer of alloy 825 consisted of Fe_3O_4 and Fe_2O_3, with the predominant distribution of NiO in the inner parts.

As it is difficult to detect the limited amount of Ni oxides by XRD and LRS, XPS was employed to further analyze the chemical state of the sample surfaces. Figure 6.59 shows the XPS Ni 2p 2/3 core-level spectra, with data analysis using software CasaXPS and application of the Shirley background correction. Compared with the NIST XPS database and published data [37, 42, 163, 164], both the alloy 625 and 825 samples did contain NiO. In addition, $Ni(OH)_2$ was only detected by XPS, which was possibly formed during the heating and cooling process.

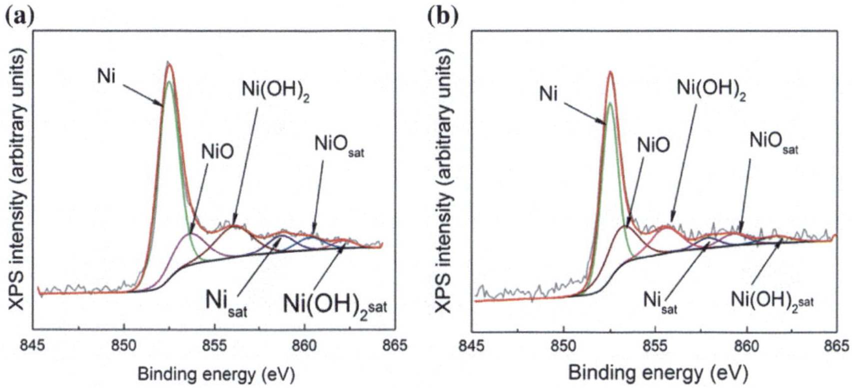

Fig. 6.59 XPS Ni 2p3/2 spectra of the samples after exposure in oxidizing SCW at 500 °C: **a** alloy 625, **b** alloy 825. Peaks corresponding to NiO and $Ni(OH)_2$ were found

6.4.2 Coupling Corrosion Behavior of Phosphate, Chloride, and Oxygen

In this work, the experiments for decomposing the pesticide wastewater were selected to investigate the corrosion behaviors of Ni-based alloy under the subcritical condition (350 °C and 25 MPa) and supercritical condition (450 °C and 25 MPa). The subcritical test without oxygen and the supercritical test with oxygen were used to simulate actual operating conditions of the preheater and the reactor, respectively. The corrosion morphology, surface chemical property, corrosion rate, and corrosion mechanism were presented in this part.

(1) **Surface Morphologies of Tested Alloys**

Figure 6.60 shows the morphologies' images of the three kinds of alloys before and after the corrosion test. The corrosion conditions of tested alloys in experiments 1, 2, 3, and 4 are presented in Fig. 6.60b–e, respectively. The surface colors of all the tested alloys showed metallic luster before corrosion, and they changed to black after corrosion, indicating lots of corrosion products fully covered the surface of tested alloys and the metal would suffer a severe corrosion under supercritical water oxidation condition. Many pits were only find on the surface of 316 SS after exposure to experiment 2, confirming 316 SS had the worst corrosion resistance in all tested alloys and oxygen coexisted with chlorine would lead to the severest corrosion.

Figure 6.61 shows the scanning electron microscope images of five tested alloys, and the conditions of experiments 1, 2, 3, and 4 are presented in Fig. 6.61 panels a, b, c, and d, respectively. 316 SS displayed the severest corrosion and spallation of five alloys under the condition of coexistence of oxygen and chlorine, as shown in Fig. 6.61 316(a). Incoloy 600, Inconel 625, and Hastelloy C-276 were also

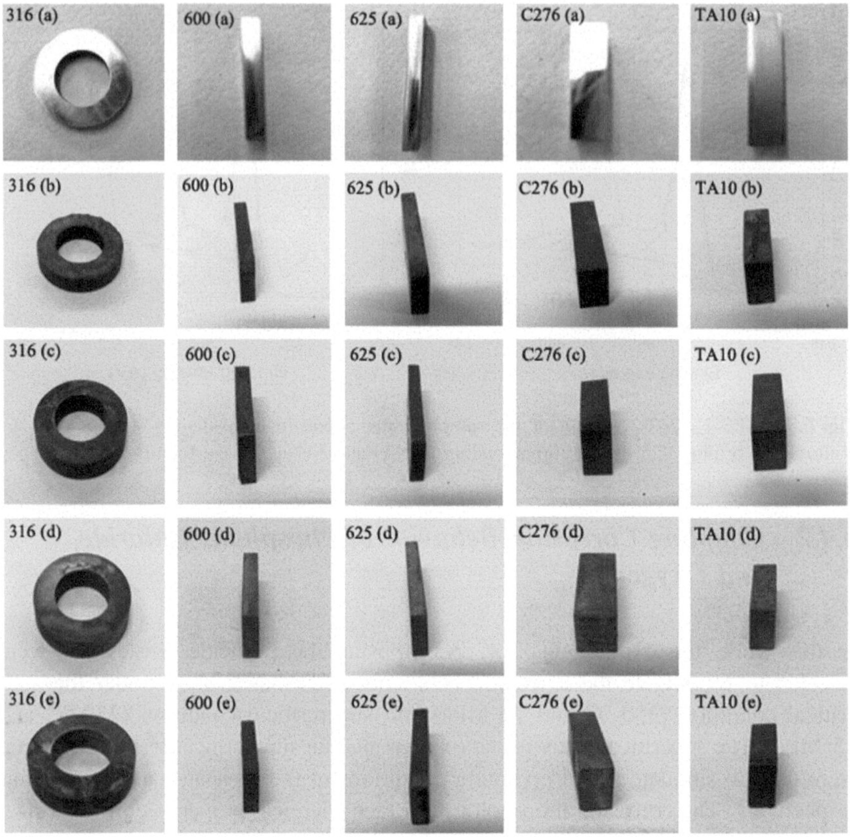

Fig. 6.60 Microscopy images of alloys before and after corrosion: **a** before corrosion, **b** experiment 1, **c** experiment 2, **d** experiment 3, and **e** experiment 4 [165]

presented the most severe corrosion in experiment 1 among four experiments, indicating the strong synergistic effect between oxygen and chlorine (see Fig. 6.61a). At the experiment of coexistence of phosphate and oxygen (Fig. 6.61b), the least corrosion on the surfaces of tested alloys in four experiments, and corrosion products were observed obviously only on the surfaces of 316 SS and Incoloy 600. By contrasting the corrosion morphologies in Fig. 6.61a, c, the alloys exposure to experiment 1 suffered from severer corrosion than that in experiment 3, indicating that the addition of phosphate was helpful for inhibiting corrosion under supercritical water. 316 SS displayed the worst corrosion resistance in all tested alloys, and the pitting corrosion was observed on 316 SS in supercritical water with chlorine. The development of pitting corrosion would lead to alloys perforated, because the concentrated corrosion mediums would cause severer corrosion in the hole. By comparing the different corrosion surface morphologies, the oxygen and chloride ion had a synergistic effect on corrosion, leading to severer corrosion.

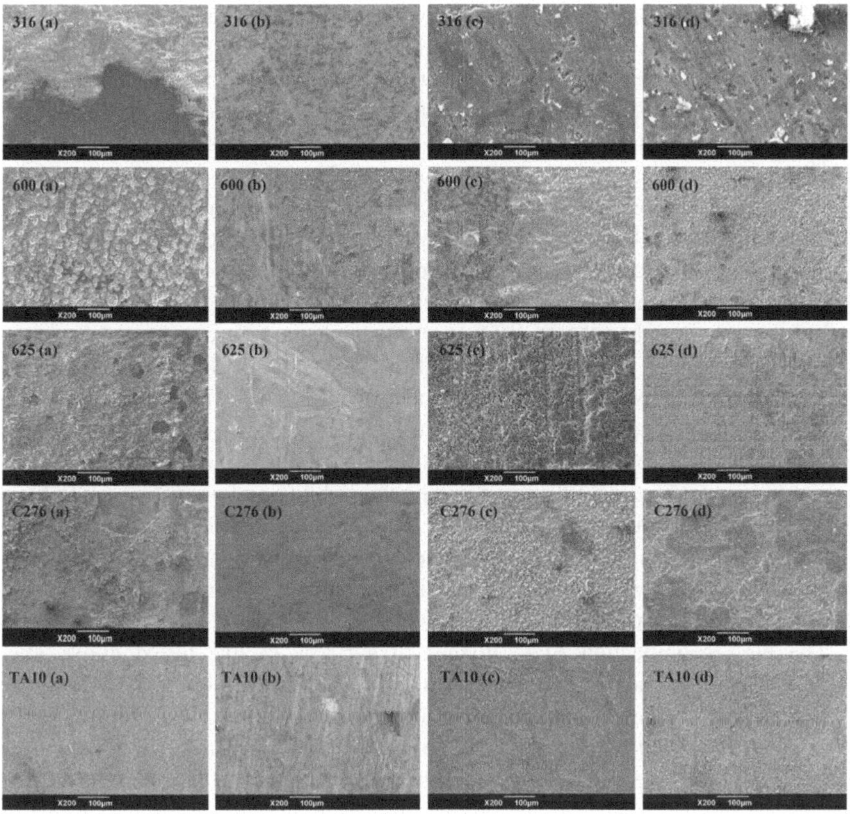

Fig. 6.61 Surface morphologies of the tested alloys after exposure to **a** experiment 1, **b** experiment 2, **c** experiment 3, and **d** experiment 4

Under all experiments, TA 10 presented best corrosion resistance in all tested alloys and the corrosion products were hardly found on surface TA 10.

As Fig. 6.61 316(a) shows, the severest oxide spallation of 316 SS was also observed, as 316 SS contained a great deal of iron which would change to loose oxide at condition of the coexistence of salt and oxidant. The loose iron oxide split into supercritical water and the matrix exposure to corrosion medium, leading to severer corrosion. The results of morphology in Fig. 6.61 316(a) are in accord with that of Fig. 6.60.

Zhang et al. [37] found that the oxide of Hastelloy C-276 also had a strong tendency to spalling under the harsh condition. The significant oxide spallation of Incoloy 800H was also observed, and using the grain boundary engineering (GBE) technology could enhance the resistance to spallation [41]. The clearly grinded scratches indicated that these three Ni-based alloys had outstanding corrosion resistance in experiment 4, as shown in Fig. 6.60c. It is well known that the Ni-based alloys are the candidate metals for constructing the SCWO system [29, 30,

36]. Incoloy 825 took on the weakest corrosion resistance in all the three tested alloys under the three conditions, with Ni-based alloys suffering a severer corrosion by the synergistic effect between salt and oxygen than that by the individual effect. The aggressive ions can break the protective oxide layer, and the matrix will be exposed to the corrosion environment [7, 18, 20, 28, 30, 37]. The pressure-bearing capacity of tube decreased as the corrosion accelerated and the safety accidents would be caused beyond the burst limit. Results showed that the coexistence of oxidant and salt should be avoided effectively so as to reduce corrosion damage on SCWO system. Therefore, the oxidant and wastewater feed should be separated in the preheat process, and the inorganic salt should be removed from the wastewater before the reactor, when the SCWO system was planned for construction.

Kim et al. [166] proposed that the stress corrosion cracking could occur in supercritical water by direct chemical attack of HCl. The oxidant and salt had a synergistic effect on corrosion, since the oxidant can deepen the depth of oxygen penetration and the ion like Cl^- can lead to the decomposition of oxides [9, 27]. As shown in Fig. 6.67, it is difficult to generate the stability and protective oxide films when the aggressive ions present in SCW. Mitton et al. [29] also observed that the Hastelloy C-276 exhibited dealloying and stress corrosion cracking in oxidizing supercritical water with chloride. The significant stress corrosion cracking and dealloying were found in stainless steel owing to the corrosion of chloride [23, 26].

The high-magnification SEM images of surfaces of five tested alloys after exposure to experiment 3 are shown in Fig. 6.62. Even though the phosphate existed in corrosion environment, 316 SS also suffered from severe pitting corrosion under the condition of coexistence of oxygen and chlorine. However, lots of anomalous oxides were observed on the surface of Incoloy 600, Inconel 625, and

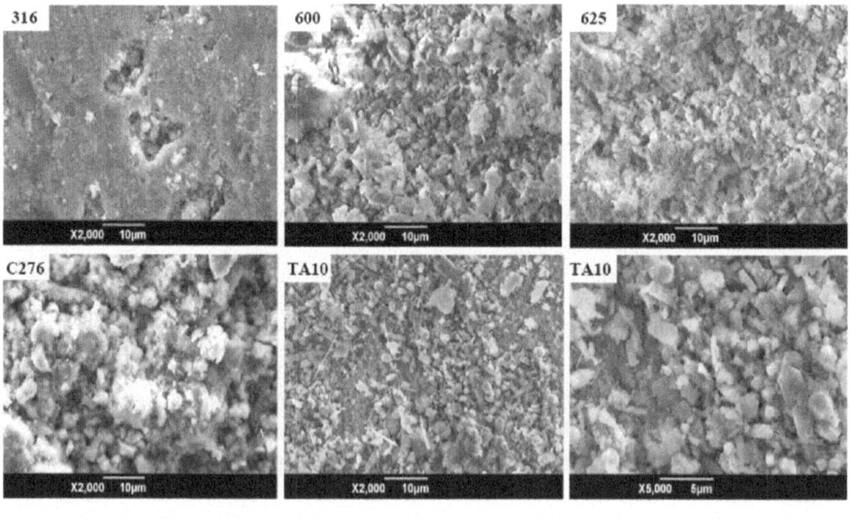

Fig. 6.62 Surface morphologies of 316 SS, Incoloy 600, Inconel 625, Hastelloy C-276, and TA 10 after exposure to experiment 3

Hastelloy C-276, indicating the metal oxide was unstable and corrosive. While, by comparing the unstable oxide generated on the surface of three Ni-based alloys, TA 10 owned the regular and stable oxide which would prevent the matrix from more corrosion at the condition of experiment 3.

The high-magnification (×5000) SEM images of surfaces of Incoloy 825 after exposure to different conditions are shown in Fig. 6.67. In oxidizing supercritical water, many oxide crystals developed into platelets fully covered the surface of Incoloy 825, as shown in 825(b). Sun et al. [28] found that the rods-like and platelets-like oxide crystals grew on the surface of Inconel 625 in oxidizing supercritical water. The 825(c) illustrates that the alloy can hardly form a stable corrosion product in supercritical saline aqueous solution. This result could be explained by the fact that the attack anions like Cl^- with high potential would gain electrons from alloying element, and then the element converted into soluble high valence form resulting in pitting corrosion. The salt and corrosion products were almost insoluble in supercritical water, which would result in less corrosion at supercritical condition than at subcritical condition [14, 33].

Incoloy 800, Incoloy 825, Inconel 625, and Hastelloy C-276 were selected for the test at high subcritical temperature. The surface of the corroded metal was analyzed by the enlarged surface image magnified 1000 × using scanning electron microscope, as displayed in Fig. 6.62. Under the subcritical condition, corroded alloys exhibited the corrosion morphologies which were different from those of specimens exposed to the supercritical condition, and Incoloy 800 exhibited the severest pitting corrosion in all the coupons in experiment 1. This difference implies that corrosion mechanisms under the subcritical and supercritical conditions are different, as a result of the variations of electrolyte dissociation and water density. Kritzer et al. [34] reported that Inconel 625 exhibited pitting corrosion at temperatures above 130–215 °C, and stress corrosion cracking occurred near the critical temperature water containing chloride. Because of the great decrease of ionic product constant, the aggressive ions will cause severer pitting corrosion under a subcritical condition than under a supercritical condition. The alloys would suffer seriously dealloying corrosion at high subcritical temperatures, which was due to free aggressive ions and high temperature [8–11, 14, 20]. The selective dissolution of alloying element was the most probably reason for pitting corrosion and stress crack corrosion. Somerday et al. [167] observed that the severe dealloying of Ni and cracking resulted in the failure of alloy C276 burst disks in a subcritical temperature water with acidic chlorinated. Boukis et al. [35] found that alloy 625 exhibited

Table 6.5 Chemical compositions of coupons (wt%)

	Ni	Cr	Mo	C	Fe	Ti
Incoloy 800	32.0	21.0	–	0.1	46.5	0.4
Incoloy 825	42.0	21.5	3.0	0.02	28.5	0.9
Inconel 625	58.3	22.0	9.0	0.01	5.0	0.4
Hastelloy C-276	45.1	16.0	16.0	0.01	5.5	–

severe pitting and intercrystalline corrosion in the preheater and cooling for the entire system at the approximately critical temperature. Under the supercritical condition, the non-polar organic compounds and gases became completely miscible with water, and the conductivity of the solution was extremely poor. The non-ionic mechanism can explain the phenomenon that small granules comprised of metal salts overspread the metal alloys. As shown in Fig. 6.67, the SEM micrographs of the coupons exposed to the experiment 1 can better support the corrosion mechanism mentioned above.

Incoloy 800 had the weakest resistance among the tested alloys under experiment 1, as a result of the severest pitting corrosion was observed. Chemical compositions of all the tested Ni-based alloys are in Table 6.5, with the Incoloy 800 lacks of molybdenum. The aggressive ions would cause a severer pitting corrosion and stress cracking corrosion on stainless steel in the subcritical water at high temperature (T = 350 °C) than in the supercritical water at low temperature (T = 450 °C). The corrosion process was that the insoluble oxide scale would change to the soluble metal chloride under the attack of chloride ion [12, 30, 31, 40]. Many researchers have found that the corrosion resistance greatly increases as molybdenum content increases with a large proportion of the chromium existing in the metals [12, 18, 25].

As shown in Fig. 6.63, the phosphorus element was detected on the surface of the tested alloys under the conditions of experiments 2 and 4, indicating insoluble phosphate could deposit on alloys stably in supercritical water with the single oxygen or chlorine condition. The phosphorus element was hardly detected on the surface of 316 SS, Incoloy 600, Inconel 625, and Hastelloy C-276 in experiment 3, due to the significant oxide spallation as a result of the synergism between chlorine and oxygen. However, the phosphorus element was observed on TA 10, as the oxide of TA 10 could exist stably in experiments 2, 3, and 4. There was no chlorine

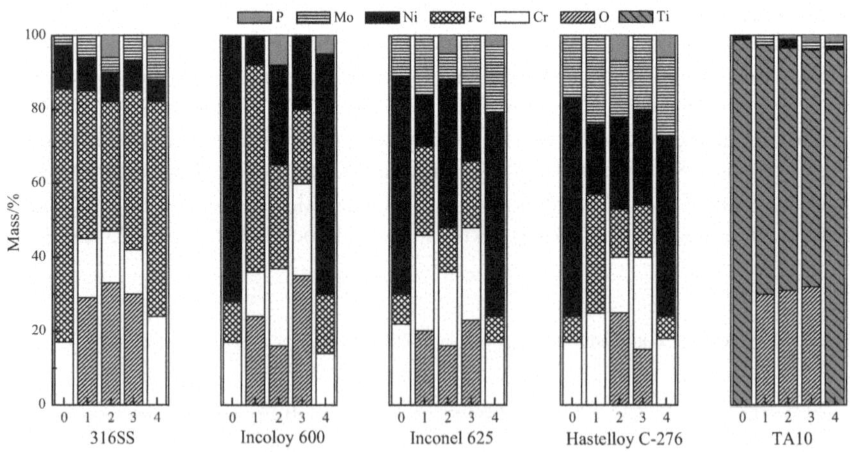

Fig. 6.63 Surface composition of corroded alloys exposed to four medium conditions, (0) matrix, (1) experiment 1, (2) experiment 2, (3) experiment 3, and (4) experiment 4

element detected under all experiments, indicating chloride ion was hard to form stable compound deposited on tested alloys. Under the condition of supercritical water with oxygen and choline, the insolubility phosphate could isolate the matrix and aggressive medium, leading to the effective reducing of corrosion. In addition, the large mass percentage of oxygen element was detected on the surface of tested alloys under the experiments 1, 2, and 3, indicating that the metal would change to oxide easily in supercritical water with oxygen. However, the mass percentage of nickel in outer layer was decreased significantly, indicating the serious depletion of nickel was observed in oxidation environment. The content of molybdenum was increased in supercritical chlorine water without oxygen, indicating molybdenum had a good corrosion resistance to chlorine. The chromium and iron could change to stable oxide in supercritical oxidation water, leading to the corrosion resistance of alloys improved. A small amount of phosphorus was detected on the surface of TA 10, as TA 10 had the best corrosion resistance of all tested alloys and little titanium ion was offered. Under the oxidation condition, the stable oxide TiO_2 generated on the surface of TA 10 at experiments 1, 2, and 3. Due to titanium and titanium oxide stably existing in supercritical water with chlorine and oxygen, titanium alloy was the best candidate metals of all tested alloys for constructing the SCWO system.

(2) **Surface Chemical Analysis of Tested Alloys**

Under the condition of high concentration of salt and oxygen, different corrosion types were observed in different specimens. Experiment 2, experiment 3, and experiment 4 have been described in Table 6.6. The mass percentages of main alloying elements were detected through the energy diffraction spectrum (EDS), as shown in Fig. 6.63. Results of (a) and (b) indicated that oxygen dominated at the surfaces of metal alloys because of the oxidizing condition. The intensities of nickel decreased more significantly than the other detected elements. Then Ni departed from the surface of corroded alloy and flowed into the reaction fluid. The intensity of carbon increased on the surfaces of corroded Ni-based alloys, which suggested that the organic matters in pesticide wastewater were decomposed not only in the fluid but also on the metal wall. A comparison of the carbon contents in three conditions showed that because of the incomplete decomposition of organic matters without oxygen causing the severest carbon contamination, the intensity of carbon increased significantly on the surface in (c). The intensities of chromium, iron, and molybdenum had a smaller diminution than that of nickel. These results indicated that unstable corrosion products of nickel can dissolve in supercritical water easily, causing the pitting corrosion.

The existence of oxygen made it dominated at the surfaces of metal alloys in (a) and (b). As shown in Fig. 6.63, the oxygen intensity in (a) was weaker than that in (b), which could be explained by the fact that chloride ion caused the dissolution of oxide in the supercritical saline aqueous solution. As shown in 825(a) and 825 (b), the intensity of iron had a lesser degree of decrease in the oxide film than that of nickel or chromium when oxygen was presented. It indicated that iron could generate stable oxide which isolated alloy and corrosive medium. However, the intensity of the iron signal decreased more than that of nickel and chromium in 825

Table 6.6 Conditions used for corrosion tests

Expt run no.	Alloy	Dissolved Oxygen (mg/L)	Dissolved Cl^- (mg/L)	Dissolved PO_4^{3-} (mg/L)	Temperature (°C)	Pressure (MPa)	Exposure Time (h)
1	316 SS	78,200	9200	0	450	25	60
	Incoloy 600	78,200	9200	0	450	25	60
	Inconel 625	78,200	9200	0	450	25	60
	Hastelloy C-276	78,200	9200	0	450	25	60
	TA 10	78,200	9200	0	450	25	60
2	316 SS	78,200	0	8600	450	25	60
	Incoloy 600	78,200	0	8600	450	25	60
	Inconel 625	78,200	0	8600	450	25	60
	Hastelloy C-276	78,200	0	8600	450	25	60
	TA 10	78,200	0	8600	450	25	60
3	316 SS	78,200	9200	8600	450	25	60
	Incoloy 600	78,200	9200	8600	450	25	60
	Inconel 625	78,200	9200	8600	450	25	60
	Hastelloy C-276	78,200	9200	8600	450	25	60
	TA 10	78,200	9200	8600	450	25	60
4	316 SS	0	9200	8600	450	25	60
	Incoloy 600	0	9200	8600	450	25	60
	Inconel 625	0	9200	8600	450	25	60
	Hastelloy C-276	0	9200	8600	450	25	60
	TA 10	0	9200	8600	450	25	60

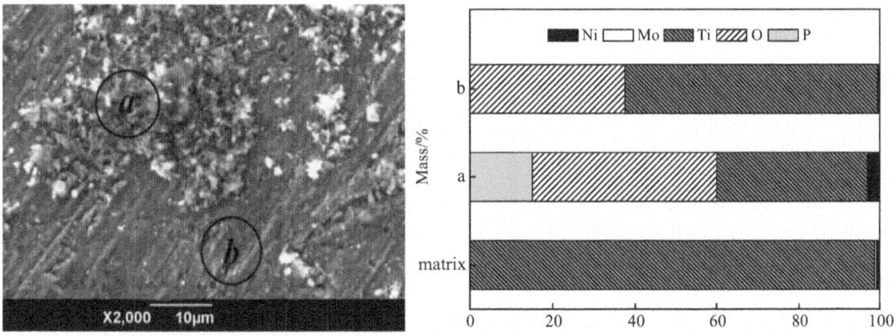

Fig. 6.64 Surface composition of corroded TA10 after exposure to experiment 2

(c), which was an indication that iron changed to ionic forms more easily under the influence of aggressive anions. As shown in Fig. 6.63(IV), oxide film was generated on the matrix of Incoloy 825 under experiment 2. The chemical composition of oxide layer can reveal the corrosion mechanism.

In order to investigate the corrosion behavior, the cross section of Incoloy 825 was analyzed by SEM and EDS after being exposed to experiment 2. The oxidative depth of Incoloy 825 was clearly observed, and distribution densities of alloying elements were also detected, as presented in Fig. 6.64. The density of white point represented the concentration of element in the X-ray mapping. The X-ray mappings revealed that Ni was depleted seriously at surface, and the distribution densities of Fe, Mo, and Cr were almost unchanged [20, 21, 26]. Kim et al. [40] also observed the selective dissolution of Ni and selective oxidation of Cr in Inconel 625. It indicates that Fe, Mo, and Cr can form stable products in order to avoid dissolving in supercritical water. The depletion of nickel promoted by acidic chlorinated was also observed in Hastelloy C-276 [39]. The Ni-based alloys exhibited different degrees of dealloying of Ni, which would cause pitting corrosion and stress corrosion cracking. In addition, the high concentration of oxygen in the outer layer indicated the formation of stable oxide and the corrosion depth.

As shown in Fig. 6.64, TiO_2 composed the oxide which covered the surface of TA 10 uniformly and the clearly grinded scratches indicated the thin oxide layer. TA 10 showed the outstanding corrosion resistance under the tested conditions, due to the thin and uniform oxide layer. As a result of deposits of phosphorus was observed on the surface of TA 10, the stable phosphate could prevent the aggressive medium form attacking the matrix. However, the chloride ion can hardly form stable chlorate to protect the metal in supercritical water with chlorine and oxygen.

Figure 6.65 depicts the XRD spectra of three Ni-based alloys after being exposed to experiment 2, experiment 3, and experiment 4. The different peaks of XRD represent the different configurations of oxides on the Ni-based alloys surface. The intensities of matrix signals were much stronger than those of the oxide signals, indicating that the oxide surface film was thin.

According to the XRD analysis (Fig. 6.65), Fe_3O_4, which was generated on the surface at the characteristic peaks, could prevent the metal from further corrosion.

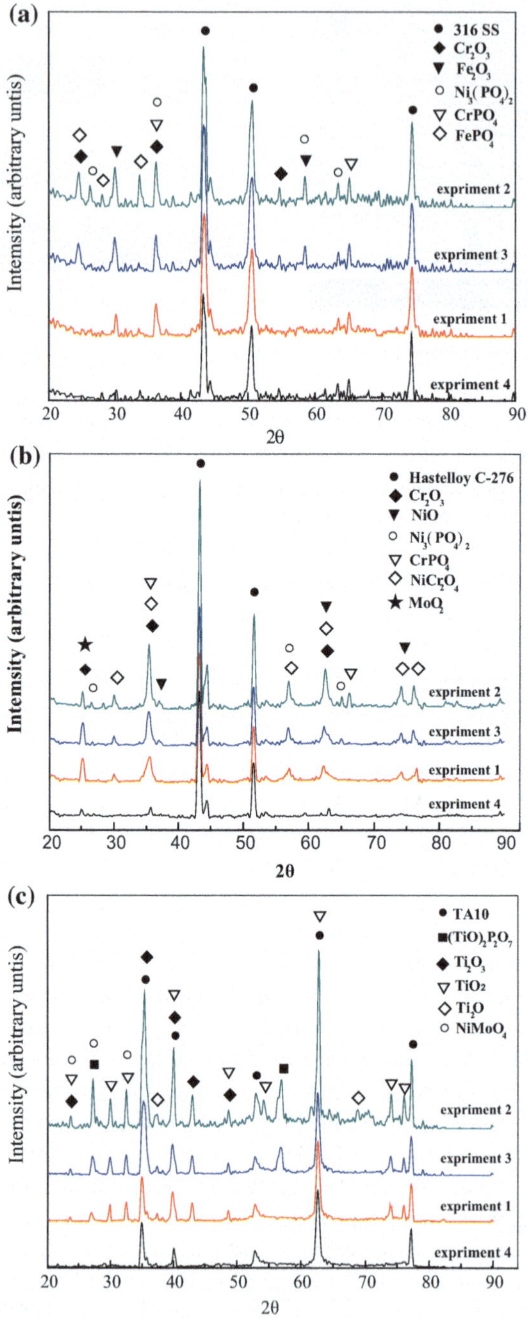

Fig. 6.65 XRD spectra for tested alloys after exposure to experiment 1, experiment 2, experiment 3, and experiment 4. **a** 316 SS, **b** Hastelloy C-276, and **c** TA10

The thickness of oxide film on tested alloy was dependent on the exposed time if the alloy and corrosion environment were definite. The intensities of MoO_2 and Cr_2O_3 were slight in XRD of three alloys in subcritical water. Under the single oxygen condition, the oxides Cr_2O_3, $NiCr_2O_4$, and NiO were found in oxide layers of all Ni-based alloys [21, 37]. The peaks of Fe_3O_4, $Ni(OH)_2$, and MoO_2 phases were found in Incoloy 825, Inconel 625, and Hastelloy C-276, respectively.

All the intensities of oxide signals except for MoO_2 were weaker under the condition of coexistence of oxygen and salt than under the condition of single oxygen, which was attributed to that the aggressive salt could cause the decomposition of oxide. Little difference in MoO_2 signals indicated that the MoO_2 was stable under the attack of salt, and the resistance to pitting corrosion would increase as the Mo content. Sun et al. [28] found that the stable Cr_2O_3 formed a protective film, and $Ni(OH)_2$, NiO, MoO_2, and $NiCr_2O_4$ were also observed in the oxide layer of Inconel 625. Ren et al. [74] also observed that the oxide layer of Inconel 625 was made up of $NiCr_2O_4$, Cr_2O_3, and NiO in the oxidizing supercritical water. Zhang et al. [37] pointed out that Hastelloy C-276 formed the dual-layer structure oxide scale consisting of outer NiO layer and inner $Cr_2O_3/NiCr_2O_4$-mixed layer. These results indicated that Cr content can improve the corrosion resistance of Ni-based alloy, and the chromizing treatment for alloy was applied to decrease the corrosion rate [23]. These finds confirmed that the metal oxide stability was the main factor for the corrosion resistance. Adschiri et al. [62] found that the solubility of chromium oxide was lower than that of iron and nickel oxide, which was an important factor affecting the corrosion resistance of alloy. Sue et al. [59] reported that oxide solubilities of Fe, Ni, and Cr were in this order: Fe > Ni > Cr, and the test was conducted in oxidizing supercritical aqueous solution of HCl.

The oxide scales of nickel-based alloys were examined by XPS to determine the chemical states of Ni, Cr, Fe, and Mo under experiment 2, experiment 3, and experiment 4, respectively. Figure 6.66 shows the core-level spectra of Ni 2p, Cr 2p, Fe 2p, and Mo 3d. The Ni 2p peaks were decomposed into three components: NiO located at a binding energy (BE) of 853.2 eV, $Ni(OH)_2$ located at the BE of 862.1 eV, and $NiCr_2O_4$ located at the BE of 880.1 eV. The signal at the BE of 576.2 eV was assigned to Cr^{3+} in Cr_2O_3. The signal located at BE of 710.5 eV was assigned to Fe_3O_4 and 227.4 eV was assigned to MoO_2. Figure 6.66.825(a), 825 (b), and 825(c) show the Ni 2p, Cr 2p, and Fe 2p core-level spectra for Incoloy 825, respectively. Using XPS techniques, NiO, $NiCr_2O_4$, Cr_2O_3, and Fe_3O_4 were observed in oxide film of Incoloy 825. Figure 6.66.625(a), 625(b), and 625(c) also show the Ni 2p, Cr 2p, and Mo 3d core-level spectra for Inconel 625, respectively. The oxide film of Inconel 625 was composed of NiO, $Ni(OH)_2$, $NiCr_2O_4$, Cr_2O_3, and MoO_2 were observed by XPS. Figure 6.66-C276(a), C276(b), and C276(c) also show the Ni 2p, Cr 2p, and Mo 3d core-level spectra for Hastelloy C-276, respectively. The oxide film of Inconel 625 composed of NiO, $NiCr_2O_4$, Cr_2O_3, and

Fig. 6.66 XPS of oxide films grown on Incoloy 825, Inconel 625, and Hastelloy C-276 after exposure to experiment 2, experiment 3, and experiment 4

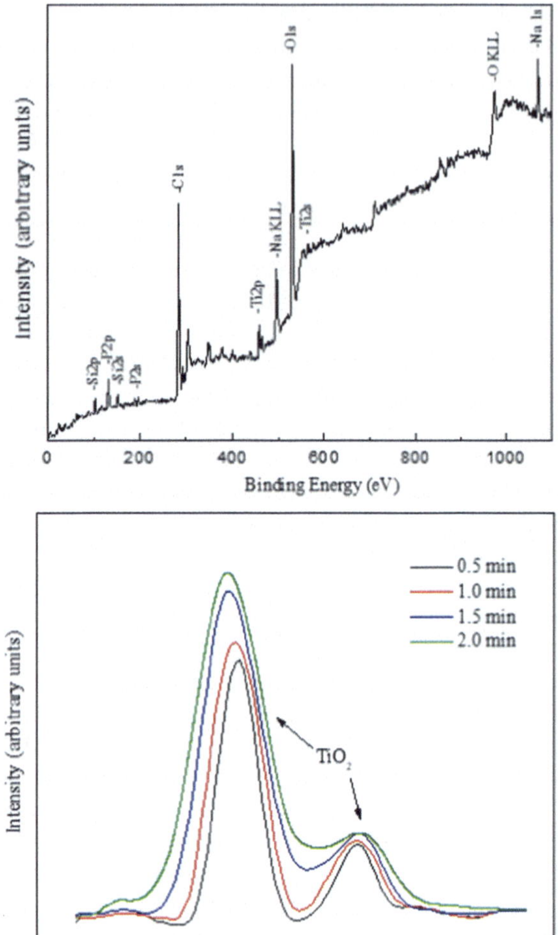

MoO_2 was observed by XPS. The aggressive ions might cause the decomposition of oxides which were helpful for preventing further corrosion, and the severe corrosion would be observed as a result of the synergism between salt and oxygen. On the whole, the results of XPS analysis are in accord with those of XRD and SEM/EDS analysis, confirming the composition of oxide scales on nickel-based alloys in the three different conditions.

6.5 Typical Corrosion Mechanisms of Alloy Steels in Supercritical Water

Lots of studies have been carried out on the mechanism of oxide scales grown on metal materials in SCW. Two typical mechanisms, such as solid-state growth mechanism and mixed model (formation of the inner layer by solid-state growth process and formation of the outer layer by a metal dissolution-oxide precipitation mechanism) depending on the water density below or above 100–200 kg/m^3 [115–118], have been proposed [1, 98, 101, 103, 109, 131, 168–170]. Solid-state growth mechanism has been used successfully in corrosion problems of various materials such as alloys and ceramics in high-temperature gaseous environments such as single gas (O_2, N_2, CO_2), air, high-temperature steam, and series of mixture gases. According to the diffusion rate of metal ions involved in the substrate, some alloying elements with relative slow rate are retained and enriched in the inner layer, while the outer layer grows by outward diffusion of metal ions, especially irons, along oxide grain boundaries and then reacts with gas species at scales/environment interface [131]. However, the mixed model, obtaining satisfying application in condensated aqueous systems, emphasizes that the outer layer is formed by a metal dissolution-oxide precipitation mechanism, and the precipitated metal ions are released from the corroding metal itself or originate elsewhere in the system. The dissolved metal cations combine with anions such as OH^- in aqueous environments to form oxides or hydroxides; they then precipitate on specimen surface to form and/or thicken the outer layer [171].

Yi et al. indicate that the ionic dissolution in the 500 °C decreased in supercritical water with a density less than 100 kg/m^3 was negligible and oxide formed by oxygen absorption without any metallic dissolution [172]. This implies that the corrosion behavior of the T91 steel in low density SCW is similar to that in gaseous environment where only solid growth takes place without any metallic dissolution. According to Macdonald [173], two corrosion mechanisms can be postulated depending on the water density. On the one hand, in lower density, the dominant mechanism seems to be a chemical oxidation mechanism based on the molecular interaction between the metal and the oxygen in diffusion process of metal ions within scales; on the other hand, when density increases, the mechanism changes to an electrochemical mechanism (definitely involved the metal dissolution) with charge transfer [115].

6.5.1 Formation and Evolution Mechanisms of Oxide Scales in Near-Pure SCW

Heat-resistant steels, such as ferritic–martensitic (F-M) steels and austenitic steels, have been extensively employed as structural materials in the high-temperature regions in fossil-fired thermal and nuclear power plants [132, 137], and the corrosion behavior of which in high-temperature supercritical water (SCW) has been widely investigated in recent years [98, 100, 117, 126, 132, 136, 140, 172, 174, 175]. Most of the previous experiments were typically carried out with exposure times above 100 h, focused primarily on cross-sectional structure of the oxide scales on heat-resistant steels in SCW, and so as to define the thickening mechanism of the scales. However, these experiments were incapable of exploring the early oxidation behavior of the steels, because the film generally formed at over times of less than 100 h. Studies over <100 h are very important because they may provide valuable guidance to further understand the formation process of scales and to improve the oxidation resistance of the steels. Here, we will discuss the early oxidation process of ferritic–martensitic (F-M) steels and austenitic steels in SCW in detail. In addition, some personal ideas about cracking and spallation mechanisms of the scales formed on austenitic steel under cyclic oxidation conditions are also included in this part.

(1) **Evolution and formation of scales on heat-resistant alloys**

F-M steels: For F-M steel, T91, exposed to SCW at 540 °C within a few hours, the relatively complete inner and outer layers almost form simultaneously [169], as shown in Fig. 6.67. The outer layer primarily comprises large columnar Fe_3O_4 grains with the longitudinal axis parallel to the oxide growth direction (i.e., perpendicular to the surface), while the fine equiaxed Fe_3O_4 and $FeCr_2O_4$ particles

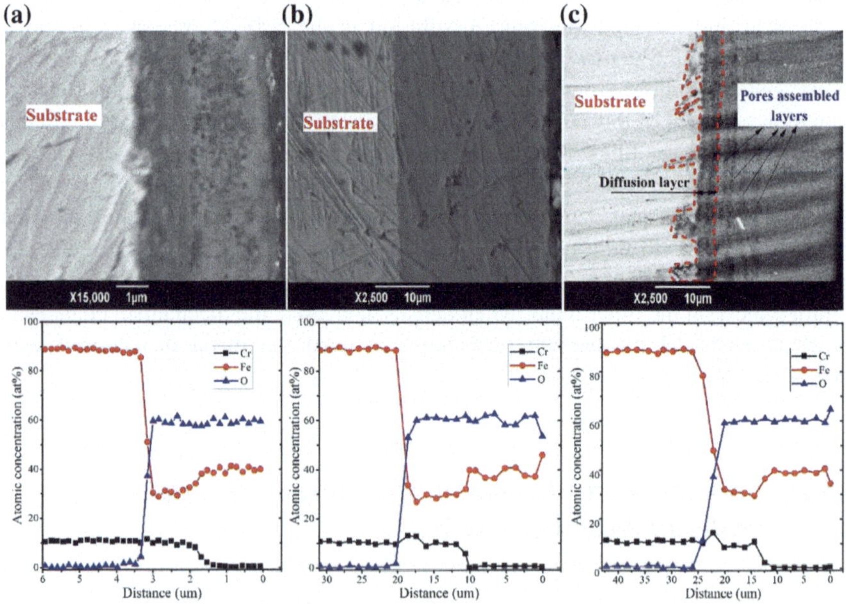

Fig. 6.67 Cross-sectional images (left: substrate, right: scales) and their corresponding elemental depth profiles of the T91 samples exposed to SCW at 540 °C for **a** 1 h, **b** 40 h, and **c** 120 h [169]

constitute the inner layer [135, 136]. Previous studies suggest that the outward migration of iron forms the outer layer and the inner layer is thickened by the inward transport of oxygen carriers [98, 100].

Once oxygen derived from SCW molecules reaches the alloy surfaces, they will rapidly react with some metal atoms located at surface activation sites, generating multiple fine oxides such as Cr_2O_3, Fe_2O_3, and Fe_3O_4 [98]. Afterward, oxygen ions permeate inward along the grains boundaries and result in the formation of chromium-rich oxides by oxidizing chromium carbides along the grain boundaries [100, 175] and/or reacting with chromium derived from adjacent grains due to the high oxygen affinity of Cr and its relatively rapid diffusion within the substrate [99]. Thus, the surface metal grains are generally outlined by the chromium-rich precipitates on either sides of the grain [175].

With the oxygen potential increasing at the outlined metal grain boundaries, ultra-fine chromium-rich spinel particles will form within the outlined metal grains because of negligible solubility of oxygen in the iron and high oxygen affinity of chromium [98]. The residual metal grain is divided by the newly generated chromium-rich oxides, into large amount of iron-rich pieces with nearly equal sizes. The oxygen potential within the residual metal grain will continue to increase during the chromium-rich particle generation continuously until it reaches the critical oxygen partial pressure sufficient to form Fe_3O_4. Therefore, the near isometric Fe-rich fine metal pieces become oxidized to partially transform equiaxed Fe_3O_4 grains, which jointly comprise the inner fine oxide layer with the preformed Cr-rich oxides. Coinciding with the inward permeation of oxygen, some Fe^{2+} from oxidized Fe-rich metal phase diffuse outward along the direction perpendicular to the original alloy surface and combine with oxygen at the scale/environment interface to form and/or augment columnar Fe_3O_4 grains [98, 100]. Finally, these columnar oxide grains spread laterally, forming the continuous, polycrystalline outer Fe-rich layers. Meanwhile, the early oxidation process also involves the outward diffusion of small amount of molybdenum and chromium, which results in their incorporation into the outer layer, as well as the formation of Fe_2O_3. However, the transformation of Fe_2O_3 into other phases, such as Fe_3O_4 and $Fe_xCr_{3-x}O4$, occurs gradually with increasing exposure time due to the presence of boundary layer near the scale/environment interface and accumulation of generated hydrogen on the inside of the boundary layer [103, 130].

The entire formation process of the two-layer scale may proceed rapidly and may be completed within several hours for F-M steels exposed to SCW at 500 °C and above, as shown in Fig. 6.67a. Tan et al. [126] also reported that when the exposure time was only 1 h, the thicknesses of the inner and outer layers for the HCM12A sample exposed to SCW at 500 °C were already 0.8 and 1.0 μm, respectively. At 540 °C, in spite of the formation of the relatively complete duplex-layer scale within several hours, the oxidation process of T91 steels did not reach the diffusion-controlled stage until the exposure time extended to hundreds of hours. The additional diffusion layer generally occurred later than the formation of the inner and outer layer [126]. It may be presumed that the interfacial reaction may be the rate-controlled process for the oxidation of F-M alloys in SCW during the early stages of less than tens of hours. Strong evidence for this presumption is that the diffusion layer at the oxide-metal interface was not observed during the initial oxidation, as shown in Fig. 6.67a, b. Extensive investigation has indicated that the rate-limiting step in the long-term oxidation of F-M alloys exposed to SCW is the outward diffusion of iron rather than the inward transport of oxygen carriers [98, 126, 136, 175]. Both transport rate of oxygen carrier inward and iron outward decrease with an increase in the thickness of the oxide scale, but the hindering effects of the oxide scales on irons may be more effective than those on the oxygen carriers, resulting in the outward diffusion of iron becoming the rate-limiting process. Thus, extra oxygen carriers can diffuse ahead of the oxide and further produce oxides along the grain boundaries [98], forming the diffusion layer with a mixture of oxides and residual metal grains, as Fig. 6.67c shows. In addition, the higher temperature may promote the appearance of the iron diffusion outward as rate-limiting step in advance [98, 126].

The complete outer layer consists primarily of magnetite and an inner Cr-containing layer form almost simultaneously on F-M steel T91 after only a 1-h exposure in SCW at approximately 540 °C, but the required exposure time to form the protective inner Cr-rich layer and diffusion layer is at least 100 h. Relative to the complete outer layer, the lack of formation priority of protective inner Cr-rich layer may result in surface treatments exhibiting little or no effect on improving the oxidation resistance of 9 wt% F-M steels in SCW. However, surface treatments, such as shot-peening, laser melting, and mechanical working, can generally decrease the corrosion rate of relative high-Cr steels with a Cr content of not less than 12 wt% [123, 128, 176–178].

For austenitic steel: However, for high-Cr steels such as austenitic stainless steels in SCW, taking austenitic stainless steel TP347H in SCW, for example, relative to the complete outer Fe-rich oxide layer, the formation priority of a continuous inner Cr-rich layer was observed [128]. This satisfying process is obviously different from that occurs on F-M steel T91. During the rapid oxidation stage, while the simultaneous nucleation of Cr_2O_3, Fe_2O_3, and Fe_3O_4 occurs at the alloy/SCW interface, the predominant formation and growth of Cr_2O_3 are necessarily due to the higher oxygen affinity of Cr and its relatively rapid diffusion within the substrate [99, 112, 143]. The nucleation number and size increase as the exposure time increases. Simultaneously, the formation and growth of oxides will lower the oxygen potential at the scale/SCW interface because of the consumption and accumulation of the oxygen and hydrogen generated by the oxidation process, respectively. The nucleation and growth of the Fe_2O_3 are suppressed, and subsequently, it may be forced to convert into Fe_3O_4, or fuses with Cr_2O_3 by a solid solution reaction to form $(Cr, Fe)_2O_3$. A production path of Cr-rich spinels resulting from a favorable reaction between Fe and Cr_2O_3 may also occur [112]. A large number of the oxide grains collide laterally with the increasing exposure time, which then excites the composite reactions of $(Cr, Fe)_2O_3$ and Fe_3O_4 as the two major oxides present on the alloy surface, to generate $(Cr, Fe)_3O_4$ spinels [103]. This process may represent the beginning of the oxidation transitory stage. The transitory stage terminates once a complete inner Cr-rich layer predominantly

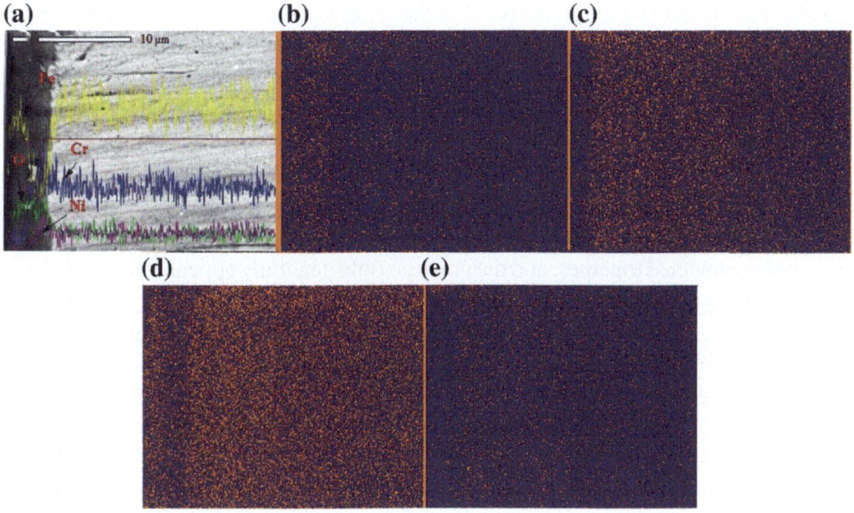

Fig. 6.68 SEM image and corresponding EDS maps of the cross-sectional TP347H exposed to the SCW at 540 °C and 25 MPa for 120 h. **b**: O, **c**: Cr, **d**: Fe, **e**: Ni [101]

consists of the (Cr, Fe)$_3$O$_4$ spinel forms. During the transitory stage, the number and size of the isolated Fe$_3$O$_4$ particles that constitute the loose outer layer also increase, but they fail to form a continuous outer layer due to insufficient time. With the growth of the inner Cr-rich layer, which acts as a barrier against the inward transport of oxygen carriers and the outward diffusion of iron ions rather than chromium ions because the latter is too tightly bound to diffuse successfully in the spinel structure [98, 105], it subsequently hampers, but does not prevent, the further development of the scale [124]. Some iron diffuses outward through the inner layer to the scale/gas interface, promoting the accumulation and growth of Fe$_3$O$_4$ particles at the scale surface. As these Fe$_3$O$_4$ particles collide laterally, merge with each other and eventually spread to cover the entire surface, a common porous, but continuous outer layer is finally formed after a longer time (Fig. 6.68) [96, 109].

(2) **Cracking and spallation**

In order to explore the cracking and spalling process of oxide scales formed on a typical austenitic steel Super304H, The cyclic oxidation test was conducted at 540 °C/25 MPa and 500 °C/23 MPa, respectively, for each cycle of 24 h, adding up 5 cycles [131]. After cyclic oxidation between 540 °C/25 MPa and 500 °C/23 MPa, cross-sectional morphology of scales formed on Super304H surface and its partial spallation is depicted in Fig. 6.70. Figure 6.68a indicates that after a 240-h cyclic oxidation, parts of surface scales peeled off. Electron microscopic observations and EDS analyses showed that the spalling area predominantly consisted of chromium-rich oxides, which were covered by a series of longitudinal cracks developing along the grain boundaries and few magnetite particles, suggesting that the exfoliated parts may be likely for parts of the outer layer of scales. As can be seen in Fig. 6.68b, the cross-sectional morphology of the scales also exhibited the presence of cracks within the outer layer and between inner/outer layer interface. For latter, the occurrence of cracks was largely attributed to presence of a large number of cavities at the interface between inner and outer layer [115]. With the corrosion process, these cavities concentrated along the magnetite/Cr-rich spinel interface and converged increasingly with each other [96], so interfacial bonding strength between the inner and outer layer became weak. With the extension of exposure times, these cavities eventually connected together, and the cracks would generally appear at the inner layer and outer layer interface as shown in Figure 6.69.

Due to the volume difference of oxides (V_{OX}) and corresponding consumed metals (V_M) in this study, or more accurately $V_{OX}/V_M > 1$, the emergence of mechanical stress within the scales is inevitable during oxidation except for the scale growth only by outward metal diffusion [96]. The compressive stress presents within the scales, being balanced by a tensile one in the alloys [179]. The compressive stress hindered inner oxides growth. These results imply the mechanical stress point that the scales growth may involve in the inward oxygen transport. If the scale grows only by outward metal diffusion, new oxides will form at the unconstrained oxide–gas interface, and no stress results. Assuming that the scales behave elastically, the corresponding growth stress would be [96].

$$\sigma_{\text{ox}} = \frac{-E_{\text{ox}}}{1 - \nu_{\text{p}}} \varepsilon_{\text{ox}} \tag{6.4}$$

where ε_{ox} represented volume changes, while E_{ox} and ν_{p} were the elastics modulus and Poisson's ratio, respectively. As the scales thicken, the volume changes aggrandized, and the growth stress will increase. When the scale stress is larger than interfacial bonding intensity between the scales and substrate, the scales will crack or even spall from the substrate. In addition to the scales' plastic deformation,

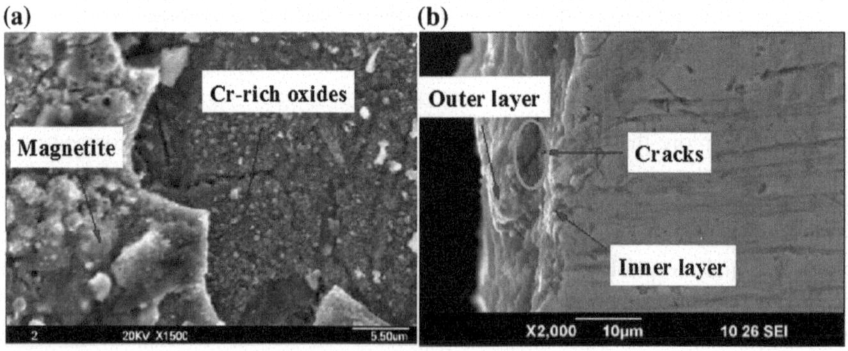

Fig. 6.69 **a** Surface morphology and **b** cross-sectional characteristics of scales formed on Super304H specimens after cyclic oxidation for total 240 h [131]

plastic deformation of the alloy substrate is also an available stress relieving mechanism. The plastic deformation of alloys increases with a decrease in grain size [180]. Therefore, the Super304 steels with smaller grain size generally own higher plastic formation ability to adjust the mentioned volume difference between oxides and its consumed metals. Moreover, chromium-rich oxides layer usually forms quickly on Super304H surface due to its fine grains during early oxidation, to a large extent, which results in a relative smaller thickening rate of the scales. Therefore, the growth stress may increase slowly. Thus, under constant temperature oxidation without external loading, the long-term stability of the scales can often maintain as tens of thousands of hours, or even longer [96, 181].

Additionally, due to significantly different thermal expansion coefficients of metals and oxides, noted as α_M and α_{ox}, respectively, a concentrated stress occurs within the scales as the environment temperatures changes sharply. For thin scales on thick substrates, the stress produced in an intact, adherent scale by a temperature change, ΔT, can be adequately approximated by [135].

$$\sigma_{ox} = \frac{-E_{ox}\Delta T(\alpha_M - \alpha_{ox})}{1 - v_p} \tag{6.5}$$

A rapid cooling of the Super304H specimens from 540 to 500 °C would further put the scale in compression. Because the resulting stress was high enough and plastic deformation rate of the scales and substrate was insufficient to relieve the

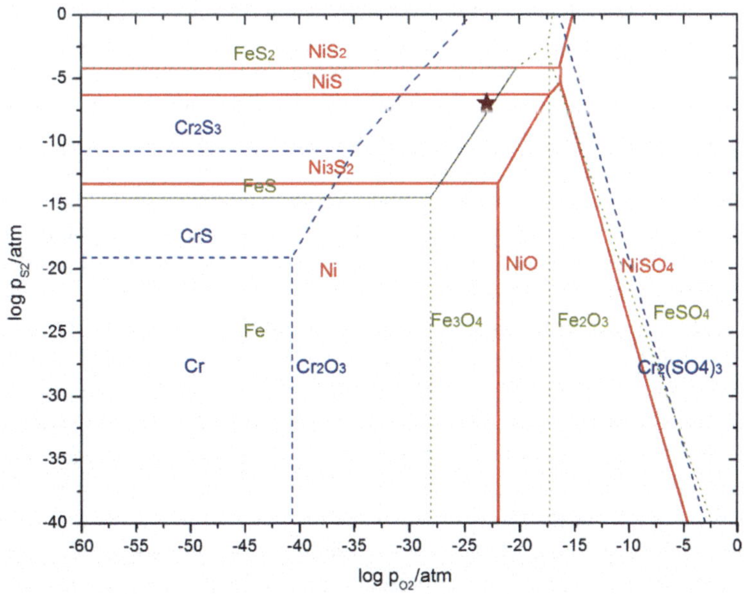

Fig. 6.70 Isothermal cross section of the simplified Ni–Cr–Fe–O–S phase at 520 °C, formed by superimposition for the Ni–O–S, Cr–O–S, and Fe–O–S phase diagrams [191]

stresses, the scales suffered mechanical failure such as multiple micro-cracks within the scales, and partial spallation of the outer layer along the inner layer and outer layer interface, as shown in Fig. 6.70. Because of wave-type scales/substrate interfaces, parts of the inner layer were firmly embedded in the substrate [131], i.e., pinning effect, which may enhance the adhesion of scales on the substrate [182]. However, the presence of plenty of cavities and micro-cracks at the inner layer and outer layer interface, to a large extent, can weaken the interfacial bonding forces. Thus, there is no doubt that the separation and ejection of fragments from the scale occur along the inner and outer interface. The phenomenon is of great significance to keep lower oxidation rate of the alloys in its subsequent service. Despite of partial spallation of the outer layer, the protective inner layer still exists continuing to restrict outward diffusion of the metals cations, which also effectively avoids the direct contact of the corrosive gases with the alloys substrate. However, a series of issues such as piping blockage, and abrasion triggered by exfoliated scales still need extensive attention, when a plant is heated up or closed down, by thermal or mechanical shock [131].

6.5.2 Effect of Inorganic Salts in Reducing SCW

(1) **Reducing sulfides**

A large amount of organic wastewater containing sulfides has been produced from industrial production, in addition to large numbers of organic pollutants, which also contains some reducing inorganic sulfur-containing species, predominantly of S^{2-}, HS^{-1}, and/or H_2S, causing diverse impacts on human health and natural ecological environment. If these wastewaters with reducing sulfides are PubMed and treated into the high-temperature supercritical aqueous systems, for the corrosion resistance of serving alloys, what will happen? In this section, effects of sulfides on the corrosion process, scale morphology, and surface microstructure of one Ni-based alloy Inconel 600 in reducing supercritical water (no dissolved oxygen) are discussed. Furthermore, an explanation is given to the corrosion mechanism of Inconel 600 in SCW containing sulfides.

The scales formed on Inconel 600 exposed to sulfide-containing SCW presented significant difference with those formed on general nickel-based alloys corroded in SCW with/without oxygen [109, 164, 183–186]. The latter scales generally showed a duplex oxide structure with Ni and Fe rich in the outer layer, and Cr rich in the inner layer. While, in present condition, the scale grown on Inconel 600 was obviously divided into the inner and outer layers: The outer layer of the formed scale was essentially of the Ni_3S_2 phase, while the inner layer was predominantly of chromium-rich spinel-type phases with a bright Ni_3S_2 network inside. The discrepancy of the scale gained in the present work to the previous studies could be attributed to the existence of sulfur-bearing species in SCW environment, which derived from sulfides present in initial aqueous solutions. In present condition of

520 °C and 25 MPa, the evaluated water density was less than 100 kg/m^3, so gaseous oxidation should be predominant corrosion process.

$$Na_2S + 2H_2O = H_2S(g) + NaOH \quad \Delta G^0 = -9.8\,kJ \tag{6.6}$$

So it can be concluded that the practical exposure environment could be regarded as H_2S-SCW atmosphere with higher steam partial pressure. Among these gases, H_2S could further dissociate into sulfur and hydrogen. Sulfur reacted with metal to form sulfides. The defect structures and transport properties theory deduced that growth mechanisms of most metal sulfides and oxides are analogous, because they exhibit similar predominant defects. However, the sulfides of common metals, such as iron, nickel, and chromium, generally have higher defects concentration due to their much larger deviations from stoichiometry than individual corresponding oxides [144, 187, 188], so that in present SCW environment containing H_2S, the Inconel 600 and Incoloy 825 showed poor corrosion resistance.

The superimposed isothermal phase diagram of the ternary Ni–O–S, Cr–O–S, and Fe–O–S systems with an indication of partial pressure of oxygen (p_{O_2}) and sulfur (p_{S_2}), calculated at 520 °C and 0.1 MPa using appropriate thermodynamic data [121], is shown in Fig. 6.70. The current corrosion tests were carried out at a pressure of 25 MPa, but the built phase diagram was at 0.1 MPa because of the absence of thermodynamic data at higher pressure. While the high pressure was expected to make difference on the fugacity of oxygen (f_{O_2}) and sulfur (f_{S_2}), the fugacity coefficients of oxygen (φ_{O_2}) were less than 1.2 at approximative present condition (500 °C and 25 MPa) [189]. Based on the above data, Tan et al. [132] reported that the difference between the phase diagrams before and after correction by the fugacity of oxygen ($f_{O_2} = \varphi_{O_2} p_{O_2}$) at 500 °C and 25 MPa was found to be negligible. Along with the consideration that similarity of the fugacity coefficients of oxygen (φ_{O_2}) and sulfur (φ_{S_2}), therefore, the isothermal phase diagram calculated at 0.1 MPa was a reasonable estimation of the expected higher pressure behavior. The diagram was schematic in that all the oxide phase and the sulfide phase for nickel, chromium, and iron had not been included. Even though the alloys (the activities of both Ni, Cr, and Fe etc. are less than unity) were exposed to the mixed gas during corrosion, the above phase diagram can also be used to a first approximation to predict the phase stability. Based on the assumption that the C/H atomic ration of organic compounds present in current feed equated approximately one-fourth and hydrogen yield (the ratio of hydrogen moles present in hydrogen divided by that inherent in original organic compounds) was 1.2 [190], hydrogen partial pressure was estimated. Then, oxygen and sulfur partial pressure of the present gas composition were estimated to be $\sim 1 \times 10^{-23}$ and $\sim 2 \times 10^{-7}$ atm based on the decomposition equilibrium of H_2O [117] and H_2S at 520 °C, respectively, which marked as a pentagram depicted in Fig. 6.70.

Also for the Ni–Cr–Fe–O–S system, the mixed gas composition was located in the Ni_3S_2 and Cr_2O_3 stability region. In addition, the partial pressure of oxygen was always lower than the dissociation pressure of both NiO and Fe_2O_3. Thus, there is no kinetics mechanism which may allow the formation of NiO and Fe_2O_3 during the corrosion of Inconel 600 in the present mixed species. Since sulfur partial pressure of the atmosphere was also much higher than CrS dissociation pressure, once the Inconel 600 came in contact with H_2S and H_2O at 520 °C, the simultaneous nucleation of chromium sulfide, Cr_2O_3 and Ni_3S_2 [141] happened at the alloy/gas interface, respectively, as the following reactions:

$$Cr + H_2S = CrS + H_2 \tag{6.7}$$

$$2Cr + 3H_2O = Cr_2O_3 + 3H_2 \tag{6.8}$$

$$3Ni + 2H_2O = Ni_3S_2 + 2H_2 \tag{6.9}$$

However, due to the less thermodynamic instability of CrS phase than that of Cr_2O_3 phase in present gas composition, the former would react with oxygen to further convert into Cr_2O_3 phase as the chromium activity decreased. This well explained why only Cr_2O_3 and Ni_3S_2 were observed on the Inconel 600 surface after initial exposure about 3 h [191]. The related investigation on sulfidation–oxidation behaviors of alloys indicated that the presence of sulfur in gas composition generally increased the critical oxygen partial pressure of the Cr_2O_3 phase [192], which resulted in a decrease of the generation driving force and stability of the Cr_2O_3 phase. Furthermore, due to the chromium activity falling rapidly with corrosion process proceeding, the continued growth of formed Cr_2O_3 particles was also hampered. As for Inconel 600, the facts that its chromium content was not enough to form continuous Cr_2O_3 protective film, coupled with the slow growth of Cr_2O_3 particles, predestinated that the chromium-rich oxide film was hard to take precedence over Ni_3S_2 layer formation. In addition, during the corrosion of the Ni-rich alloy at high temperature, combining with the facts that Ni_3S_2 grows very rapidly in view of its highly defective nature, it was reasonable to expect the preferential generation of Ni_3S_2 film as the outer subscale. The formation of the outer sulfide subscale decreased the sulfur potential and thereby increased the oxygen potential underneath. It was worth mentioned that as the Ni diffused outward continuously from the alloy substrate to formed Ni_3S_2 at the scale/gas interface [141], the chromium activity increased between the Ni_3S_2 sublayer and substrate due to the chromium enrichment. The above-mentioned points together resulted in the formation of chromium-rich oxides (mainly spinel-type phases) and sulfides in the inner scale. It was well known that the oxygen partial pressure needed for a continuous Cr_2O_3 layer to develop on an alloy surface was usually at least 10^2–10^3 times higher than the Cr_2O_3–CrS/Cr_2S_3 equilibrium value [193]. The formation of the inner scale layer consisted of mostly chromium-rich oxides was supported by the above standpoint, because the present gas composition was quite far from the Cr_3S_2–Cr_2O_3 equilibrium line with the oxygen partial pressure about 10^6 times higher than the equilibrium value (Fig. 6.70). The presence of the inner layer rich in spinel-type phases in contact with the alloy was able to reduce the growth rate of the external Ni_3S_2 due to its barrier action to the nickel outward diffusion. But the reduced degree was limited, because the inner layer also contained a network of Ni_3S_2 particles, which provided a rapid transport path for inward diffusion of oxygen and outward diffusion of metal ions [194].

For the compact Ni_3S_2 sublayer in Inconel 600, it grew essentially by means of an outward diffusion of nickel, because the diffusion coefficient of nickel in Ni_3S_2 was several orders of magnitude bigger than that of sulfur [187, 195]. The formation of oxides and sulfides in the inner layer suggested that the inward diffusion of oxygen and sulfur had occurred. The inward diffusion of oxygen through the outer Ni_3S_2 layer could proceed by two means: oxygen ions diffusion and oxygen-bearing molecule transport. Rahmel et al. pointed out that the diffusion coefficient of oxygen ion in sulfide was rather small [195]. So the oxygen-bearing molecules (i.e., H_2O) might be the mainly oxygen provider for the inner layer oxides, which were able to penetrate through the external Ni_3S_2 layer, probably through discontinuities of the outer layer such as pores and cracks, specifically observed on the scale surface [191]. The water molecules reached the inner surface of the external Ni_3S_2 generally dissociated into oxygen and hydrogen. The oxygen in the form of O^{2-} diffused inward continuously and reacted with metal to produce the metal oxides, which composited to chromium-rich spinel-type phase [109, 196–198]. The permeated SCW molecules could increase the oxygen partial pressure near substrate/scale surface, and interacted with the scale interior to dissolve hydrogen in the oxide generating defects or even produce harmfully volatile hydrated species [144, 199–201]. The inward transport resistance of H_2S was significantly larger than that of H_2O, due to the former with a larger molecular diameter, which well explained the cause that the inner layer was predominantly of oxides, whereas the external layer consisted of mostly sulfides.

(2) **Ammonium chloride**

Corrosion behavior of Ni-based alloys has been investigated in salt solutions such as Na_2SO_4 [202] and NaCl [203] at subcritical and supercritical temperature, which showed that the corrosive ions can result in pitting corrosion and intergranular corrosion. Moreover, Asselin et al. [9, 148, 150, 151] used ammoniacal sulfate solution as the corrosion medium to test the corrosion behavior of Ni-based alloys and stainless steels and concluded that sulfate and ammonia, which was not corrosive separately in supercritical water, showed severe corrosion attack to test samples. Cr (III) in the alloy oxide was dissolved, and Ni was also preferentially dissolved and formed ammines with ammonia, which both led to the unstable of oxides. However, chloride ion is generally more aggressive than sulfate ion. So, what will happen for corrosion behavior of materials of interest if coexistent of chloride and ammonium ions? The following section will aim to figure out the synergism effect of ammonium and chloride on the oxide structures and corrosion rates of typical Ni-based alloy and stainless steel in supercritical water.

Ni-based alloys and stainless steels were tested in supercritical water containing 6000 ppm NH_4Cl at 400 °C to investigate the corrosion mechanism and to evaluate the corrosion resistance. The migration of alloying elements was used to understand the corrosion behavior. Depletion of metal ions and formation of corrosion products were principal behaviors of alloys in SCWG systems [149].

The coexistence of ammonium and chloride ion in supercritical water built a severe corrosion environment for candidate structural materials. Ammonium chloride may produce acids in supercritical water environment [145], as shown in Eq. (6.10).

$$NH_4Cl + O_2 \longrightarrow HNO_3 + HCl + H_2O \tag{6.10}$$

Nitric acid and hydrochloric acid may form with the presence of chloride ion and oxygen in supercritical water. Selective dissolution of Fe and Ni, and the transpassive dissolution of Cr were the prominent steps for Ni-based alloys and stainless steels in acidic high-temperature water [155]. However, most of the oxygen in the reactor space had been eliminated by nitro gas bubble; dissolved oxygen (DO) content in solutions was little. Hence, the concentration of acid would be limited. In a word, the serious corrosion of samples in this study may not be caused by the formation of acid. The obvious weight loss of tested alloys may be due to the formation of ammine $Ni(NH_3)_n^{2+}$ by the following equation [147, 148]:

$$Ni + \frac{1}{2}O_2 + nNH_3 + H_2O \longrightarrow Ni(NH_3)_n^{2+} + 2OH^- \tag{6.11}$$

Ni and Cr are the main elements in Ni-based alloy. In this study, Ni was depleted from the substrate of both Ni-based alloys and stainless steels after exposing in supercritical water containing NH_4Cl. Based on the research by Asselin et al. [9, 148, 150, 151], Ni was not stable in high-temperature water with the presence of ammonium ion, forming nickel ammines which were soluble in hot water [204]. In high-density supercritical water, Ni was also unstable in some cases. $NiCl_2$ was found on the oxide films of alloy 625 which were exposed in oxidizing SCW with HCl and NaCl at 500 °C and 46.5 MPa for 160 h, and high concentration of Ni was found in the effluent of the experiment system [205]. Ni oxide was also unstable in pure supercritical water. Zhang et al. [37] indicated that Ni^{2+} were able to diffuse through the NiO and Cr_2O_3 oxide which formed on the surface of Hastelloy C-276 and kept forming NiO with oxygen diffusing inward to the alloy. Many researches had confirmed that Cr had excellent corrosion resistance by forming Cr_2O_3 films, which were relatively compact and were able to resist the inward diffusion of oxygen. Kim et al. [206] found that Cr oxide formed on Ni-based alloy was stable at pH 2 in supercritical water while Ni and Fe were selectively dissolved. However, Kritzer et al. [207, 208] suggested that transpassive dissolution would occur to Cr when exposing in SCW with high dissolved oxygen content, forming soluble chromate. Though the SCWG process was operated in reducing environment where oxygen was limited, structural materials may also form oxides by reacting with the water at the SCW/oxide surface [209].

Figure 6.71 is the schematic of the corrosion mechanism of samples in SCW containing NH_4Cl at 400 °C. The mechanism of stainless steels and Ni-based alloys

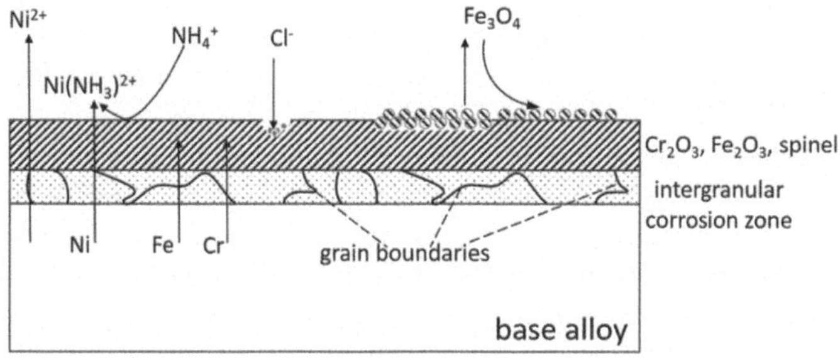

Fig. 6.71 Schematic of the corrosion mechanism of samples in SCW containing NH_4Cl at 400 °C [149]

is similar: Selective dissolution and formation of nickel ammines occurred to Ni, pitting phenomena caused by the concentrating of Cl^- in defects, dissolution and re-deposition of Fe_3O_4. Remarkably, intergranular corrosion phenomena were more distinct in 316 SS for the reason that the material mainly formed Fe oxides, which were unstable and had weak resistance against ion diffusion.

6.5.3 *Corrosion Mechanisms in Oxidizing SCW*

The supercritical water environment in SCWO process always contains aggressive components such as oxygen and salts, as the intrinsic species within the organic wastes to be treated or reaction products from SCWO [210–212]. In the above text, the corresponding mechanisms of alloys in supercritical water environments with a single corrosion medium like salt or oxidant are provided. In the next moment, under the actual corrosion condition of the coexistence of salt (Cl^{-1} and PO_4^{3-}) and oxidant (dissolved oxygen), the complicated corrosion problems will be discussed. It is well known that the chemical stability of corrosion products determines the corrosion resistance of construction materials for SCWO. In previous studies, many researchers focused on the complex formation mechanism of corrosion products due to the special properties of the supercritical water and various aggressive species. In high-density aqueous systems (density above 100–200 kg/m^3), the oxide film growth mechanism, metal dissolution mechanism, and metal salt precipitation mechanism are typical mechanisms investigated and accepted by researchers [196]. In the following parts, the corrosion mechanism of austenitic stainless steels, nickel-based alloys, and titanium alloys in two supercritical water environments, (1) single dissolved oxygen and (2) coexistence of chloride and phosphate ions and dissolved oxygen, are discussed. The migration and transformation of Cr and Ni were employed to interpret the corrosion behavior. The multilayered structure, depletion of metal ion, and formation of corrosion product are prominent corrosion feature under SCWO condition.

(1) **Effect of dissolved oxygen**

Generally, the corrosion resistance of a metal alloy is dependent on the chemical stability, arrangement, compactness, strength, plastic, and thermal expansion coefficient of the corrosion products. The corrosion mechanism under SCWO is more complex than that of water at high temperature due to the special properties of the supercritical water and presence of dissolved oxygen. In the same corrosion mediums, the corrosion resistance of the alloy depends on the contents and types of elements in alloy [68]. The major constructional alloys in SCWO systems are Ni-based alloys, which are mainly composed of chromium and nickel. The migrations and transformations of Cr and Ni are employed to explain the corrosion behavior [68].

Firstly, let us talk about the corrosion behaviors of Ni-based alloys in oxidizing SCW with dissolved oxygen of 5000 ppm, but without various additional inorganic salts. The oxygen concentration in the experiment, 5000 ppm, is relatively less than that in the reactors of SCWO systems in which the oxygen concentration can reach 40,000 ppm [208] or even 70,000 ppm [213]. However, compared with the concentration of oxygen in the working fluids of power generation systems, which contain less than 8.3 ppm of DO [184, 214–216], the oxygen concentration in this experiment is sufficiently high to result in a high corrosion potential and the ensuing high corrosion rates of alloys.

The growth rates of the oxide films were affected by the DO content in SCW. Some studies have reported the oxidation of Ni-based alloys in SCW at 500 °C with different content of DO. However, the duration of their studies was different. For comparing the results, the "oxidation rate" was used to summarize the results in previous studies, as shown in Table 6.7.

The oxidation rate was evaluated as follows:

$$\text{Oxidation rate} = (K * \xi)/t \ \mu\text{m/year} \tag{6.12}$$

where K is a constant equal to 8760, ξ is the thickness of the oxide film in μm, and t is time of exposure in h. The values of ξ and t were obtained from other literatures.

Table 6.7 summarizes the corrosion rates of Ni-based alloy 625 in neutral SCW as a function of the DO content. The corrosion rates of the alloys clearly increased with the increasing oxygen concentration, possibly related to the control of the

Table 6.7 Summary of oxidation rates of Ni-based alloy 625 in different oxygen concentrations [217]

Oxygen concentration (ppm)	Temperature (°C)	Oxidation rate (μm/year)	References
46,080	500	500	[23]
9400	500	49.7	[22]
5000	500	28.0	This work
8.3	500	17.2	[17]
0.15	700	14.0	[18]

oxidation rate via the diffusion of ions through the oxide scale. The diffusion rate was affected by the steady-state oxygen partial pressures at the interface between the oxide and water, which can be described by Fick's first law of diffusion:

$$j = -D\frac{\mathrm{d}C}{\mathrm{d}x} \tag{6.13}$$

where j is the ion flux, D is the diffusion coefficient, C is the ion concentration, and x is the scale thickness.

Furthermore, the scale growth rate can be described by the following equation.

$$\frac{\mathrm{d}x}{\mathrm{d}t} = j * \mathrm{V}_{\mathrm{ox}} \tag{6.14}$$

where V_{ox} is the molar volume of the formed oxide.

Oxygen can be dissolved almost completely in SCW because of its non-polar nature. In addition, the DO content affects the oxide structures via changes in the corrosion potential of the environment. Tan et al. [184] reported that NiO was mainly formed on the alloy 625 surface after exposure to SCW containing 25 ppb oxygen, and a similar conclusion was reported by Kritzer et al. [208] for the corrosion behavior of alloy 625 in neutral SCW containing 46,080 ppm oxygen. However, Sun et al. [160] conducted an experiment in SCW containing 9410 ppm of DO and reported a duplex oxide structure for Ni-based alloy 625 comprising an inner layer of fine-grain Cr_2O_3 and an outer layer of large-grain NiO and $Ni(OH)_2$. Moreover, Chang et al. [161] have examined the corrosion of Ni-based alloy 625 in SCW containing 150 ppb and 8.3 ppm of DO at 700 °C. They reported that Ni (Cr, $Fe)_2O_4$ was formed on the outer sample layer, and the inner layer mainly consisted of Cr_2O_3 for SCW with a low oxygen concentration (150 ppb). Nevertheless, the only oxide at high DO content (8.3 ppm) experiment was NiO.

As the concentration of oxygen in SCW clearly affects the corrosion potential, potential pH diagrams for nickel and chromium at elevated temperatures were constructed to estimate the chemical stability of the elements in Ni-based alloy under SCW conditions with different oxygen concentrations [218]. Cook and Olive [219] have calculated the pH value of water to be approximately 9.4 at 500 °C and 24 MPa. In the E-pH diagrams, NiO and $HCrO_4^-$ were mainly formed from nickel and chromium, respectively. The corrosion potentials of the Ni-based alloy increased because of the relatively high oxygen concentration (5000 ppm); consequently, the transformation from a stable Cr (III) oxide to soluble Cr (VI) oxide rapidly occurred. However, because of the low ion product and density of water at 500 °C, the Cr (VI) oxide was maintained in a loose state in alloys, which dissolved during the cooling of the systems [207, 220]. Hence, a Cr_2O_3 layer was not detected in this experiment. Conversely, NiO was stable in strongly oxidizing SCW, forming protective films on samples.

(2) Synergistic effects of chloride, phosphate, and dissolved oxygen in oxidizing SCW

Generally, the chloride ion can react with metal oxide to generate the dissolved metal chloride, inducing the destruction of protective oxide scales, so the severe corrosion will occur when chlorides coexist with oxygen [68], while the addition of phosphate is generally helpful for inhibiting corrosion under high-density (>100–200 kg/m^3) supercritical aqueous systems [221].

Figure 6.72 illustrates the possible corrosion mechanism of 316 SS, Ni-based alloys, and TA10 under coexistence of chloride and phosphate ions and dissolved oxygen. 316 SS is mainly composed of iron and chromium, and the iron oxides depart from the corroding surface into fluid phase. As 316 SS lacked enough molybdenum, the severe pitting corrosion occurred on the surface. The rapid dissolution of metal ions occurred in the corrosion pit. The dissolved metal ions reacted with phosphate, to form insoluble $CrPO_4$ and $FePO_4$ on the surface of 316 SS; however, an insoluble iron chloride species cannot be formed because the chloride ion was hard to form stable chlorate in SCW. This finding confirms that the dominant corrosion mechanism in subcritical water and high-density supercritical water is an electrochemical mechanism which becomes of lower importance in high-temperature supercritical water with a density less than 100 kg/m^3. The spallation of loose scales could reduce the protection of phosphate because the phosphate deposit would spall with scales. The spallation of scales would expose the alloy to the corrosive environment, so the further corrosion would occur at the unprotected areas. Ni was depleted from the substrate of Ni-based alloys significantly, and $CrPO_4$ and $Ni_3(PO_4)_2$ formed on the oxide film. Kim reported that the diffusion rate of nickel determined the growth rate of alloyed oxide layer [205].

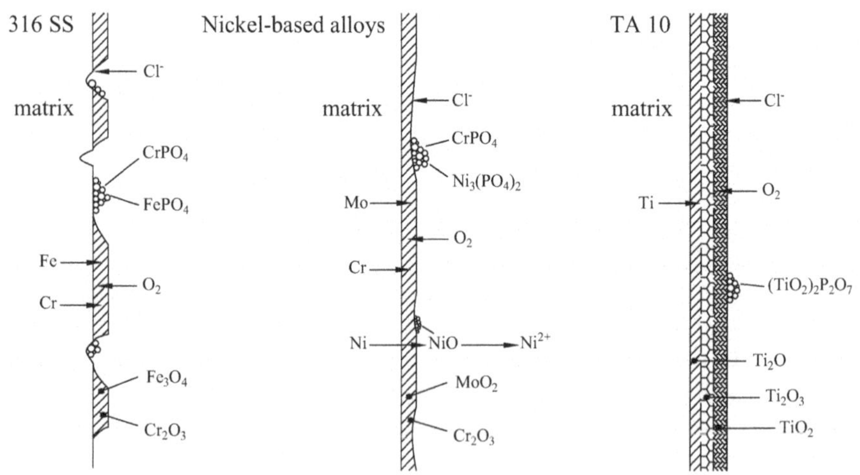

Fig. 6.72 Mechanism of the corrosion of 316SS, nickel-based alloys, and TA10 after exposure to supercritical water with chlorine salt and phosphate coexisting with oxygen [221]

Hara et al. found that nickel showed the largest weight gain among all tested metals and the corrosion rate decreased with increasing chromium content for Ni-based alloys in oxidizing supercritical water (500 °C, 29.4 MPa) [207, 208]. The molybdenum in nickel-based alloys contributed to enhance the resistance to pitting corrosion and the matrix suffered less pitting corrosion. The chromium could form more stable oxide Cr_2O_3 than iron and nickel. Adschiri et al. reported that chromium oxide solubility was lower than iron oxide in high-temperature water and metal oxide solubility increased as temperature increased [210]. Sue et al. found that iron oxide solubility was the highest among the tested metal oxides in supercritical HCl aqueous solutions with oxygen [211]. The titanium alloy TA10 showed the best corrosion resistance of the materials tested, and the titanium oxide film was the most uniform of all surface films formed. A three-layer oxide film was formed on the surface of TA10, and the oxygen content in oxide increased gradually from inner to outer. Due to the low release of titanium into the solution, titanium phosphate was not formed to a significant extent on TA10. However, $(TiO)_2P_2O_7$ was detected on the surface of TA10 in the presence of oxygen and phosphate, implying the titanium would react with phosphate. Boukis et al. observed that titanium oxide phosphate $Ti_2P_3O_{20}$ formed on the surface of titanium grade 2 in subcritical water with H_3PO_4 and O_2 for 200 h [208]. Lu proposed a reaction mechanism for titanium oxide phosphate formation ($-Ti_2O(PO_4)_2 \cdot 2H_2O$) in Ti–$H_3PO_4$ system solution at 250 °C after 24-h immersion [202]. Xiao Hua also found that $Ti_5O_4(PO_4)$ generated on the surface of Ti in the treatment of chlorpyrifos ($C_9H_{11}OSPNCl_3$) by SCWO at 450 °C and 25 MPa [206].

6.6 Suggestions for Material Selection

6.6.1 Suggestions for Construct Materials

As shown in Fig. 6.73, the material selection of a conventional SCWO process was divided into seven parts according to the different reaction conditions and mediums. In this flow diagram, oxygen and wastewater both entered reactor after they were preheated by different preheating pipes, which could avoid synergistic effect of oxidant and aggressive feeds effectively. The material selection of a whole SCWO process includes seven parts, and the guideline of material selection is provided in detail.

The high-carbon alloys are not allowed to build transport pipe for oxygen, as a certain rate of oxygen could react with the carbon to cause explosion. The copper tube, brass tube, and Ni–Cu alloy can be used for transporting the high-speed oxygen, and the stainless steel can be used for transporting the low-speed oxygen among −183 to 300 °C at supercritical pressure.

The low-temperature preheating process is used for preheating feed from normal temperature to 300 °C. There are two situations should be taken into consider for

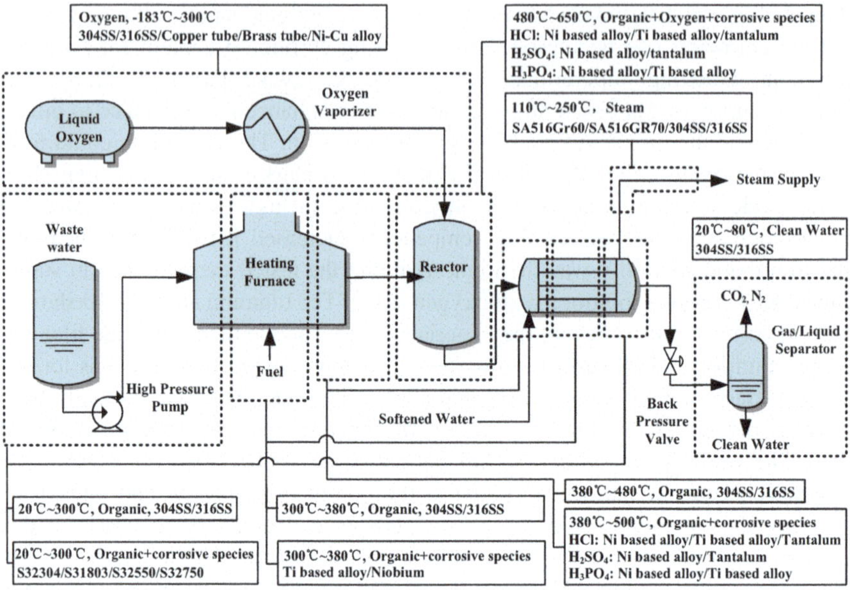

Fig. 6.73 Material selection for SCWO system in flow diagram

materials selection in this low-temperature process. In the first case, 304 SS and 316 SS are recommended at the weak corrosive condition that the feeds did not contain halide. In the second case, dual-phase steel can endure the severe corrosion due to the feeds contain halide. The dual-phase steels like S32304, S31803, S32550, and S32750 own an excellent corrosion resistance to halide; however they cannot be used at temperature over 300 °C due to the decline of mechanical strength. The materials which are applied to the low-temperature preheating pipe for wastewater can also be used in the low-temperature zone of vapor generator.

The high-temperature heating process is used for heating feed from 300 to 380 °C, and this high-temperature subcritical process is considered as the severest corrosion part of SCWO system. When the feed do not contain corrosive species like halogenated hydrocarbons or halide ion, 304 SS or 316 SS will be selected as candidate materials. When the feed containing corrosive species in this condition, the titanium-based alloys, niobium, and tantalum are the suitable materials for constructing the heating pipe. The medium-temperature zone of vapor generator could use the same materials which applied to the high-temperature heating pipe for feed.

This is high-temperature supercritical area from 380 to 480 °C in heating furnace. The 304 SS and 316 SS will also be applied for constructing the supercritical heating pipe in absent of halide. When the feed contains the aggressive species like HCl, H_3PO_4, and H_2SO_4, the nickel-based alloys are usable materials and the corrosion rates should be measured through corrosion tests under different conditions. However, the titanium-based alloys showed a good resistance to corrosion in the oxidizing supercritical water contained aggressive species, especially chloride ion. The material selection for this process is as same as the high-temperature zone of vapor generator.

The temperature of reactor is usually among 480–650 °C, and the hazardous waste decomposes immediately in this process. The reactor would suffer a severer corrosion from the synergistic effect between oxygen and aggressive species than that from the individual effect. The materials which used in the high-temperature supercritical heating area could also be used for constructing reactor. Moreover, the wall thickness of reactor should be thick enough for reaching the strength requirement under this high-temperature condition.

The work temperature of pipe for transporting steam is usually among 150–250 °C, and the softened water is heated to create clean steam which is a weak corrosive medium. The low-alloy steels like SA516Gr60 and SA516GR70 can be used for constructing this weak corrosion part of system. The high-alloy steels like 304 SS and 316 SS are also recommended as the constructing materials under this condition.

The heat exchanger cooled the treated wastewater from supercritical temperature to room temperature in this unit. The work temperature of this unit is among 20–60 °C, and the common stainless steel (304 SS, 316 SS) is the suitable material in this low-temperature area. The selection of materials depends on corrosion resistance and mechanical strength of materials for constructing SCWO plant, especially at supercritical condition. In this section, a guidance of candidate materials selection is supplied for SCWO system. According to the above suggestion, the kinds of candidate materials will be determined. Then, the corrosion behavior and rate of selected material should be investigated through the corrosion test in specific condition, which is the foundation work for building SCWO system.

6.6.2 Several Points on Further Research

In conclusion, no single material can withstand all kinds of SCWO conditions; however, the suitable material can be selected for a certain condition considering for corrosion resistance, strength, and economy. The appropriate corrosion control approaches can also reduce the corrosion effectively, especially in the highly corrosive feed with high-pressure and high-temperature requirements. In order to accomplish the commercialization of SCWO in a more safe and economical way, in our opinions, some potentially valuable researches are necessary and are summarized as follows.

(1) Online corrosion monitoring methods. The corrosion monitoring technologies such as electric resistivity technology, pulsed magnetic flux leakage technology, and ultrasonic technology can be applied in SCW system. The effective monitoring methods can predict the service life of the equipment, and the appropriate corrosion control technology can be used for reducing

corrosion timely. The corrosion monitoring and corrosion control methods can prevent dangerous occurrence from happening under condition of high temperature and high pressure.

(2) Corrosion control approaches. The effective control approaches should be applied for reducing the corrosion in the whole SCWO system. Based on the analyzing of various reasons for the corrosion, the control approaches consist of three aspects: the equipment structure optimization, reducing the corrosion of reaction medium, and optimizing operation parameters. The effective corrosion control approaches could reduce the manufacturing cost and the service life of equipment will be elongated.

(3) Corrosion test under the real condition. The high pressure and flow must be considered as the important factors in the further corrosion test. In the previous corrosion researches, these two important factors were rarely taken into account. The high pressure and flow which can accelerate corrosion significantly are consistent with the real situation.

(4) Develop and determine calculation model. Based on corrosion test results and theoretical analysis, the exact calculation model should be established for predicting the tendency of corrosion. The good model will make up for the shortage of monitoring methods under high-temperature and high-pressure condition. The effective factors and causes can be analyzed according to the accurate simulation calculation results. The classical parabolic law is still used somewhat uncritically since they "seem to work" and they, at the phenomenological level at least, can relate the growth rate of oxide film to other measurable transport properties of the oxide, such as diffusion coefficients. But these models cannot account for fundamental quantities (e.g., the "jump distance") occurring at the metal/scale interface, neglecting the existence of space-charge regions near the metal/scale and scale/environment interfaces, and fail to account for the behavior at limiting small times (finite rate) and limiting long times (steady-state). Now, at the atomic level, we are developing a point defect model for the growth of corrosion scales of steels and alloys in supercritical aqueous systems with/without oxygen [222], which aims to quantificationally describe the interfacial corrosion reactions and defects transport within the scales and explains the corrosion kinetics in a way with clear physical connotations compared to the classical parabolic law.

References

1. R. Viswanathan, J. Sarver, J. M. Tanzosh, Boiler materials for ultra-supercritical coal power plants-steamside oxidation. Journal of Materials Engineering and Performance, 2006, 15: 255–274.
2. t. U. S. N. E. R. A. Committee, N. a. t. G. I. I. F. (GIF), A technology roadmap for generation IV nuclear energy systems. 2002.

3. P. Kritzer, Corrosion in high-temperature and supercritical water and aqueous solutions: a review. Journal of Supercritical Fluids, 2004, 29: 1–29.
4. P. A. Marrone, G. T. Hong, Corrosion control methods in supercritical water oxidation and gasification processes. The Journal of Supercritical Fluids, 2009, 51: 83–103.
5. G. S. Was, S. S. Teysseyre, J. S. McKinley, Corrosion and Stress Corrosion Cracking of Iron and Nickel-base Austenitic Alloys in Supercritical Water. CORROSION 2004, 2004.
6. H. Inoue, Y. Maeda, T. Mizuno, Corrosion Potential Fluctuations and Polarization Resistance Measurements on High Subcritical and Supercritical Water Oxidation Environment. CORROSION 2004, 2004.
7. G. S. Was, Z. Jiao, S. S. Teysseyre, Corrosion of austenitic alloys in supercritical water. CORROSION 2005, 2005.
8. X. Gao, X. Wu, Z. Zhang, H. Guan, E.-h. Han, Characterization of oxide films grown on 316L stainless steel exposed to H-containing supercritical water. The Journal of supercritical fluids, 2007, 42: 157–163.
9. E. Asselin, A. Alfantazi, S. Rogak, Corrosion of nickel–chromium alloys, stainless steel and niobium at supercritical water oxidation conditions. Corrosion Science, 2010, 52: 118–124.
10. M. Sun, X. Wu, Z. Zhang, E.-H. Han, Oxidation of 316 stainless steel in supercritical water. Corrosion Science, 2009, 51: 1069–1072.
11. T. M. Hayward, I. M. Svishchev, R. C. Makhija, Stainless steel flow reactor for supercritical water oxidation: corrosion tests. The Journal of supercritical fluids, 2003, 27: 275–281.
12. H. Kim, J. Yoon, J. Han, B. Mitton, R. Latanision, Y. Kim, Influence of chromizing treatment on the corrosion behavior of AISI 316 stainless steel in supercritical water oxidation. Metals and Materials International, 2004, 10: 83–88.
13. H.-C. Lee, S.-H. Son, K.-Y. Hwang, C.-H. Lee, Surface chemical analysis on the corrosion of alloys in the supercritical water oxidation of halogenated hydrocarbon. Industrial & engineering chemistry research, 2006, 45: 3412–3419.
14. V. S. Rao, H. Kwon, Reactor corrosion in ceria production by hydrothermal synthesis under supercritical conditions. Corrosion, 2007, 63: 359–365.
15. S.-H. Son, J.-H. Lee, S.-H. Byeon, C.-H. Lee, Surface chemical analysis of corroded alloys in subcritical and supercritical water oxidation of 2-chlorophenol in continuous anticorrosive reactor system. Industrial & Engineering Chemistry Research, 2008, 47: 2265–2272.
16. R. Fujisawa, K. Nishimura, M. Sakaihara, Y. Watanabe, Y. Kurata, T. Nishida, Corrosion Behavior of Ni Base Alloys and 316 Stainless Steel in Less Oxidizing or Reducing SCW Containing HCl. Corrosion 2005, 2005.
17. M. Son, Y. Kurata, Y. Ikushima, Corrosion behavior of metals in SCW environments containing salts and oxygen. Corrosion 2002, 2002.
18. Y. Watanabe, T. Adschiri, T. Kobayashi, Significance of Water Density on Corrosion Behavior of Alloys in Supercritical Water. Corrosion 2001, 2001.
19. R. Fujisawa, M. Sakaihara, Y. Kurata, Y. Watanabe, Corrosion behaviour of nickel base alloys and 316 stainless steel in supercritical water under alkaline conditions. Corrosion engineering, science and technology, 2005, 40: 244–248.
20. M. Drews, M. Williams, M. Barr, The corrosion of sol-gel-coated type 316 SS in chlorinated SC water. Industrial & engineering chemistry research, 2000, 39: 4772–4783.
21. Y. Daigo, Y. Watanabe, K. Sugahara, T. Isobe, Compatibility of nickel-based alloys with supercritical water applications: Aging effects on corrosion resistance and mechanical properties. Corrosion, 2006, 62: 174–181.
22. R. Fujisawa, T. Nishida, Y. Kurata, Y. Watanabe, M. Sakaihara, K. Nishimura, Cracking Susceptibility of Ni Base Alloys and 316 Stainless Steel in Less Oxidizing or Reducing SCW. CORROSION 2005, 2005.
23. C. Bosch, D. Delafosse, Stress Corrosion Cracking of Ni-Based and Fe-Based Alloys in Supercritical Water. Corrosion 2005, 2005.
24. H. Kim, An Investigation of Corrosion Mechanisms of Constructional Alloys in Supercritical Water Oxidation (SCWO) Systems. 2004, Massachusetts Institute of Technology.

25. Y. Daigo, Y. Watanabe, K. Sue, Effect of chromium ion from autoclave material on corrosion behavior of nickel-based alloys in supercritical water. Corrosion, 2007, 63: 277–284.
26. J. H. Y. D. B. Mitton, J. A. Cline, H. S. Kim, N. Eliaz and R. M. Latanision, Corrosion behavior of Nickel-Based Alloys in Supercritical Water Oxidation Systems. Ind. Eng. Chem. Res, 2000, 39: 4689–4696.
27. J. Konys, S. Fodi, J. Hausselt, H. Schmidt, V. Casal, Corrosion of high-temperature alloys in chloride-containing supercritical water oxidation systems. Corrosion, 1999, 55: 45–51.
28. M. Sun, X. Wu, Z. Zhang, E.-H. Han, Analyses of oxide films grown on Alloy 625 in oxidizing supercritical water. The Journal of Supercritical Fluids, 2008, 47: 309–317.
29. J.-H. L. Sang-Ha Son, Sang-Hoon Byeon and Chang-Ha Lee, Surface Chemical Analysis of Corroded Alloys in Subcritical and Supercritical Water Oxidation of 2-Chlorophenol in Continuous Anticorrosive Reactor System. Ind. Eng. Chem. Res, 2008, 47: 2265–2272.
30. H. Kim, D. Mitton, R. Latanision, Corrosion behavior of Ni-base alloys in aqueous HCl solution of pH 2 at high temperature and pressure. Corrosion Science, 2010, 52: 801–809.
31. D. B. M. a. R. M. L. H. Kim, Stress Corrosion Cracking of Alloy 625 in pH 2 Aqueous Solution at High Temperature and Pressure. Corrosion, 2011, 67: 035002–1-035002-8.
32. E. Asselin, A. Alfantazi, S. Rogak, Thermodynamics of the corrosion of alloy 625 supercritical water oxidation reactor tubing in ammoniacal sulfate solution. Corrosion, 2008, 64: 301–314.
33. P. Kritzer, N. Boukis, E. Dinjus, Corrosion of alloy 625 in high-temperature, high-pressure sulfate solutions. Corrosion, 1998, 54: 689–699.
34. P. Kritzer, N. Boukis, E. Dinjus, Corrosion of alloy 625 in aqueous solutions containing chloride and oxygen. Corrosion (Houston, TX, U S), 1998, 54: 824–834.
35. N. Boukis, Corrosion phenomena on alloy 625 in aqueous solutions containing hydrochloric acid and oxygen under subcritical and supercritical conditions. N. Boukis, and P. Kritzer (ITC, Forschungszentrum Karlsruhe, Germany), Paper, 1997.
36. V. Kolarik, B. Michelfelder, M. Wagner, Corrosion of Alloys 625 and Pure Chromium in Cl-Containing Fluids During Supercritical Water Oxidation (SCWO). Corrosion 99, 1999.
37. Q. Zhang, R. Tang, K. Yin, X. Luo, L. Zhang, Corrosion behavior of Hastelloy C-276 in supercritical water. Corros Sci, 2009, 51: 2092–2097.
38. J.-H. L. Sang-Ha Son, Chang-Ha Lee, Corrosion phenomena of alloys by subcritical and supercritical water oxidation of 2-chlorophenol. The Journal of Supercritical Fluids, 2008, 44: 370–378.
39. B. Somerday, K. Wiggans, R. Bradshaw, Environment-assisted failure of alloy C-276 burst disks in a batch supercritical water oxidation reactor. Engineering Failure Analysis, 2006, 13: 80–95.
40. K. HATAKEDA, Y. IKUSHIMA, N. SAITO, O. SATO, T. AIZAWA, C. RYU, Corrosion of Supercritical Water Oxidation Systems for PCBs Destruction. Nippon Kagakkai Koen Yokoshu, 1999, 76: 625.
41. L. Tan, K. Sridharan, T. Allen, The effect of grain boundary engineering on the oxidation behavior of INCOLOY alloy 800H in supercritical water. Journal of nuclear materials, 2006, 348: 263–271.
42. Z. Zhang, J. Wang, E.-H. Han, W. Ke, Influence of dissolved oxygen on oxide films of Alloy 690TT with different surface status in simulated primary water. Corros Sci, 2011, 53: 3623–3635.
43. J. Konys, A. Ruck, J. Novotny, J. Hausselt, Corrosion of Nickel Based Alloy G-30 Between 100 and 420? C, pH Values From 2 to 6 and Pressures Ranging From 250–480 Bar Under Conditions of Supercritical Water Oxidation. Corrosion 2001, 2001.
44. J. Konys, S. Fodi, A. Ruck, Comparison of Corrosion of Nickel and Chromium Based Alloys Under Conditions of Supercritical Water Oxidation. Corrosion 99, 1999.
45. C. Schroer, J. Konys, J. Novotny, J. Hausselt, Aspects of the corrosion kinetics for binary nickel-chromium alloys under conditions found in supercritical water oxidation plants processing chlorinated substances. Corrosion, 2006, 62: 444–459.

46. S. Iimura, M. Miyasaka, H. Yakuwa, Effect of Cr and Mo Contents on the Corrosion Properties of Ni-Base Alloys in a Supercritical Water Including Chloride. CORROSION 2002, 2002.
47. N. Hara, S. Soma, K. Sugimoto, S. Tanaka, Corrosion Resistance of Fe-Cr and Ni-Cr Alloys in Oxidizing Supercritical HCl Solution. Corrosion 2002, 2002.
48. N. Boukis, C. Friedrich, E. Dinjus, Titanium as Reactor Material for SCWO Applications-First Experimental Results. CORROSION 98, 1998.
49. Y. Tsuchiya, K. Hiruta, N. Saito, Y. Akai, Corrosion behavior of metals for SCWO reactors for organic waste processing plants. CORROSION 2001, 2001.
50. Y. Tsuchiya, N. Saito, Y. Akai, K. Yamada, T. Takada, Corrosion Behavior of SCWO Reactor Materials for Organic Waste Processing Plants. Corrosion Engineering, 2003, 52: 829–840.
51. M. Schacht, N. Boukis, E. Dinjus, Corrosion of alumina ceramics in acidic aqueous solutions at high temperatures and pressures. Journal of Materials Science, 2000, 35: 6251–6258.
52. M. Nagae, T. Yoshio, K. Oda, Corrosion Behaviour of Structural Ceramics in Supercritical Water. Advances in Science and Technology, 2006, 45: 173–177.
53. M. Nagae, Y. Koyama, S. Yasutake, T. Yoshio, K. Oda, Corrosion behavior of sialon ceramics in supercritical water. Journal of the American Ceramic Society, 2006, 89: 3550–3553.
54. N. Boukis, N. Claussen, K. Ebert, R. Janssen, M. Schacht, Corrosion screening tests of high-performance ceramics in supercritical water containing oxygen and hydrochloric acid. Journal of the European Ceramic Society, 1997, 17: 71–76.
55. K. M. Garcia, R. Mizia, Corrosion investigation of multilayered ceramics and experimental nickel alloys in SCWO process environments, in Other Information: PBD: Feb 1995. 1995. p. Medium: ED; Size: 200 p.
56. R. Wang, a, Z. D., a, J. L., a, W. C., a, W. Z., b, D. G., c, Stability of MgO-, CeO2- and SiO2-doped Cr2O3 ceramics in high-temperature supercritical water. Corrosion Science, 2014.
57. P. Kritzer, N. Boukis, G. Franz, E. Dinjus, The Corrosion of Niobium in Oxidizing Sub- and Supercritical Aqueous Solutions of HCl and H_2SO_4. Journal of Materials Science Letters, 1999, 18: 25–27.
58. C. Friedrich, P. Kritzer, N. Boukis, G. Franz, E. Dinjus, The corrosion of tantalum in oxidizing sub-and supercritical aqueous solutions of HCl, H_2SO_4 and H_3PO_4. Journal of materials science, 1999, 34: 3137–3141.
59. K. Sue, K. Arai, Y. Watanabe, N. Tsujinaka, T. Adschiri, Relationship between corrosion rate and metal oxide solubility in supercritical water. CORROSION 2002, 2002.
60. Y. Ikushima, M. Son, Y. Kurata, H. Kim, K. Hatakeda, Corrosion behavior of metals in supercritical water solutions in the presence of salts. CORROSION 2001, 2001.
61. N. Hara, S. Tanaka, K. Sugimoto, Corrosion Behavior of Constituent Metals of Stainless Alloys in SCWO Environments. Corrosion, 2001.
62. T. Adschiri, K. Sue, K. Arai, Y. Watanabe, Estimation of metal oxide solubility and understanding corrosion in supercritical water. Corrosion 2001.
63. S. A. Wood, L. L. Baker, Experimental Study of Metal Corrosion in Supercritical Brines: Application to Supercritical Water Oxidation of Hazardous Wastes. 2000, DTIC Document.
64. P. Kritzer, N. Boukis, E. Dinjus, Transpassive dissolution of alloy 625, chromium, nickel, and molybdenum in high-temperature solutions containing hydrochloric acid and oxygen. Corrosion, 2000, 56: 265–272.
65. A. M. Olmedo, a, M. G. A., b, G. D. *, a, R. B., a, Corrosion behavior of T91 and type AISI 403 stainless steel in supercritical water Procedia Materials Science 2012, 1: 543–549.
66. M. Sun, X. Wu, E.-H. Han, J. Rao, Microstructural characteristics of oxide scales grown on stainless steel exposed to supercritical water. Scripta Materialia, 2009, 61: 996–999.

67. D. Mitton, S. Zhang, K. Hautanen, J. Cline, Evaluating stress corrosion and corrosion aspects in supercritical water oxidation systems for the destruction of hazardous waste. D. B. Mitton, S-H. Zhang, K. E. Hautanen, J. A. Cline, E-H. Han, and R. M. Latanision (Massachusetts Inst. of Technology, Cambridge, MA), Paper, 1997.
68. X. Tang, S. Wang, D. Xu, Y. Gong, J. Zhang, Y. Wang, Corrosion behavior of Ni-based alloys in supercritical water containing high concentrations of salt and oxygen. Ind Eng Chem Res, 2013, 52: 18241–18250.
69. D. D. Macdonald, S. N. Lvov, Development of Advanced In-Situ Techniques for Chemistry Monitoring and Corrosion Mitigation in SCWO Environments, in Other Information: PBD: 31 Mar 2000. 2000. p. Medium: ED; Size: vp.
70. T. Mizuno, H. Inoue, Y. Maeda, Corrosion–Potential Fluctuation and Polarization Resistance Method for Corrosion Monitoring in Subcritical and Supercritical Water Environments. NACE International, 2004.
71. L. B. K. a. D. D. Macdonald, Corrosion in Supercritical Water Oxidation Systems: A Phenomenological Analysis. The Electrochemical Society, 1995, 142: 4069–4073.
72. L. J.-s. M. Z.-y. Z. J.-y. M. C.-a. M. X.-b. L. Xiao-hua, Corrosion of titanium in supercritical water oxidation environments. Transactions of Nonferrous Metals Society of China, 2002, 12: 1054–1057.
73. D. Mitton, N. Eliaz, J. Cline, R. Latanision, An overview of the current understanding of corrosion in SCWO systems for the destruction of hazardous waste products. Technology, 2001, 16: 30–53.
74. X. Ren, K. Sridharan, T. Allen, Corrosion behavior of alloys 625 and 718 in supercritical water. Corrosion (Houston, TX, U S), 2007, 63: 603–612.
75. K.-H. Chang, a, J.-H. H., a, C.-B. Y., b, T.-K. Y., b. a, F.-R. C. *, b,, J.-J. Kai, Corrosion behavior of Alloy 625 in supercritical water environments. Progress in Nuclear Energy, 2011, 57: 20–31.
76. K. Hatakeda, Y. Ikushima, N. Saito, C. Liew, T. Aizawa, Corrosion on continuous supercritical water oxidation for polychlorinated biphenyls. International Journal of High Pressure Research, 2001, 20: 393–401.
77. N. Boukis, E. D. E, G. Franz, W. Habicht, Corrosion Resistant Materials for SCWO-Applications: Experimental Results From Long Time Experiments. Corrosion 2001.
78. Y. Kurata, M. Son, Corrosion Problem of Supercritical Water Oxidation Constructional Materials. Review of High Pressure Science and Technology, 2001, 11: 324–331.
79. T. Oe, T. Iwamori, S. Kawasaki, A. Suzuki, H. Daimon, K. Fujie, Corrosion resistance in highly concentrated hydrochloric acid and creep rupture strength of a Ti-Pd alloy for a supercritical water oxidation reactor. Corrosion, 2007, 63: 793–798.
80. N. Saito, Y. Tsuchiya, Y. Akai, H. Omura, T. Takada, N. Hara, Corrosion performance of metals for supercritical water, oxidation-utilized organic waste-processing reactors. Corrosion, 2006, 62: 383–394.
81. H. G. Wheat, M. Schmerling, E. F. Gloyna, T. Danielson, L. G. Li, Apparatus for Studying Corrosion in Sub and Supercritical Water Oxidation Environments. Corrosion 2000.
82. C. Frayret, P. Botella, T. Jaszay, M.-H. Delville, Titanium Dissolution-Passivation in Highly Chloridic and Oxygenated Aqueous Solutions Reaction Mechanism Extended to Supercritical Water Conditions. Journal of The Electrochemical Society, 2004, 151: B543-B550.
83. J.-S. Lu, Corrosion of titanium in phosphoric acid at 250° C. Transactions of Nonferrous Metals Society of China, 2009, 19: 552–556.
84. X. Guan, D. Macdonald, Determination of Corrosion Mechanisms and Estimation of Electrochemical Kinetics of Metal Corrosion in High Subcritical and Supercritical Aqueous Systems. Corrosion, 2009, 65: 376–387.
85. T. Richard, J. Poirier, C. Reverte, C. Aymonier, A. Loppinet-Serani, G. Iskender, E.-B. Pablo, F. Marias, Corrosion of ceramics for vinasse gasification in supercritical water. Journal of the European Ceramic Society, 2012.

86. Z. Dong, a, W. C., a, W. Z., b, D. G., c, Corrosion behavior of chromium oxide based ceramics in supercritical water (SCW) environments. Corrosion Science, 2012, 65: 461–471.
87. Z. Fang, Nano-Structured Coatings. Rapid Production of Micro-and Nano-particles Using Supercritical Water, 2010: 57–62.
88. F. Khatkhatay, L. Jiao, J. Jian, W. Zhang, Z. Jiao, J. Gan, H. Zhang, X. Zhang, H. Wang, Superior corrosion resistance properties of TiN-based coatings on Zircaloy tubes in supercritical water. Journal of Nuclear Materials, 2014, 451: 346–351.
89. Y. Watanabe, T. Adschiri, K. Shoji, Effects of Oxygen Concentration on Corrosion Behavior of Alloys in Acidic Supercritical Water. Corrosion 2002.
90. S. Penttilä, a, A. T., a, J. L., b, W. Z., b, R. N., Effect of surface modification on the corrosion resistance of austenitic stainless steel 316L in supercritical water conditions. The Journal of Supercritical Fluids, 2013, 81: 157–163.
91. B. Stellwag, The mechanism of oxide film formation on austenitic stainless steels in high temperature water. Corros Sci, 1998, 40: 337–370.
92. Y. Daigo, Y. Watanabe, K. Sue, Corrosion Mitigation in Supercritical Water with Chromium Ion. Corrosion, 2007, 63: 1085–1093.
93. J. A. Cline, D. B. Mitton, R. M. Latanision, P. Marrone, J. W. Tester, Corrosion Mechanisms of Alloy N10276 in Hydrothermal HCL Solutions: Failure Analysis and Exposure Studies. CORROSION 2001, 2001.
94. X. Y. Tang, S. Z. Wang, L. L. Qian, J. M. Lu, Selection of material with high salt concentration in preheating section of supercritical water oxidation system, in Advanced Research on Material Engineering, Chemistry And Environment, H. Zhang, D. Jin, and X. J. Zhao, Editors. 2013, Trans Tech Publications Ltd.: Stafa-Zurich. p. 440–443.
95. R. Fujisawa, K. Nishimura, T. Nishida, M. Sakaihara, Y. Kurata, Y. Watanabe, Corrosion behavior of nickel-based alloys and type 316 stainless steel in slightly oxidizing or reducing supercritical water. corrosion, 2006, 62: 270–274.
96. R. Dooley, I. Wright, P. Tortorelli, Program on technology innovation: oxide growth and exfoliation on alloys exposed to steam, in Palo Alto, CA: EPRI. 2007.
97. G. S. Was, S. Teysseyre, Z. Jiao, Corrosion of austenitic alloys in supercritical water. Corrosion, 2006, 62: 989–1005.
98. J. Bischoff, A. T. Motta, Oxidation behavior of ferritic-martensitic and ODS steels in supercritical water. Journal of Nuclear Materials, 2012, 424: 261–276.
99. H. Helong, Z. Zhangjian, L. Ming, Z. Lefu, W. Man, L. Shaofu, G. Changchun, Study of the corrosion behavior of a 18Cr-oxide dispersion strengthened steel in supercritical water. Corrosion Science, 2012, 65: 209–13.
100. J. Bischoff, A. T. Motta, C. Eichfeld, R. J. Comstock, G. Cao, T. R. Allen, Corrosion of ferritic-martensitic steels in steam and supercritical water. Journal of Nuclear Materials, 2013, 441: 604–611.
101. Y. Li, S. Wang, P. Sun, D. Xu, M. Ren, Y. Guo, G. Lin, Early oxidation mechanism of austenitic stainless steel TP347H in supercritical water. Corrosion Science, 2017, 128: 241–252.
102. Y. Behnamian, A. Mostafaei, A. Kohandehghan, B. S. Amirkhiz, D. Serate, Y. Sun, S. Liu, E. Aghaie, Y. Zeng, M. Chmielus, W. Zheng, D. Guzonas, W. Chen, J. L. Luo, A comparative study of oxide scales grown on stainless steel and nickel-based superalloys in ultra-high temperature supercritical water at 800 °C. Corrosion Science, 2016, 106: 188–207.
103. D. Rodriguez, D. Chidambaram, Oxidation of stainless steel 316 and Nitronic 50 in supercritical and ultrasupercritical water. Applied Surface Science, 2015, 347: 10–16.
104. D. Rodriguez, A. Merwin, D. Chidambaram, On the oxidation of stainless steel alloy 304 in subcritical and supercritical water. J Nucl Mater, 2014, 452: 440–445.
105. K. I. Choudhry, S. Mahboubi, G. A. Botton, J. R. Kish, I. M. Svishchev, Corrosion of engineering materials in a supercritical water cooled reactor: Characterization of oxide scales on Alloy 800H and stainless steel 316. Corrosion Science, 2015, 100: 222–230.
106. S. J. Rothman, L. J. Nowicki, G. E. Murch, Self-diffusion in austenitic Fe-Cr-Ni alloys. Journal of Physics F: Metal Physics, 1980, 10: 383–398.

107. G. C. Allen, S. J. Harris, J. A. Jutson, J. M. Dyke, A study of a number of mixed transition metal oxide spinels using X-ray photoelectron spectroscopy. Applied Surface Science, 1989, 37: 111–134.
108. S. Mischler, H. J. Mathieu, D. Landolt, Investigation of a passive film on an iron chromium alloy by AES and XPS. Surface & Interface Analysis, 1988, 11: 182–188.
109. Z. Xiangyu, H. Enhou, W. Xinqiang, Corrosion behavior of Alloy 690 in aerated supercritical water. Corrosion Science, 2013, 66: 369–79.
110. T. P. Moffat, R. M. Latanision, An Electrochemical and X-Ray Photoelectron Spectroscopy Study of the Passive State of Chromium. Journal of the Electrochemical Society, 1991, 139: 1869–1879.
111. G. C. Allen, P. M. Tucker, Multiplet splitting of X-ray photoelectron lines of chromium complexes. The effect of covalency on the 2p core level spin-orbit separation. Inorganica Chimica Acta, 1976, 16: 41–45.
112. L. Tan, T. R. Allen, Y. Yang, Corrosion behavior of alloy 800H (Fe-21Cr-32Ni) in supercritical water. Corros Sci, 2011, 53: 703–711.
113. P. Marcus, J. M. Grimal, The anodic dissolution and passivation of NiCrFe alloys studied by ESCA. Corrosion Science, 1992, 33: 805–814.
114. J. C. Langevoort, I. Sutherland, L. J. Hanekamp, P. J. Gellings, On the oxide formation on stainless steels AISI 304 and incoloy 800H investigated with XPS. Applied Surface Science, 1987, 28: 167–179.
115. D. G. Briceno, F. Blazquez, A. S. Maderuelo, Oxidation of austenitic and ferritic/martensitic alloys in supercritical water. Journal of Supercritical Fluids, 2013, 78: 103–113.
116. T. Darko, I. H. Hillier, J. Kendrick, A theoretical study of the 1s photoelectron spectrum of O2. The Journal of Chemical Physics, 1977, 67: 1792–1793.
117. N. Zhang, H. Xu, B. Li, Y. Bai, D. Liu, Influence of the dissolved oxygen content on corrosion of the ferritic-martensitic steel P92 in supercritical water. Corrosion Science, 2012, 56: 123–128.
118. Q. Zhang, K. Yin, R. Tang, L. Zhang, X. Luo, Corrosion behavior of Hastelloy C-276 in supercritical water. Corrosion Science, 2009, 51: 2092–7.
119. X. Zhong, X. Wu, E.-H. Han, Effects of exposure temperature and time on corrosion behavior of a ferritic–martensitic steel P92 in aerated supercritical water. Corrosion Science, 2015, 90: 511–521.
120. S. Penttilä, I. Betova, M. Bojinov, P. Kinnunen, A. Toivonen, Oxidation model for construction materials in supercritical water—Estimation of kinetic and transport parameters. Corrosion Science, 2015, 100: 36–46.
121. A. Roine, HSC Chemistry Thermo-chemical Database. 2007, Version.
122. Y. toguro, M. Sakakibara, T. Saito, H. Ito, Y. Inoue, Oxidation Behavior of Austenitic Heat-Resisting Steels in a High Temperature and High Pressure Steam Environment. Transactions of the Iron & Steel Institute of Japan, 1988, 28: 761–768.
123. A. N. Hansson, H. Danielsen, F. B. Grumsen, M. Montgomery, Microstructural investigation of the oxide formed on TP 347H FG during long-term steam oxidation. Materials and Corrosion, 2010, 61: 665–675.
124. Y. Behnamian, A. Mostafaei, A. Kohandehghan, B. S. Amirkhiz, J. Li, R. Zahiri, E. Aghaie, W. Zheng, D. Guzonas, M. Chmielus, W. Chen, J. L. Luo, Internal oxidation and crack susceptibility of alloy 310S stainless steel after long term exposure to supercritical water at 500 °C. The Journal of Supercritical Fluids, 2017, 120, Part 1: 161–172.
125. J. Yuan, X. Wu, W. Wang, S. Zhu, F. Wang, The Effect of Surface Finish on the Scaling Behavior of Stainless Steel in Steam and Supercritical Water. Oxidation of Metals, 2013, 79: 541–551.
126. L. Tan, X. Ren, T. R. Allen, Corrosion behavior of 9–12% Cr ferritic-martensitic steels in supercritical water. Corrosion Science, 2010, 52: 1520-1528.
127. Y. Li, S. Wang, P. Sun, J. Yang, X. Tang, D. Xu, Y. Guo, J. Yang, D. D. Macdonald, Investigation on early formation and evolution of oxide scales on ferritic–martensitic steels in supercritical water. Corros Sci, 2018.

128. Y. Li, s. Wang, P. Sun, D. Xu, M. Ren, Y. Guo, G. Lin, Early oxidation mechanism of austenitic stainless steel TP347H in supercritical water. Corrosion Science, 2017.
129. Y. Yang, Q.-z. Yan, Y.-f. Yang, L.-f. Zhang, C.-c. Ge, Corrosion Behavior of Ferritic/Martensitic Steels CNS-I and Modified CNS-II in Supercritical Water. Journal of Iron and Steel Research International, 2012, 19: 69–73.
130. K. Yin, S. Qiu, R. Tang, Q. Zhang, L. Zhang, Corrosion behavior of ferritic/martensitic steel P92 in supercritical water. Journal of Supercritical Fluids, 2009, 50: 235–239.
131. Y. Li, S. Wang, P. Sun, Z. Tong, D. Xu, Y. Guo, J. Yang, Early oxidation of Super304H stainless steel and its scales stability in supercritical water environments. International Journal of Hydrogen Energy, 2016, 41: 15764–15771.
132. L. Tan, Y. Yang, T. R. Allen, Oxidation behavior of iron-based alloy HCM12A exposed in supercritical water. Corrosion Science, 2006, 48: 3123–3138.
133. B. J. Tan, K. J. Klabunde, P. M. A. Sherwood, XPS studies of solvated metal atom dispersed (SMAD) catalysts. Evidence for layered cobalt-manganese particles on alumina and silica. Journal of the American Chemical Society, 1991, 113: 855–861.
134. J. N. Wanklyn, The role of molybdenum in the crevice corrosion of stainless steels. Corrosion Science, 1981, 21: 211–225.
135. X. Huang, Y. Shen, Oxidation behavior of an 11Cr ferritic-martensitic steel after Ar-ions irradiation in supercritical water. Journal of Nuclear Materials, 2015, 461: 1–9.
136. L. Tan, M. T. Machut, K. Sridharan, T. R. Allen, Corrosion behavior of a ferritic/martensitic steel HCM12A exposed to harsh environments. Journal of Nuclear Materials, 2007, 371: 161–170.
137. P. Ampornrat, G. S. Was, Oxidation of ferritic-martensitic alloys T91, HCM12A and HT-9 in supercritical water. Journal of Nuclear Materials, 2007, 371: 1–17.
138. N. Hiramatsu, F. H. Stott, The Effects of Molybdenum on the High-Temperature Oxidation Resistance of Thin Foils of Fe–20Cr–5Al at Very High Temperatures. Oxidation of Metals, 2000, 53: 561–576.
139. Y. Chen, K. Sridharan, T. Allen, Corrosion behavior of ferritic–martensitic steel T91 in supercritical water. Corrosion Science, 2006, 48: 2843–2854.
140. J. Bischoff, A. T. Motta, R. J. Comstock, Evolution of the oxide structure of 9CrODS steel exposed to supercritical water. Journal of Nuclear Materials, 2009, 392: 272–279.
141. W. Kai, C. H. Lee, T. W. Lee, C. H. Wu, Sulfidation behavior of Inconel 738 superalloy at 500 900 degrees C. Oxidation of Metals, 2001, 56: 51-71.
142. D. S. Williams, R. Moller, H. J. Grabke, The High-temperature Corrosion of Alloy-800 in Carburizing, Oxidizing, and Sulfidizing Environments. Oxidation of Metals, 1981, 16: 253–266.
143. Y. H. Li, S. Z. Wang, X. Y. Tang, D. H. Xu, Y. Guo, J. Zhang, L. L. Qian, Effects of Sulfides on the Corrosion Behavior of Inconel 600 and Incoloy 825 in Supercritical Water. Oxid Met, 2015, 84: 509–526.
144. J. C. Lee, M. J. Kim, D. B. Lee, High-Temperature Corrosion of Aluminized and Chromized Fe-25.8%Cr-19.5%Ni Alloys in N-2/H2S/H2O-Mixed Gases. Oxidation of Metals, 2014, 81: 617–630.
145. S. Baur, H. Schmidt, A. Krämer, J. Gerber, The destruction of industrial aqueous waste containing biocides in supercritical water—development of the SUWOX process for the technical application. J Supercrit Fluids, 2005, 33: 149–157.
146. P. Kritzer, N. Boukis, E. Dinjus, Transpassive Dissolution of Alloy 625, Chromium, Nickel, and Molybdenum in High-Temperature Solutions Containing Hydrochloric Acid and Oxygen. Corrosion -Houston Tx-, 2000, 56: 265–272.
147. S. Morioka, K. Shimakage, Studies on the Ammonia Pressure Leaching of Metallic Nickel and Copper Powders. Materials Transactions Jim, 1971, 12: 197–205.
148. E. Asselin, A. Alfantazi, S. Rogak, A Polarization Study of Alloy 625, Nickel, Chromium, and Molybdenum in Ammoniated Sulfate Solutions. Corrosion -Houston Tx-, 2005, 61: 579–586.

149. J. Yang, S. Wang, D. Xu, Y. Guo, C. Yang, Y. Li, Effect of ammonium chloride on corrosion behavior of Ni-based alloys and stainless steel in supercritical water gasification process. Int J Hydrogen Energy, 2017, 42: 19788–19797.
150. E. Asselin, G. Nazari, A. Alfantazi, S. Rogak. Corrosion of Alloy 316 SCWO Reactor Tubing In Ammoniacal Sulfate Solution. in Nace Conference. 2009.
151. E. Asselin, A. Alfantazi, S. Rogak, Thermodynamics of the Corrosion of Alloy 625 Supercritical Water Oxidation Reactor Tubing in Ammoniacal Sulfate Solution. Corrosion - Houston Tx-, 2008, 64: 301–314.
152. P. Kritzer, Corrosion in high-temperature and supercritical water and aqueous solutions: a review. J Supercrit Fluids, 2004, 29: 1–29.
153. P. Kritzer, N. Boukis, E. Dinjus, Corrosion of alloy 625 in aqueous solutions containing chloride and oxygen. Corrosion the Journal of Science & Engineering, 1998, 54: 824–834.
154. Y. Behnamian, A. Mostafaei, A. Kohandehghan, B. S. Amirkhiz, D. Serate, Y. Sun, S. Liu, E. Aghaie, Y. Zeng, M. Chmielus, A comparative study of oxide scales grown on stainless steel and nickel-based superalloys in ultra-high temperature supercritical water at 800 °C. Corros Sci, 2016, 106: 188–207.
155. H. Kim, D. B. Mitton, R. M. Latanision, Corrosion behavior of Ni-base alloys in aqueous HCl solution of pH 2 at high temperature and pressure. Corrosion Science, 2010, 52: 801–809.
156. R. Dieckmann, T. O. Mason, J. D. Hodge, H. Schmalzried, Defects and cation diffusion in magnetite (III.) tracer diffusion of foreign tracer cations as a function of temperature and oxygen potential. Ber Bunsenges Phys Chem, 1978, 82: 778–783.
157. J. Q. Yang, S. Z. Wang, J. Lei, H. T. Xu, Y. S. Zhang, D. H. Xu. Corrosion Behavior of Candidate Materials for Supercritical Water Oxidation Reactor for Sewage Sludge Processing Plants. in Solid State Phenomena. 2018. Trans Tech Publ.
158. E. Dinjus, P. Kritzer, N. Boukis, Review of the corrosion of nickel-based alloys and stainless steels in strongly oxidizing pressurized high-temperature solutions at subcritical and supercritical temperatures. Corrosion (Houston, TX, U S), 2000, 56: 1093–1104.
159. S. Teysseyre, Z. Jiao. Corrosion of austenitic alloys in supercritical water. J Volcanol Geotherm Res, 2005, 120: 235–247.
160. M. Sun, X. Wu, Z. Zhang, E.-H. Han, Analyses of oxide films grown on Alloy 625 in oxidizing supercritical water. J Supercrit Fluids, 2008, 47: 309–317.
161. K. H. Chang, S. M. Chen, T. K. Yeh, J. J. Kai, Effect of dissolved oxygen content on the oxide structure of Alloy 625 in supercritical water environments at 700 °C. Corros Sci, 2014, 81: 21–26.
162. K. H. Chang, J. H. Huang, C. B. Yan, T. K. Yeh, F. R. Chen, J. J. Kai, Corrosion behavior of Alloy 625 in supercritical water environments. Prog Nucl Energy, 2012, 57: 20–31.
163. N. McIntyre, D. Zetaruk, D. Owen, X-Ray photoelectron studies of the aqueous oxidation of Inconel-600 alloy. J Electrochem Soc, 1979, 126: 750–760.
164. A. Machet, A. Galtayries, S. Zanna, L. Klein, V. Maurice, P. Jolivet, M. Foucault, P. Combrade, P. Scott, P. Marcus, XPS and STM study of the growth and structure of passive films in high temperature water on a nickel-base alloy. Electrochim Acta, 2004, 49: 3957–3964.
165. X. Tang, S. Wang, L. Qian, Y. Li, Z. Lin, D. Xu, Corrosion behavior of nickel base alloys, stainless steel and titanium alloy in supercritical water containing chloride, phosphate and oxygen. Chem Eng Res Des, 2015, 100: 530–541.
166. H. Kim, D. Mitton, R. Latanision, Stress corrosion cracking of Alloy 625 in pH 2 aqueous solution at high temperature and pressure. Corrosion, 2011, 67: 35002–35002.
167. K. T. W. B.P. Somerday, R.W. Bradshaw, Environment-assisted failure of alloy C-276 burst disks in a batch supercritical water oxidation reactor. Engineering Failure Analysis, 2006, 13: 80–95.
168. X. Gao, X. Wu, Z. Zhang, H. Guan, E. Han, Characterization of oxide films grown on 316L stainless steel exposed to H2O2-containing supercritical water. Journal of Supercritical Fluids, 2007, 42: 157–163.

169. Y. Li, S. Wang, P. Sun, J. Yang, X. Tang, D. Xu, Y. Guo, J. Yang, D. D. Macdonald, Investigation on early formation and evolution of oxide scales on ferritic–martensitic steels in supercritical water. Corrosion Science, 2018, 135: 136–146.
170. Y. Li, S. Wang, J. Yang, D. Xu, Y. Guo, L. Qian, W. Song, Corrosion characteristics of a nickel-base alloy C-276 in harsh environments. Int J Hydrogen Energy, 2017, 42: 19829–19835.
171. Y. H. Li, S. Z. Wang, L. S. Wang, P. F. Yu, J. Q. Yang, Corrosion Behavior of Austenitic Steel 304 in Nearcritical Aqueous Solutions. Advances in Engineering Research, 2015, 45: 189–192.
172. Y. Yi, B. Lee, S. Kim, J. Jang, Corrosion and corrosion fatigue behaviors of 9cr steel in a supercritical water condition. Materials Science and Engineering: A, 2006, 429: 161–168.
173. D. D. Macdonald, X. Guan, Volume of Activation for the Corrosion of Type 304 Stainless Steel in High Subcritical and Supercritical Aqueous Systems. Corrosion, 2009, 65: 427–437.
174. N. Q. Zhang, Z. L. Zhu, H. Xu, X. P. Mao, J. Li, Oxidation of ferritic and ferritic-martensitic steels in flowing and static supercritical water. Corrosion Science, 2016, 103: 124–131.
175. J. Bischoff, A. T. Motta, EFTEM and EELS analysis of the oxide layer formed on HCM12A exposed to SCW. Journal of Nuclear Materials, 2012, 430: 171–180.
176. Z. Dong, Z. Liu, M. Li, J.-L. Luo, W. Chen, W. Zheng, D. Guzonas, Effect of ultrasonic impact peening on the corrosion of ferritic-martensitic steels in supercritical water. Journal of Nuclear Materials, 2015, 457: 266–272.
177. L. Tan, X. Ren, K. Sridharan, T. R. Allen, Effect of shot-peening on the oxidation of alloy 800H exposed to supercritical water and cyclic oxidation. Corrosion Science, 2008, 50: 2040–2046.
178. Y. H. Li, S. Z. Wang, P. P. Sun, X. D. Li, Research on a Surface Shot Peeling Process for Increasing the Anti-Oxidation Property of Super304H Steel in High-Temperature Steam. Advanced Materials Research, 2014, 908: 77–80.
179. Y. H. Li, S. Wang, X. D. Li, M. M. Ren, Analysis on the Structure and Mechanism of Cracking and Spalling for Super304H Steel Oxide Films in High-Temperature Steam. Advanced Materials Research, 2014, 908: 72–76.
180. P. Kofstad, High Temperature Corrosion. (London and New York, NY, USA), 1988, 40: 1331–1347.
181. R. Viswanathan, K. Coleman, J. Shingledecker, J. Sarver, G. Stanko, M. Borden, W. Mohn, S. Goodstine, I. Perrin, Boiler materials for ultrasupercritical coal power plants, in US Department of Energy, Report No. DE-FG26-01NT41175. 2006.
182. R. F. Wang, Z. Q. Dong, J. L. Luo, W. X. Chen, W. Y. Zheng, D. Guzonas, Stability of MgO-, CeO2- and SiO2-doped Cr2O3 ceramics in high-temperature supercritical water. Corrosion Science, 2014, 82: 339–346.
183. Z. Shen, L. Wu, L. Zhang, Z. Duan, Y. Sun, Corrosion Behavior of Nickel Base Alloy 800H in High-temperature and High-pressured Water. Corrosion Science and Protection Technology, 2014, 26: 113–118.
184. L. Tan, X. Ren, K. Sridharan, T. R. Allen, Corrosion behavior of Ni-base alloys for advanced high temperature water-cooled nuclear plants. Corrosion Science, 2008, 50: 3056–3062.
185. X. Ren, K. Sridharan, T. R. Allen, Corrosion Behavior of Alloys 625 and 718 in Supercritical Water. Corrosion, 2007, 63: 603–612.
186. P. Xu, L. Y. Zhao, K. Sridharan, T. R. Allen, Oxidation behavior of grain boundary engineered alloy 690 in supercritical water environment. Journal of Nuclear Materials, 2012, 422: 143–151.
187. W. Znamirowski, F. Gesmundo, S. Mrowec, M. Danielewski, K. Godlewski, F. Viani, THE SULFIDATION OF MANGANESE AT LOW SULFUR PRESSURES AT 700-950-DEGREES-C. Oxidation of Metals, 1991, 35: 175-198.
188. K. Przybylski, T. Narita, W. W. Smeltzer, Kinetics and Mechanism of High-Temperature Sulfidation of an Fe-23.4Cr-18.6Al Alloy in H_2S-H_2 Atmospheres. Oxidation of Metals, 1992, 38: 1–32.

189. B. V. Gamson, K. M. Watson, National Petroleum News 1944, 36: 623.
190. A. Kruse, Hydrothermal biomass gasification. The Journal of Supercritical Fluids, 2009, 47: 391–399.
191. Y. Li, S. Wang, X. Tang, D. Xu, Y. Guo, J. Zhang, L. Qian, Effects of Sulfides on the Corrosion Behavior of Inconel 600 and Incoloy 825 in Supercritical Water. Oxidation of Metals, 2015, 84: 509–526.
192. W. Kai, M. T. Chang, C. D. Liu, T. W. Lee, Corrosion behavior of Y-Al alloys in a H-2/H2S/H2O gas mixture at 800–950 degrees C. Oxidation of Metals, 2002, 57: 523-547.
193. H. Xu, M. G. Hocking, P. S. Sidky, Sulfidation Oxidation Behavior of Alloy-800H in SO2-O2 and H2-H2S-CO-CO2 Atmospheres. Oxidation of Metals, 1994, 41: 81–101.
194. H. Li, W. Chen, Effect of Sulfur Partial Pressures on Oxidation Behavior of Fe-Ni-Cr Alloys. Oxidation of Metals, 2012, 78: 103–122.
195. M. Schulte, A. Rahmel, M. Schutze, The sulfidation behavior of several commercial ferritic and austenitic steels. Oxidation of Metals, 1998, 49: 33–70.
196. D. Rodriguez, A. Merwin, D. Chidambaram, On the oxidation of stainless steel alloy 304 in subcritical and supercritical water. J Nucl Mater, 2014, 452: 440–445.
197. C. Kai Hsiang, C. Shih Ming, Y. Tsung Kuang, K. Ji Jung, Effect of dissolved oxygen content on the oxide structure of Alloy 625 in supercritical water environments at 700 degC. Corrosion Science, 2014, 81: 21–6.
198. H. L. Hu, Z. J. Zhou, L. Liao, L. F. Zhang, M. Wang, S. F. Li, C. C. Ge, Corrosion behavior of a 14Cr-ODS steel in supercritical water. Journal of Nuclear Materials, 2013, 437: 196–200.
199. S. Guillou, C. Cabet, C. Desgranges, L. Marchetti, Y. Wouters, Influence of Hydrogen and Water Vapour on the Kinetics of Chromium Oxide Growth at High Temperature. Oxidation of Metals, 2011, 76: 193–214.
200. D. M. Gorman, R. L. Higginson, H. Du, G. McColvin, A. T. Fry, R. C. Thomson, Microstructural Analysis of IN617 and IN625 Oxidised in the Presence of Steam for use in Ultra-Supercritical Power Plant. Oxidation of Metals, 2013, 79: 553–566.
201. P. Berthod, L. Aranda, S. Mathieu, M. Vilasi, Influence of Water Vapour on the Rate of Oxidation of a Ni-25wt.%Cr Alloy at High Temperature. Oxidation of Metals, 2013, 79: 517–527.
202. D. J. Kim, H. C. Kwon, P. K. Hong, Effects of the solution temperature and the pH on the electrochemical properties of the surface oxide films formed on Alloy 600. Corros Sci, 2008, 50: 1221–1227.
203. Z. Szklarska-Smialowska, D. Grimes, J. Park, The kinetics of pit growth on alloy 600 in chloride solutions at high temperatures. Corros Sci, 1987, 27: 859–867.
204. K. Osseo-Asare, Application of Activity-Activity Diagrams to Ammonia Hydrometallurgy II. The Copper-, Nickel-, Cobalt-Ammonia-Water Systems at Elevated Temperatures,". Journal of Macro-molecular Science Part B Physics, 1981: 359–369.
205. V. Kolarik, B. Michelfelder, M. Wagner, Corrosion of Alloys 625 and Pure Chromium in Cl-Containing Fluids During Supercritical Water Oxidation (SCWO). Eighteenth-Century Life, 1999, 23: 30–45.
206. H. Kim, D. B. Mitton, R. M. Latanision, Effect of pH and Temperature on Corrosion of Nickel-Base Alloys in High Temperature and Pressure Aqueous Solutions. J Electrochem Soc, 2010, 157: C194.
207. P. Kritzer, N. Boukis, E. Dinjus, Factors controlling corrosion in high-temperature aqueous solutions: a contribution to the dissociation and solubility data influencing corrosion processes. J Supercrit Fluids, 1999, 15: 205–227.
208. P. Kritzer, N. Boukis, E. Dinjus, The corrosion of nickel-base alloy 625 in sub- and supercritical aqueous solutions of oxygen: A long time study. J Mater Sci Lett, 1999, 18: 1845–1847.
209. Z. Zhu, H. Xu, D. Jiang, G. Yue, B. Li, N. Zhang, The role of dissolved oxygen in supercritical water in the oxidation of ferritic-martensitic steel. Journal of Supercritical Fluids the, 2015, 108: 17–21.

210. J. W. Tester, H. R. Holgate, F. J. Armellini, P. A. Webley, W. R. Killilea, G. T. Hong, H. E. Barner, Supercritical water oxidation technology, in Emerging Technologies in Hazardous Waste Management III. 1993, American Chemical Society: Washington. p. 35–76.
211. V. Vadillo, M. B. García-Jarana, J. Sánchez-Oneto, J. R. Portela, E. J. M. de la Ossa, Supercritical water oxidation of flammable industrial wastewaters: economic perspectives of an industrial plant. J Chem Technol Biotechnol, 2011, 86: 1049–1057.
212. D. Xu, S. Wang, J. Zhang, X. Tang, Y. Guo, C. Huang, Supercritical water oxidation of a pesticide wastewater. Chem Eng Res Des, 2015, 94: 396–406.
213. X. Tang, S. Wang, L. Qian, Y. Li, Z. Lin, D. Xu, Y. Zhang, Corrosion behavior of nickel base alloys, stainless steel and titanium alloy in supercritical water containing chloride, phosphate and oxygen. Chem Eng Res Des, 2015, 100: 530–541.
214. W. Kuang, X. Wu, E.-H. Han, J. Rao, The mechanism of oxide film formation on Alloy 690 in oxygenated high temperature water. Corros Sci, 2011, 53: 3853–3860.
215. S. Bsat, X. Huang, Corrosion behaviour of IN625 in superheated steam at 800 °C. Corros Eng, Sci Technol, 2016, 51: 321–328.
216. W. Kuang, E.-H. Han, X. Wu, J. Rao, Microstructural characteristics of the oxide scale formed on 304 stainless steel in oxygenated high temperature water. Corros Sci, 2010, 52: 3654–3660.
217. J. Yang, S. Wang, X. Tang, Y. Wang, Y. Li, Effect of low oxygen concentration on the oxidation behavior of Ni-based alloys 625 and 825 in supercritical water. The Journal of Supercritical Fluids, 2018, 131: 1–10.
218. T. Takada, H. Omura, Y. Akai, N. Hara, N. Saito, Y. Tsuchiya, T. Takada, H. Omura, Y. Akai, N. Hara, Corrosion performance of metals for supercritical water, oxidation-utilized organic waste-processing reactors. Corrosion (Houston, TX, U S), 2006, 62: 383–394.
219. W. G. Cook, R. P. Olive, Pourbaix diagrams for chromium, aluminum and titanium extended to high-subcritical and low-supercritical conditions. Corros Sci, 2012, 58: 291–298.
220. P. Kritzer, N. Boukis, E. Dinjus, Corrosion of alloy 625 in high-temperature, high-pressure sulfate solutions. Corrosion (Houston, TX, U S), 1998, 54: 689–699.
221. X. Y. Tang, S. Z. Wang, L. L. Qian, Y. H. Li, Z. H. Lin, D. H. Xu, Y. P. Zhang, Corrosion behavior of nickel base alloys, stainless steel and titanium alloy in supercritical water containing chloride, phosphate and oxygen. Chemical Engineering Research & Design, 2015, 100: 530–541.
222. Y. Li, D. D. Macdonald, A Point Defect Model for the Growth of oxide Scales on Metals and Alloys in Supercritical Aqueous Systems. 2018.

Chapter 7
Study on Salt Deposition and Crystallization Properties in Sub/Supercritical Water

7.1 Salt Deposition in a Supercritical Water Oxidation System

7.1.1 Deposition and Hazard of Inorganic Salts

Above the critical point (T = 374.15 °C, P = 22.1 MPa), water exists as a single phase and has analogous transportation properties of gas and solvent properties like liquid, which is known as supercritical water (SCW) [1, 2]. SCW is a non-polar solvent that can be completely miscible with organic compounds and gases such as hydrogen, nitrogen, and oxygen. However, it is a poor solvent for polar materials such as inorganic salts and other ionic compounds [3, 4], because the dielectric constant of SCW sharply decreases dozens or even hundreds of times than that of the water under ambient conditions, exhibiting non-polarity [2, 5]. In relation to viscosity, it is small enough to carry out reactions at very high velocities under supercritical conditions [6]. Therefore, SCW is an admirable medium that has been utilized in SCWO processes for the degradation of wastes such as sewage sludge, chemical weapons, and other refractory organic compounds [7]. Meaningfully, original heavy metals such as Mn, Zn, and Cu are also possible to be recovered by SCWO reactions in the form of insoluble oxides and salts [8], which exhibits practical application for the treatment of refractory waste. Besides, SCWO offers some other potential advantages such as: providing easier salts separation due to their extremely low solubilities in SCW, a very short residence time (usually less than several minutes), achieving heat self-sufficiency and energy saving with a relatively low concentration of organic matter feedstock due to exothermic reactions [9]. Therefore, SCWO technology is an effective and promising destruction method for organic wastewater and sewage sludge, or an alternative to conventional methods.

However, SCWO is seriously hindered by corrosion and plugging problems of the reactor [10, 11]. Reactor materials are particularly susceptible to corrosion due to dissolved salts in water at high temperature, high density, and high oxidation

S. Wang et al., *Supercritical Water Processing Technologies for Environment, Energy and Nanomaterial Applications*,
https://doi.org/10.1007/978-981-13-9326-6_7

SCWO conditions. Methods that manage corrosion issues in SCWO systems have already been reviewed elsewhere [12]. Salt deposition is a direct consequence of poor solubilities of most salts in SCW and its negative impacts broadly include three aspects. First of all, these precipitated salts are prone to adhere to the reactor gradually forming a salt layer there, reducing the heat transfer capability across the reactor wall [13]. Secondly, a dead micro-environment is created between the salt layer and the reactor inner wall, where considerable corrosion quite possibly occurs [12]. Thirdly, the accumulation of precipitated salts may also lead to an increased pressure drop, plugging of the reactor or associated pipelines, even the shutdown of the whole expensive SCWO equipment if it is not controlled well. These drawbacks of salt deposition seriously reduce the reliability and operational economy of the SCWO apparatus [14]. In order to overcome these unfavorable influences on commercial application of SCWO, plenty of methods and reactor configurations have been developed over the past three decades. In 2004, Hodes et al. [15] and Marrone et al. [14] systematically reviewed and summarized deposition problems and solutions in commercial/full-scale applications of SCWO. However, it seems that no any method or reactor design used at the laboratory or commercial level exhibited better than all the others, or was fully and successfully applied to the operation of a long-term continuous flow system.

With the remarkable development of SCWO, the research on salt deposition and its control means have been updated in the past decade. Thus, systematical summarizations of current research status concerning salt deposition problems are significant. This present article aims to review and summarize the current research status of salt deposition problems, such as research methods, salt types, solubilities, deposition properties, separation performance, deposition mechanism, and salt deposition control techniques.

7.1.2 Salt Deposition Theory

Under supercritical water conditions, the process of salt deposition on the inner wall of the reactor can be roughly divided into three stages: saturated precipitation, crystal growth, and deposition. However, the mechanism of salt deposition is still unclear. Hodes et al. [15] reported the basic knowledge related to salt deposition in the SCWO reactor and introduced the phase diagram of the brine system under supercritical water conditions and the complex heat transfer model. LaJeunesse et al. [16] explored the deposition characteristics of salt in a supercritical water oxidation reactor and quantitatively grasped the salt deposition process by adjusting the salt formation condition. Shan and the other [17] combined water molecule simulation theory and aggregated water molecular structure model theory, suggesting that the root cause of the sharp decline of salt solubility in supercritical water with temperature rise is attributed to a strong charge dipole effect. Studies by Fedotova [18] have shown that temperature and pressure have opposite effects on ion hydration and ionic polymerization of NaCl. Hodes et al. [19] indicated that the soluble salt is transferred by molecular diffusion to the salt layer and solution interface at the wall, where it is heterogeneously nucleated, and a considerable proportion of salt deposits are found in the porous salt layer. The existing research on salt deposition theory is still not deep in salt crystallization and deposition mechanism. The study of sedimentary kinetics has been rarely reported, and the evolution of salt from micro-nucleation to macroscopic deposition in supercritical water has not been fully revealed.

7.1.3 Salt Types

Salts found in SCWO mostly originate from several sources. The overall oxidation of organic wastewater containing heteroatoms such as S, P, and Cl, results in an acidic solution [15, 16] and/or produces some inorganic salts in the presence of alkaline compounds. To minimize corrosion associated with the acidic solution above, neutralizing it with a base is adopted, so large amounts of salts are produced [20]. It can not also be ignored that the wastewater feedstock inherently contains some salts. Therefore, the salts in this article mainly focus on those formed from the overall oxidation process, neutralization reaction, and feedstock itself.

Table 7.1 summarizes the classification of salts in SCWO of wastewater or sewage sludge. Salt solids carried by the feedstock commonly have very low solubility over the entire range of from subcritical to supercritical conditions. These solids exist in the form of sand, clay, and rust, hardly sticking to reactor surfaces compared with nascent salt particles, so they are not responsible for scaling or plugging [14], and are defined as insoluble salts. Correspondingly, soluble salts refer to the salts that exhibit a high solubility in water at the ambient temperature, but undergo a dramatic decrease in the solubility when the temperature is heated to supercritical conditions, due to their small dielectric constants in SCW [21]. These salts seem to be much stickier and are prone to adhere to the hot surface of the reactor. Hence, the phenomenon of scaling or plugging easily takes place precisely because of low solubilities of these salts in SCWO.

Phase equilibrium relationships have an important effect on the morphology and size of salt particles formed in SCWO [22]. To better understand the deposition process of salt-type solids, extensive information on soluble salts is required in SCWO, including dissolution characteristics, phase behavior, and other thermal physical properties. These salts typically consist of carbonates, sulfates, nitrates and phosphates, etc. The salt-water system can be classified into the binary system, ternary system, even quaternary or higher system according to phase behavior data [15, 23]. The binary salt-water system (i.e., one salt dissolved in water) can be divided into two main categories (i.e., type 1 salt and type 2 salt), depending on whether the critical behavior is observed in saturated solutions. The type 1 salt does not exhibit critical behavior in saturated solutions [15, 24], and its diluted solution forms a saturated liquid and a saturated vapor phase under supercritical conditions. However, the type 2 salt displays critical behavior in saturated solutions, and its

Table 7.1 Classification of salts in SCWO of wastewater or sewage sludge

Category	Criteria	Characteristics	Typical examples
Insoluble salts	Solubility	These salts have very low solubility from subcritical to supercritical conditions	Sand, clay, and rust, etc. [14]
Soluble salts		These salts exhibit a high solubility at the ambient temperature, but undergo a dramatic decrease in solubility when the temperature is heated to supercritical conditions	Nascent salts: carbonates, sulfates, nitrates, and phosphates [15, 21]
Type 1 salt	Phase behavior	Type 1 salt does not exhibit critical behavior in saturated solutions, its diluted solution forms a saturated liquid and a saturated vapor phase under supercritical conditions	K_3PO_4, KCl, KNO_3, NaCl, etc. [15, 23]
Type 2 salt		Type 2 salt displays critical behaviors in saturated solutions, and its diluted solution forms a supercritical fluid phase and a solid salt phase	Na_2SO_4, Na_2CO_3, etc. [25, 30]

diluted solution (e.g., Na_2SO_4 or Na_2CO_3) forms a supercritical fluid phase and a solid salt phase [15, 25]. Notably, the type 1 salt and the type 2 salt have more detailed classifications involving subtypes a, b, c, d, and e, respectively, according to their specific characteristics of phase behavior [26, 27]. Similarly, the binary salt-water system can also be classified into two types based on solubility near the water's critical point [28]. Valyashko [27, 29] classified the ternary salt-salt-water system by the behavior of the bounding binary salt-water system. However, the ternary salt system is very complex and variable due to the mixed components of various type 1 salts, type 2 salts and their subsystems, as well as the combination of type 1 and type 2 salts [15]. It is necessary to obtain more phase behavior data to analyze the system in depth.

7.2 Research Methods

The formation of salt deposition is a transition process from micro- to macro-levels in SCWO. Considering that very extensive information relating to brine mixtures is difficult to be obtained under harsh conditions of high temperature and pressure, special methods are employed to study salt deposition behavior in SCW. Currently, salt deposition problems have been investigated by many research institutions such as the University of Valladolid in Spain [31], Stanford Institute for International Studies in California, the Boreskov Institute of Catalysis in Russia [32], the University of British Columbia in Canada [33], and Xi'an Jiaotong University in China [34]. Generally, major research methods can be divided into three categories, i.e., experimental methods, numerical simulation, and a combination of them.

7.2.1 Experimental Methods

Experiments have been performed on salt deposition by many researchers in their various devices at harsh SCWO conditions. Technological improvements in materials and measuring instruments strongly promote the study of salt deposition. Table 7.2 summarizes some experimental studies of salt deposition in SCW. It can be noticed that experiments are usually conducted in the range of 18–30 MPa and 200–600 °C. The experimental apparatus is commonly a continuous flow plant or a

Table 7.2 Summary of some experimental studies of salt deposition in SCW

Investigators	Research contents	Experimental techniques
Peterson et al. [36]	Salt precipitation of salt-water mixtures (K_3PO_4 and Na_2SO_4). Temperature: 370–450 °C (bulk fluid). Pressure: 25 MPa	Continuous reverse-flow separator, normal-phase dynamic images by neutron radiography
Makaev et al. [37]	Phase diagram and salt deposition of ternary salt mixtures. Temperature: 350–600 °C. Pressure: 26.5–33.4 MPa (270–340 kg/cm^2)	Continuous flow tubular type reactor, online digital monitoring (temperature and pressure)
Schubert et al. [20, 38, 39]	Salt precipitation and separation of type 1 salts, type 2 salts and their mixtures. Temperature: 350–450 °C (bulk fluid). Pressure: 30 MPa	Continuous reverse-flow salt separator, titanium grade 5 visual experimental devices, concentration analysis by ion chromatography
Príkopský et al. [13]	Deposition characteristics of the Na_2SO_4 solution. Temperature: 300–500 °C (bulk fluid). Pressure: 25 MPa	Transpiring wall reactor with a hydrothermal flame, salt deposition measurements by electric conductivity and ionic chromatography
Haroldsen et al. [40]	Deposition characteristics of the Na_2SO_4 solution. Temperature: 450–550 °C (bulk fluid). Pressure: 25–26 MPa	Transpiring wall reactor, platelet liner using stainless steel 304, online analysis (conductivity, pH, and absorption spectra)
Karakama et al. [35]	Deposition and transport of magnetite particles in SCW. Temperature: 200–400 °C. Pressure: 23–24 MPa	Once-through flow apparatus, stainless steel 316L, online thermal monitoring, neutron ray technology
Hegde and Hicks [23]	Salt distribution and transport in liquid, vapor, and supercritical fluid. Temperature: 103–377 °C. Pressure: 0–18.2 MPa	Batch test cell, 4.3 cm^3, Inconel 625, online imaging diagnostics by video camera
Kawasaki et al. [41]	Flow and deposition characteristics of aqueous salt solutions (e.g., type 1 salts and type 2 salts). Temperature: 350–600 °C. Pressure: 18–25 MPa	Vertical tube reactor, 7.24 cm^3, Inconel 625, horizontal tube reactor, 314 cm^3, Inconel 625 Batch system, 650 mL, Hastelloy C-276, concentration measurement by ion chromatography

batch plant, and it is typically made of high-temperature corrosion resistance material. Use of measuring and analyzing means includes visual technology, neutron ray technology, online and offline diagnostics, and a combination of two or more technologies together. Experimental experiences indicate that multiple techniques are required to characterize the deposition process if the underlying rate-limiting steps are not known before [35].

7.2.2 Numerical Simulation

Research methods, based on experiments, are seriously restricted by extremely harsh SCWO reaction conditions, so it is difficult to directly gain comprehensive fluid information in SCWO [42]. Nevertheless, numerical simulation has great advantages in the calculation of fluid properties and the salt deposition process. Consequently, detailed information can be obtained by simulation, which especially helps reveal the mechanisms of salt deposition and design a new reactor.

Different numerical models are built to calculate and predict various salt deposition processes, and obtained information involves solubilities, nucleation and growth properties of salt particles, salt separation, and salt deposition rate. For instance, Masoodiyeh et al. [43] adopted three thermodynamic models to calculate the solubilities of six inorganic salts at high temperature and pressure conditions, and the corrected model had a good agreement with experimental data. Molecular dynamics simulation seemed to be a valid method to investigate the nucleation and growth properties of nanoparticles [44, 45]. Hovland et al. [46, 47] also utilized molecular models to examine salt precipitation theory at special supercritical conditions attained by seawater at depths exceeding 2800 m. Solubilities and deposition characteristics of Na_2SO_4 and K_2SO_4 in SCW were discussed, and a natural convection model was established to determine salt deposition rates [19]. Information regarding the salt separation process near and above the critical point of water was obtained by means of numerical model [48]. However, the model neglected salt deposition probably occurring in the porous salt layer, although it took deposition rates of the salt layer covering the internal cylinder surface into account. Deposition modeling is mostly limited by unavailable information on the properties of salt solutions and mass transfer in supercritical fluids. Therefore, many scholars have no choice but to ignore this information in their models, resulting in greater differences from the actual salt deposition process. In spite of this, the simulation method is still a powerful tool to offer theoretical guidance for optimum reactor designs avoiding reactor plugging induced by salt deposition [49].

7.2.3 Combination of Experimental and Numerical Methods

In order to recognize more comprehensive information on salt deposition, several researchers attempt to investigate the salt deposition process in SCWO using the combination method of experimentation and numerical simulation. As it is expected, this integrated approach exhibits certain advantages in both of them. For research on the salt solubility, finite points were attained by experimental methods to verify and correct the theoretical model [50], and then more values can be easily calculated through this established model. Similarly, Chan et al. [51] reported that experimental tests were performed to obtain a judgment of salt formation and deposition in a SCWO reactor, and then a mathematical model was established to predict the reactor pressure rising time was a function of the reactor axial temperature and the salt solubility. In relation to the formation of salt precipitation, the microscopic process of salt crystals entering into the solution could be observed with a liquid X-ray diffractometer. Moreover, molecular dynamics methods were employed to reveal structure changes during special ion dissociation and polymerization processes and demonstrate the changes of salt micro-cluster states in the dissolution process [52]. Therefore, simple experiment techniques can obtain some necessary information requirement for deposition models and verify the validity of these models. Correspondingly, numerical simulation is able to easily and comprehensively gain some information and further supplement experiment data. These advantages are more available especially for the multi-salt mixtures encountered in SCWO processes. The combination of the experiment and simulation studies is a desired method for salt deposition, which is favored and employed by more researchers.

7.3 Study on Salt Behavior Characteristics in Sub/ Supercritical Water

7.3.1 Salt Solubility in Sub/Supercritical Water

Saturated precipitation is an initial stage of salt deposition in SCWO, largely depending on the solubility of the single salt or their mixtures in SCW. Many studies concerning the salt solubility have been performed at certain conditions relevant to a common SCWO process. Figure 7.1 elucidates the solubilities of the single salt and their mixtures verse temperature at 25 MPa. As shown in Fig. 7.1, solubilities of both the type 1 salts (represented by NaCl and KCl) and the type 2 salts (represented by Na_2CO_3 and Na_2SO_4) sharply drop down due to the breaking of substantial hydrogen bonds from subcritical to supercritical conditions [15]. Specifically, the solubility of NaCl in SCW reduces to about 200 mg/L at 500 °C and 25 MPa [53]. For the type 2 salts, solubilities of Na_2CO_3 and Na_2SO_4 decrease to 20 mg/L and 0.3 mg/L at 440 °C and 25 MPa, respectively, which

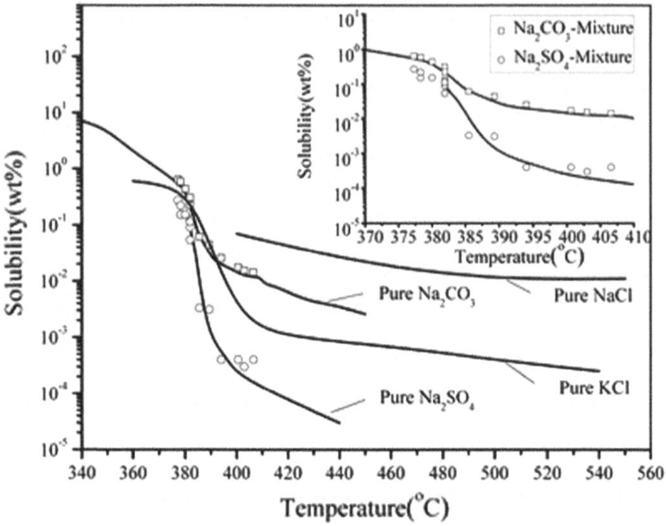

Fig. 7.1 Solubilities of the single salt and their mixtures verse temperature at 25 MPa. Na_2CO_3-mixture: solubility of Na_2CO_3 in the Na_2CO_3–Na_2SO_4 mixture; Na_2SO_4-mixture: solubility of Na_2CO_3 in the Na_2CO_3–Na_2SO_4 mixture. The plotting data come from Khan and Rogak [28] and Armellini and Tester [53]

decrease almost four orders of magnitude across the measuring temperature range of 300–440 °C. Solubilities of salts are too low to easily precipitate in SCWO. The trend of the Na_2CO_3–Na_2SO_4 mixture indicates that the solubility of Na_2CO_3 in the form of a mixture is almost same as that of the single binary salt-water system. However, Na_2SO_4 exhibits obvious difference between single salt and the mixture in the presence of Na_2CO_3 [28]. It may be explained that the addition of one type of salt leads to some subtle changes in solubilities of other salts at certain conditions [38], but this mechanism needs study in depth. Urusova and Valyashko [54] reported that solubilities of NaF and Li_2CO_3 displayed quite different trends from those of other alkali metals sulfates and carbonates. Therefore, solubilities of various salt mixtures can not be determined just by that of a given single salt.

The aforementioned solubility data are generally obtained by experimental methods, but numerical models are also useful tools to calculate relevant solubility values. Moreover, numerical simulations of some models have a good agreement with the experimental data [43, 55], which expands the investigation ranges of extreme temperatures and pressures that experimental method hard to realize.

7.3.2 Salt Deposition Properties in Sub/Supercritical Water

The salt deposition process can be divided into three stages: saturated precipitation, crystallization growth, and the deposition process. Research shows that each type of salt has its unique deposition performance in SCWO, which is mainly determined by operation conditions and the nature of the salt solution or its mixture. In respect of operation conditions, Fedotova [18] pointed out that temperature and pressure had positive effect on hydration and polymerization of NaCl ions. At the temperature of 400 °C, Kawasaki et al. [41] reported that these salts such as NaCl, KCl, K_2CO_3, $KHCO_3$, and $CaCl_2$ impossibly caused serious salt deposition problems. Additionally, except NaCl, these salts are able to flow without plugging or pressure fluctuation at 25 MPa and 450–600 °C. Diameters of salt particles in SCW vary from a few microns to several hundreds of nanometers under various pH values and heat flux conditions [35]. Meanwhile, the salt deposition process can also be quantitatively controlled by regulating salt formation conditions in SCWO [16]. At certain operation conditions, the nature of the solution containing salt concentration and component seems to be very important to the deposition process. Kawasaki et al. [41] tested flow performance of salt solutions with low flow rates of 0.8–65 cm/s and high flow rates of 45–196 cm/s in a tube apparatus at 350–600 °C and 25 MPa, and found that these ions in the NaCl solution were almost completely recovered in the effluent at 400 °C. They also noticed that NaCl tended to agglomerate, but KCl and $CaCl_2$ were prone to disperse and deposit inside the vessel at 500 °C in a batch autoclave. The salt deposition rate, as an important index describing the deposition process, was tested by Hodes et al. [56]. They argued that the salt deposition rates of both Na_2SO_4 and K_2SO_4 were on the order of 1 g/min, and the transportation of salts was mainly dominated by natural convection. Therefore, the salt deposition process is very complex and affected by various factors including heat transfer, mass transfer, and solubility, etc. These factors may have a coupling influence on salt deposition [57], which greatly increases complexity and difficulty of research. Furthermore, salt deposition problems should be discriminatingly treated for various organic wastewaters in their SCWO processes.

7.3.3 Salt Separation Properties in Sub/Supercritical Water

Great majorities of inorganic salts may be separated from SCW due to their very low solubilities. For example, copper ions can be separated in the form of oxides with the separation efficiency of 99.8% at 500 °C [58]. It is also possible to efficiently recovery silicon [59], dysprosium [60], and chromium [61] in the form of their oxides from sludge or aqueous solutions in a batch reactor, and the higher temperature plays a positive role in the SCWO process. Similarly, phosphates as type 1 salts are successfully recovered via a special reverse-flow tank separator at a relatively low temperature (390 °C), whereas the separation of other type 1 salts such as $NaNO_3$ and KNO_3 requires a higher temperature (415 °C [20]). In addition to the impact of operation parameters (e.g., temperature and pressure), the separation efficiency is also significantly influenced by the nature of initial wastewater components. For most type 1 salts, their separation efficiencies are in the range of 80–97% [20]. However, it is impossible to recover a single type 2 salt as a concentrated brine with a salt separator [38]. The observation of phosphorus migration and transformation in SCW shows that organic phosphorus almost completely converts to phosphates entering into solid residue, and the remaining phosphorus (mainly in the form of HPO_4^{2-}) in the solution is almost soluble [62]. Fauvel et al. [63] also pointed out that phosphorus could be completely recovered as phosphoric acid in effluent and no obvious corrosion occurred in the reactor.

Schubert et al. [38] presented that the ternary mixture of two type 1 salts (such as KH_2PO_4–K_2HPO_4–H_2O or $NaNO_3$–K_2CO_3–H_2O) exhibited an analogous separation characteristic to the binary mixture of type 1 salts. However, separation performance of these mixtures differs from that of the corresponding single salt solution. As for the binary type 2 salt-water mixtures of Na_2CO_3, Na_2SO_4, and K_2SO_4 eventually depositing inside the salt separator, it is feasible to recovery concentrated salts in the case of the solution containing type 1 salts. The high separation efficiency (up to 95%) may also be achieved due to the possible formation of a certain amount of type 1 salts (such as K_2CO_3). Schubert et al. [38] also investigated the separation efficiencies of both the Na_3PO_4–Na_2SO_4–H_2O mixture and the Na_2SO_4–K_2SO_4–H_2O mixture with the same concentration of potassium, sodium, sulfate, and phosphate. They found that these two kinds of mixtures had the same trend in separation performance, indicating that a certain amount of type 1 salt is actually formed in solutions. However, the amount of type 1 salts in such a mixture depends on the mixture itself. Kruse et al. [64] described that salts could be caught by other salts via the second formation of a salt-rich aqueous phase (hydrothermal brine), and their experimental results offered supports for this point.

Therefore, current research on salt deposition is mainly concentrated on salt macroscopic properties such as solubility, recovery, and separation in SCW. It is impossible to determine the separation performance of a given salt mixture just according to the corresponding single salt because complex interaction exists during different salt mixtures; however, specific mechanisms are still unclear so far.

7.4 Study on Salt Crystallization and Deposition Mechanisms in Sub/Supercritical Water

Figure 7.2 indicates the influence factors analysis on salt deposition mechanisms. As illustrated in Fig. 7.2, salt deposition, especially occurring in an extremely harsh SCWO environment, experiences complex and variable processes, which is easily influenced by coupling effect of operation conditions and the inherent nature of salts in water. Analysis based on existing research results indicates that solubility, phase

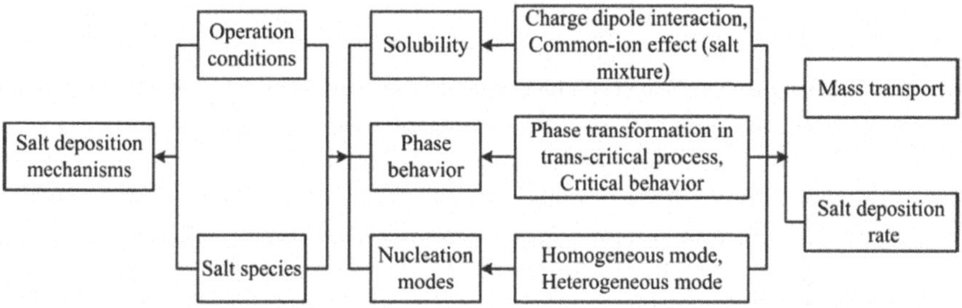

Fig. 7.2 Influence factors analysis on salt deposition mechanisms

behavior, and nucleation modes seem to be the key points to evaluate the entire salt deposition process. As mentioned before, the particular property of SCW leads to low solubilities of most salts in SCWO environments, and thus saturated precipitation and salt deposition take place inevitably. Some researchers point out that dramatic changes about the salt solubility with operation parameters (e.g., temperature and pressure) radically attribute to strong charge dipole interaction [17]. For the mixture containing various types of salt, Schubert et al. [20] proposed that adding certainly specific salts or ions into the raw salt solution probably led to some interesting shifts in solubility properties due to possible hydrolysis and polymerization relating to these new types of salts. This was called as common-ion effect, and relevant experiment data reported by them proved this finding. In any case, research on solubility properties of a single salt or mixture is fundamental to reveal salt deposition mechanisms.

Phase behavior further explains that thermo-physical properties comprising dissolution characteristics undergo significant changes in the trans-critical process. Three distinct regimes can be defined into subcritical regime, near-critical regime and supercritical regime in the investigation of salt deposition, each possessing respective characteristics. Hegde and Hicks [23] reported that both liquid and vapor phases coexist in the subcritical regime, but obvious distinctions in the vapor–liquid phase disappear gradually as the temperature exceeds the critical point. For the particular regime near the critical point, the thermo-physical properties of the two phases rapidly convert due to sharp reduction of energy required for transition between the two phases, but the phase distinction is still retained [23]. As a consequence, a slight change in temperatures and pressures possibly results in rapid phase transitions and solubility decrease. The transition of phase behavior induces the variations of thermo-physical properties, which essentially explains the evolving law of the salt solubility in the entire range. Moreover, various types of salts exhibit a relatively large gap of phase behavior, as described in Sect. 7.2.

Extremely low solubilities of most inorganic salts in SCW easily cause saturated precipitation along with the accumulation of new salts produced from oxidation reactions in SCWO. The specific precipitation process may be divided into a homogeneous mode and a heterogeneous mode based on the means that it occurs in the bulk fluid [23]. The expected homogeneous precipitation takes place uniformly in the bulk fluid and hardly adheres to the reactor's inner surface. Thus, it is possible to manage salt deposition only in the bulk fluid away from the reactor wall by homogeneous nucleation [65]. In addition, the formation of particulate deposition in the bulk fluid helps reduce the volume of pure salts compared with crystal deposition in which salts inevitably deposit on the reactor wall. It is because the flow fluid will remove parts of salts when the thickness of salt layer reaches a certain value and then continues a new deposition cycle under particulates fouling conditions [30], thus, particulates deposition helps dominate the thickness of salt deposition. The heterogeneous mode is dominated by concentration gradients in the bulk fluid. The concentration gradients are easily established in the presence of hot regions where local thermal gradients and higher temperatures make the fluid nature, such as the dielectric constant, different from that of the bulk fluid. Ultimately, the gap of concentration gradients promotes salt precipitation and results in a large proportion of salts deposition in the salt layer [19, 23]. The salt deposition process can be divided into two steps, i.e., mass transport and adhering to the wall, and the deposition rate depends on the net resistance of these two processes [35, 66]. Smith et al. [67] reported that the total salt deposition rate was the sum of the deposition rate occurring at the salt layer-solution interface and that inside the porous salt layer. The salt layer covering the inner wall is formed by that dissolved salts are transported to the interface between the salt layer and reaction fluid by molecular diffusion [48].

Apparently, the exact process of salt deposition for a given salt solution is investigated by many researchers, but no one single theory or idea is capable to provide a thorough explanation for specific deposition mechanisms. Besides, the prediction of salt deposition seems to be quite difficult and needs extensive information on salts' phase behavior in SCW. Therefore, in order to reveal the mechanisms of salt deposition, the entire evolution process of salts from microscopic nucleation to macroscopic deposition in SCW requires to be further explored from the analysis of the above influence factors.

7.5 Research on Prevention Control Technology of Salt Deposition in Sub/Supercritical Water

Salt precipitation is a direct result of the poor salt solubility in SCW. These salts are either originated from feedstock or released from the products of organics oxidation, which may precipitate and deposit inside SCWO equipments. In many traditional fields, some methods have been used to avoid or decrease salt deposition, including distillation, ions exchange, electronic dialysis, and reverse osmosis, etc. However, these methods are not suitable for the scale-up SCWO system due to the limitation of the low removal efficiency for complex feedstock components at harsh operation conditions. Measures of solving salt deposition problems in SCWO plants can be briefly summarized into special operation techniques and reactor designs. To the best of our knowledge, there are no further summarization and review regarding the progress of these two aspects after the report of Marrone et al. in 2004. Therefore, further systematical review based on previous reports is significant as a supplement of salt deposition information.

7.5.1 Source Control Technologies

In overcoming the problem of reactor clogging caused by salt deposition in SCWO, the most optimal and efficient method is to avoid the formation of solid salts from the source or to increase its solubility. One such method is to interfere with salt formation by the addition of additives. For example, Makaev et al. [37] can avoid the deposition of a second type of salt by adding a first type of salt (KCl, K_2SO_4, NaCl) to the initial solution of the second type of salt (K_2SO_4, Na_2SO_4). The study found that when the salt precipitated in SCWO is mainly calcium sulfate and aluminum phosphate, in order to prevent the precipitation of these salts, Na_2PO_4 can be added to the fluid to interfere with Ca, Al, and Mg ions to form a solid salt [68]; when the SCWO feed contains sulfur and chlorine, McBrayer et al. [69] suggested the use of $NaNO_3$, NaOH, KNO_3, or KOH as possible additives. Kawasaki et al. [41] considered that $KHCO_3$, K_2CO_3, or KOH can be used as a neutralizing agent, while NaOH is not suitable as a neutralizing agent due to its deposition characteristics. Another way to avoid solid salt formation is to increase the density of supercritical water by increasing the pressure, thereby increasing the solubility of most salts, because the increase in density repairs the hydrogen bonding and polarity effects in some supercritical waters. Therefore, many salts are partially dissolved at 69–137.9 MPa, for example, the solubility of NaCl at 70 MPa and 80 °C can be greater than 200 g/L [14]. However, considering that increasing the pressure will lead to an increase in the solubility of corrosive substances, and the solubility of protective oxides will increase, which will accelerate the corrosion rate of the reactor [11], so Kritzer and Dinjus [10] proposed that the solution in the reactor must be guaranteed. The density should be less than 200 kg/m. When salt precipitation occurred, Wofford et al. [65] controlled the operating conditions to cause the salt to reflect the homogeneous nucleation of the mainstream away from the reactor wall. When the salt is further deposited on the wall of the reactor, Khan [30] reported that particle deposition helps to reduce the amount of pure salt deposition compared to crystal deposition because of the instability of salt deposition in the presence of particulate scale. Once the salt deposit reaches a certain thickness, the flowing fluid will remove part of the salt deposit and then continue the new deposition cycle.

Therefore, the dissolution and deposition characteristics of the salt in supercritical water can be changed by the additive, which interferes with the formation of salt deposition. Although increasing the pressure helps to increase the solubility of the salt, it can increase reactor corrosion and increase equipment investment. When the salt is precipitated, it can be uniformly deposited at a position away from the wall surface of the reactor by controlling the operating conditions. Particle deposition helps to control the thickness of the salt deposit as it deposits at the wall. These methods can control the formation of salt deposits from the source or control salt deposition to the permissible range.

In Tang Xingying's study, phosphate and sulfate were added to Na_2SO_4 aqueous solution to investigate the change in dissolved behavior. After the salt tests were conducted at different SCWO conditions, the conductivities, cations, and anions were determined by conductivity meter, graphite furnace atomic absorption spectrophotometer (AAS), and ion chromatography (IC). This research will offer the results for the lack of solubility information of mixture of two binary salts, which is helpful for fabricating supercritical part in SCWO system.

(1) **Experimental**

Materials, sodium sulfate, potassium sulfate, and potassium phosphate, which were all purchased from Tianhao Chemical Co. (Tianjin, China), were of 99.5% purity and were used without further treatment. The electrical conductivity of deionized water was 15 μs/cm, which was made by water purification machine. Three kinds of Ni-based alloys (Incoloy800, Incoloy825, and dual-phase steel2205) were selected as candidate alloys and their chemical compositions are shown in Table 7.3.

The salt solution was prepared by a certain concentration for the solubility experiments. The electric mixer was used to make the salt aqueous solution mixed fully. Experiments were performed around 24 MPa and increasing temperatures. Before the experiments performed, the continuous kettle-reactor system would take 1.5 h to achieve a steady state condition (T = 320 ± 1 °C, P = 24 MPa). When the steady state was reached, the salt solution was come into the experiment system and the two salt solutions were preheated through two independent pipes. The preheating temperatures were kept at 280 ± 1 °C and the temperature of reactor was kept at 320 ± 1 °C. Then the conductivity of effluent increased to the maximum value

Table 7.3 Details of Na_2SO_4 solubility experiments

Run no.	Salt	Influent salt concentration [wt%]	Gauge pressure [MPa]	Mass flow rate [L/h]	Test item
E1	Na_2SO_4	5.0	24.2	2.0	Conductivity
E2	Na_2SO_4	8.0	24.1	2.0	Na^+
	K_2SO_4	0.4		1.2	
	K_2SO_4	0.8		1.2	
	K_2SO_4	1.2		1.2	
E3	Na_2SO_4	8.0	24.3	2.0	Na^+
	K_3PO_4	2.4		1.2	
	K_3PO_4	4.8		1.2	
	K_3PO_4	7.2		1.2	

closed to influent and the temperature of reactor was raised. The conductivity of effluent varied with the increasing temperature of reactor. The 316 stainless steel and Hastelloy C-276 were used for constructing the low-temperature regions ($T < 350$ °C) and the high-temperature regions ($T > 350$ °C), respectively. The maximum operating temperature and pressure of the reactor were 550 °C and 32 MPa, respectively. The reactor had a volume of 200 mL and type-K thermocouple was used for measuring the temperatures for the inner reactor. The pressure was monitored by a pressure gauge and the jumping-up pressure of safety valve was 30 MPa. The cooling coil was used to quenching effluent from reactor and the back pressure valve was used for keeping the pressure of system. A filter with a 2 μm Hastelloy C-276 filter was installed on the outlet in reactor, as shown in Fig. 7.3. Two pumps were 8 and 5 L/h inflows, respectively. In this experiment, the Na_2SO_4 aqueous solution was stored in tank 1, and the K_2SO_4 and K_3PO_4 were stored in tank 2.

The conductivity was determined by a sartorius professional meter (PP-50 model). Ionic chromatography (PIC-10A, Qingdao Puren Instrument Co. Ltd., China) was used for measurement of the Na^+ concentration, and the measurement error was estimated to be ±2%. The lower limit of detection of Na^+ was 7×10^{-3} μg/g.

As Fig. 7.4 shown, the conductivity of the effluent was determined when the Na_2SO_4 salt solution was introduced into the system and sample was taken every 5 min. It was noticed that the salt solution moves from entry to exit of system would take 1.5 h. The conductivity of effluent would increase to 72,200 μs/cm after the steady state achieved. Then, the temperature of kettle reactor would rise from 320 to 400 °C and the conductivity showed a sharp drop between 355 and 370 °C.

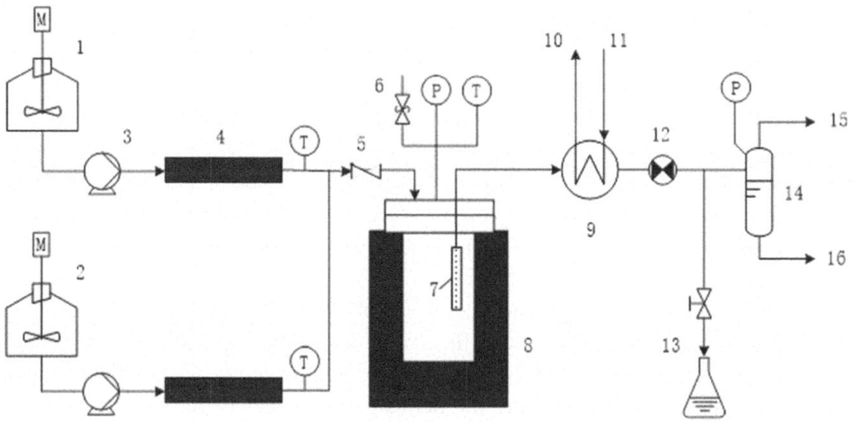

Fig. 7.3 Schematic of continuous kettle-reactor system: 1 salt solution tank 1; 2 salt solution tank 2; 3 high-pressure metering pump; 4 preheater; 5 check valve; 6 safety valve; 7 filter; 8 kettle reactor; 9 cooler; 10 cooling water inlet; 11 cooling water outlet; 12 back pressure valve; 13 liquid collector; 14 gas-liquid separator; 15 gas outlet; 16 liquid outlet

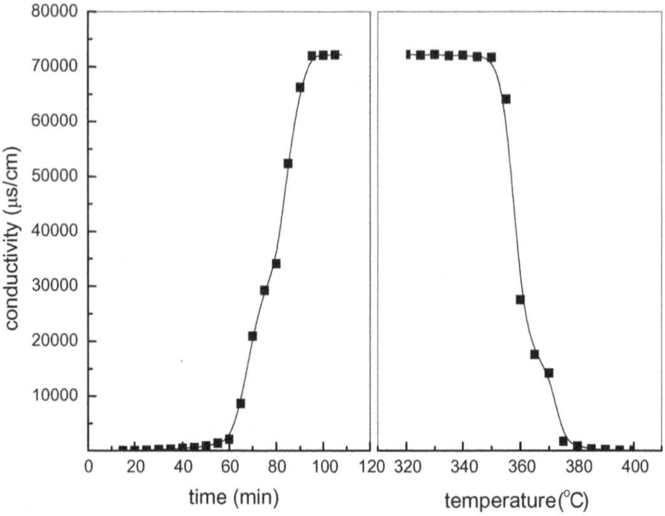

Fig. 7.4 Effluent conductivity versus time and temperature run "E1"

These results implied that almost all of Na_2SO_4 had been deposited in kettle reactor when temperature over 370 °C.

The solubility of sodium sulfate was determined after mixed with K_2SO_4 or K_3PO_4 (see Fig. 7.5). It is interesting to find that the Na_2SO_4 was less soluble in the presence of K_2SO_4 at low fluid density, and the solubility of Na_2SO_4 was decreased as the concentration of K_2SO_4 increased, as shown in Fig. 7.5a. This find can be explained by the common-ion effect, however, the solubility of Na_2SO_4 in presence of K_2SO_4 complied well with the pure Na_2SO_4 at the condition of $T \leq 355$ °C and $T \geq 380$ °C, $P = 24$ MPa [28]. As Fig. 7.5b showed, Na_2SO_4 deposit was less with the increase of K_3PO_4 addition. This result indicated that the addition of K_3PO_4 was suggested to improve the liquidity of salt and the agglomeration of salt particles was inhibited. Phosphate was always used as anti-sludging agent at normal temperature and pressure conditions. Threshold effect mechanism, lattice distortion mechanism, and dispersion mechanism were used to explain the salt deposition inhibited by phosphate. This anti-sludging feature of phosphate would be applied for controlling the salt solubility and deposit in supercritical water system effectively.

Images of salts deposition during the salt mixture experiments are shown in Fig. 7.6. In the presence of K_2SO_4, the solubility of Na_2SO_4 reduced more significantly than pure Na_2SO_4, leading to severe plugging in the supercritical area of system. The method that observing the deposited salts was depressurization at the high temperature, and then the kettle was opened and the power was cut off. The water would evaporate rapidly by the heat stored in system. Running "E3," little salt was deposited in kettle reactor which indicated the blocking was prevented from

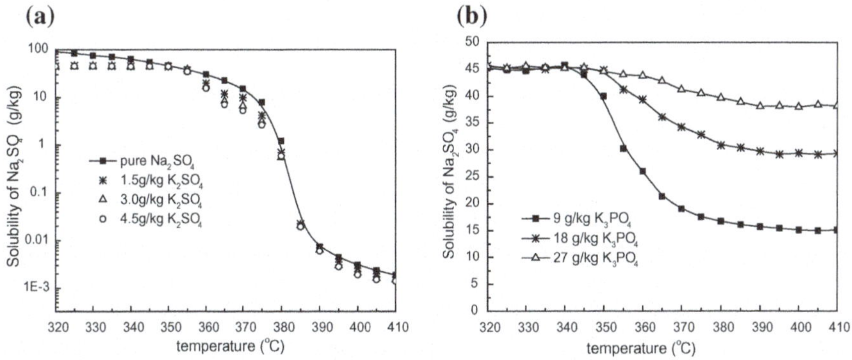

Fig. 7.5 Solubility changes of Na_2SO_4, **a** solubility of mixture of Na_2SO_4 and K_2SO_4 run "E2" and **b** solubility of mixture of Na_2SO_4 and K_3PO_4 run "E3"

(1) (2)

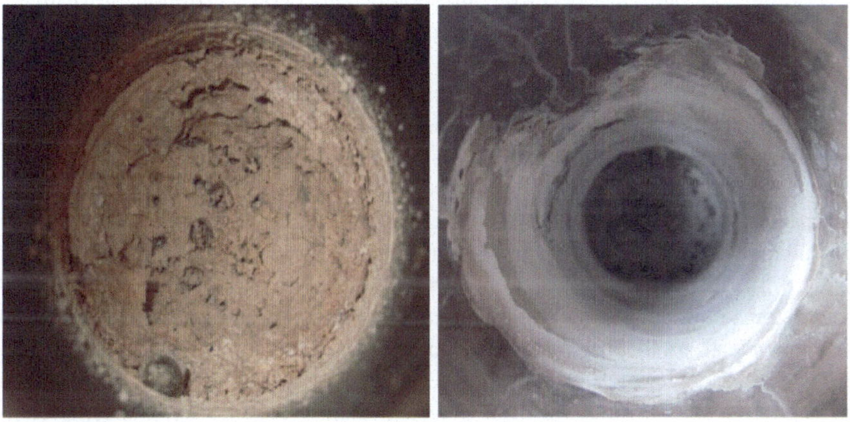

Fig. 7.6 Photos for salts deposited in kettle reactor, **1** run "E2" and **2** run "E3"

happening due to the addition of K_3PO_4. This advantage of phosphate would be used to inhibit the plugging in supercritical water system.

Salt solubility of Na_2SO_4 was investigated in presence of K_2SO_4 and K_3PO_4, respectively. The Na_2SO_4 was less soluble when it mixed with K_2SO_4 at a temperature within the range of 355–380 °C, and the common-ion effect could be used to explain this phenomenon. The K_3PO_4 was helpful for inhabiting the blocking in the supercritical water system, and the deposition of salt decreased as the addition of K_3PO_4 increased. The K_3PO_4 could be used as anti-sludging agent which would inhibit the plugging effectively.

7.5.2 Special Operation Techniques

The reliable way to overcome salt deposition is subsection control along the process of wastewater treatment. The primary consideration is to minimize the salt content entering into the reactor. The objective can be achieved through restricting the content of salts in the feedstock, and further setting a salt separator to remove salts before the reactor. Besides, cold feed injection restrains organics decomposition to salts and avoids salt deposition in the preheater, so it can prevent feed lines from salt deposition and plugging [70]. The second stage of salt control is inside the reactor, by inhibiting salt precipitation or making precipitated salts away from the inner surface of the reactor. The present techniques mainly include additive, extreme pressure, low turbulence, and homogeneous precipitation, etc. Nonetheless, salt deposition on the reactor's internal wall seems to be inevitable for a long-term running of a SCWO system in the deteriorating case, so techniques of mechanical brushing, rotating scraper, and reactor flushing can be employed to deal with the salt layer attaching to the internal wall. Eventually, in order to overcome salt deposition and plugging occurring in cooling and depressurization downstream lines, salts carried by the reactor effluent need further managements. Minimizing the salt content of the reactor effluent is one of the fundamental and effective approaches to overcoming salt deposition appearing in the cooling heat exchanger. Industrial techniques of density separation and filtration are used to remove salt particles from the reactor downstream fluid [71, 72]. Although many methods have been developed for the control of salt deposition, each method has its own characteristics, and no "right" approach exhibits better than all the others or properly works in all cases [14]. For instance, filtration is easy to be plugged and corroded, and difficult to select appropriate filtration materials for the actual heteroatoms-containing wastewater. Increasing pressure to minimize salt deposition may also lead to the increase of solubilities of both corrosive substances and protective oxides, which accelerates the corrosion rate of the reactor [11]. The appropriate means for a given SCWO system mainly depends upon specific combination of operation conditions and properties of the feedstock. It is noted that more than one of these methods can be employed in a SCWO process, and they are also generally used in conjunction with a specific reactor form. Table 7.4 systematically illustrates some special operation techniques and their fundamental features.

As mentioned above, there are a variety of operation techniques that have been developed for overcoming salt deposition, but few of them can solve the problem thoroughly. A valuable idea, source control, is proposed here to overcome or decrease salt deposition in the whole SCWO system. The primary purpose of the source control is to minimize the amount of salts in the feedstock and in the reaction fluid. Vadillo et al. [5] pointed out that wastewater treated by SCWO tubular reactors requires to satisfy certain requirements of salt concentrations (e.g., $NaCl < 200$ ppm, $Na_2SO_4 < 1$ ppm and $Na_2CO_3 < 1$ ppm, etc.). Nevertheless, the salt content of the wastewater is difficult to reduce and control on-line due to its

Table 7.4 Specific operation techniques for overcoming reactor plugging induced by salt deposition

Category	Operation	Definition	Characteristics	Application status
Approaches to decrease the salts content entering into the reactor	Salts removal before the reactor	Most salts are removed by a salt separator prior to the reactor	Separation efficiency depends upon initial wastewater features. Stability and economy of the SCWO system are hindered by the separator	The salt separator was tested with a proof of concept by Schubert et al. [20]
Techniques for overcoming salt deposition on the reactor's inner wall	Extreme pressure	The high pressure can improve the salt solubility, so decreasing the risk of salt deposition	Increasing the pressure will accelerate the reactor corrosion rate and increase the reactor investment	This approach was demonstrated by Los Alamos National Laboratory [73], and German acronym, SUWOX [74]. Related patents have been applied by Dassel et al. [75] and Barber [76]
	Additive (including fluidized particles and certain salts)	The fluidized particles provide a large nucleation surface for salts. Certain salts convert the deposition nature of the salt mixture	It is a key to determine the quantity and kind of the additive. The additive solids may aggravate valve corrosion. Extensive information on the properties of most salts and their mixture is required	The method was employed in testing stage by General Atomics, and recently issued patents were obtained by Hazlebeck [77], Modell [78] and Whiting and Mehta [79]
	Low turbulence, homogeneous precipitation	Make salts precipitate in the bulk fluid and be far away from the reactor's inner surface by controlling operation conditions	Homogeneous nucleation occurring in the bulk fluid needs harsh operation conditions and reactor structure	The technique was patented by Hydroprocessing [65]

(continued)

Table 7.4 (continued)

Category	Operation	Definition	Characteristics	Application status
	High-velocity flow	Ensure salt particles to suspend in a high-velocity flow before they are re-dissolved or removed	The appropriate velocity depends on many specific factors, such as the reactor diameter and orientation, fluid-phase physical properties, particle size, particle density, and particle concentration	This method was proved and patented by MODEC [80, 81]
	Sedimentation by gravity	Achieve sedimentation by gravity at a low fluid velocity and a sufficient density difference conditions	This technique is suitable for big particles, but the smaller particles are limited by the sedimentation velocity. A periodical cleaning is necessary	The method was still in an experimental stage and patented by Dell'Orco et al. [82]
	Setting a sacrificial or mobile surface	A sacrificial or mobile surface is set in the reactor, providing a large surface for salt deposition	The reactor can successfully be operated for a certain period of time, but it inevitably will take place plugging eventually	The method was a trial in the laboratory-scale and patented by Whiting and Mehta [79] and Nauflett et al. [83]
Salt removal after salt deposition on the reactor wall	Mechanical brushing	Transport a mechanical brushing from the reactor top to its bottom to physically remove salts sticking on the reactor's inner surface	The brushing intermittently works according to the salt deposition rate. The improved brushing is suitable for various types of reactors, but this induces a	The method was used in the MODEC pilot-scale system and patented by Gairns and Joustra [84] and Modell et al. [85]

(continued)

Table 7.4 (continued)

Category	Operation	Definition	Characteristics	Application status
			relatively complicated reactor configuration	
	Rotating scraper	Adopt mechanical rotation of a scraper to remove salt deposition on the reactor inner surface	Reactor needs to be shut down as it works, and the rotation frequency depends on the deposition rate and natural feature of salts	The method was demonstrated by GA in a SCWO reactor system built for the US Navy [86]. Another evolving scraper device was proposed and patented by Hazlebeck et al. [87]
	Reactor flushing	Use cool water or acid solution to dissolve salts and scaling on the reactor surface	The technique can work off-line (need additional downtime) or on-line (need to install two or more parallel reactors) with any type of reactor or reactor combination, and the clearing time depends on the nature of the salt layer and the flow rate of cleaning liquid	This method was used in the laboratory-scale and patented by Abitibi-Price and GA [88, 89]

Note Main information in this table is obtained from the literature [14]

complex component features and harsh operation conditions. A desired result is that these salts can be removed before entering into the reactor. Thus, the salt separation seems to be a good means of achieving salt removal, completely solving salt deposition problems both in the reactor and in downstream pipelines. Salt separation before the reactor was described by Schubert et al. [20, 39], and they designed and manufactured a continuous salt separator (see Fig. 7.7) for the first time. The separator evolves from the MODAR reverse-flow reactor used for SCWO [90–92], however, it seems more slender due to the need for an external heating of the vessel.

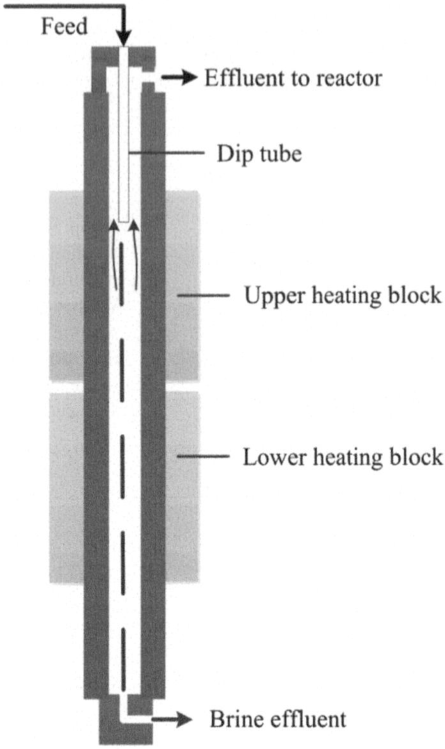

Fig. 7.7 Sketch of continuous salt separator (adapted from Refs. [20, 38])

Two insulated parts of upper high heating block and bottom low block form two distinct thermal zones, i.e., supercritical zone and subcritical zone. The upper keeps a supercritical temperature; salt particles precipitate from the bulk fluid and re-dissolve in the bottom subcritical zone. Ultimately, salts are discharged as brines from the bottom of the separator vessel to achieve salt separation. Experimental results show that the separation efficiency of the type 1 salts is in the range of 80–97% [20]. This successful utilization of the salt separator provides a feasible choice for the SCWO treatment of the high-salt wastewater. Salt separation in the reactor can also be accomplished by a centrifuge reactor [93], and high-speed rotation makes salt particles separate from the reaction fluid by centrifugal force. Nevertheless, the reactor eventually occurs plugging because of the deposition of sticky salt [14]. The improving approach is that salt separation is applied in combination with other techniques such as filtration, sacrificing liner, and rotary scraper.

In 2009, in order to realize the removal of inorganic salts dissolved in subcritical water, the large difference in solubility of inorganic salts in subcritical and supercritical regions was skillfully utilized, and the desalination and slag removal device in the supercritical water oxidation system was newly developed (see Fig. 7.8). By combining with the countercurrent process of the countercurrent tank reactor, the salt is poured into the bottom, and through precise temperature control, a

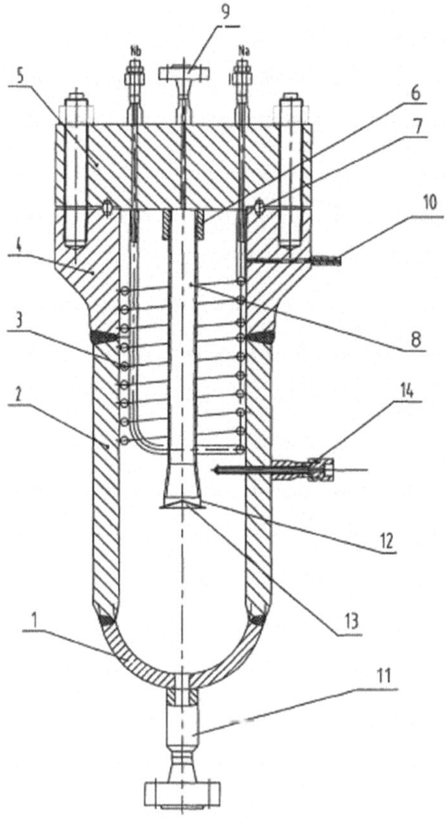

Fig. 7.8 Desalination and slag removal device for supercritical water treatment of waste organic matter [94]. 1 Spherical head; 2 straight tube of cylinder; 3 heat exchange coil; 4 upper end of cylinder; 5 end cover; 6 casing; 7 octagonal pad; 8 central conduit; 9 inlet pipe; 10 outlet pipe; 11 salt pipe; 12 connecting rod; 13 reflecting baffle; 14 temperature sensor; 15 row salt pipe; 16 first valve; 17 second valve; 18 third valve

subcritical zone is formed at the bottom of the reactor to dissolve and discharge the inorganic salt, and then the subcritical brine is heated by heat exchange to supercritical state to achieve separation and removal of inorganic salts.

In 2010, in order to alleviate the salt deposition and plugging pressure of the reactor, a predesalting device for use in a supercritical water oxidation system was developed, as shown in Fig. 7.9. By utilizing the characteristic that the inorganic salt is almost insoluble in supercritical water, the precipitated inorganic salt is removed by gravity sedimentation and a porous baffle, and a subcritical region is formed at the bottom of the device by precise temperature control, so that the inorganic salt is re-dissolved and discharged. It is an efficient separation and removal method.

In order to overcome technical problems of heating high-salt and high-concentration wastewater to supercritical conditions, such as severe corrosion, salt deposition, and coking blockage, Wang Shuzhong et al. innovatively developed a supercritical DC heating furnace in 2012, as shown in Fig. 7.10. By integrating partial oxidation anti-coking technology, positive and negative rinsing technology, subcritical water rinsing technology, and continuous online desalination process, it

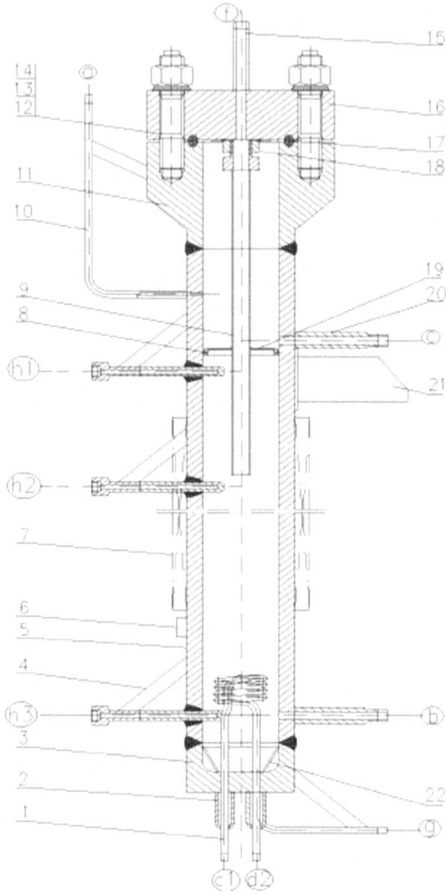

Fig. 7.9 Predesalter for supercritical water treatment of waste organic matter [95]. 1 Cooling coil; 2 casing; 3 lower flat cover; 4 rib plate; 5 simplified; 6 attached temperature sensor; 7 electric heater; 8 support ring; 9 center tube; 10 zero take-up; 11 simplified end; 12 head stud; 13 nut; 14 spherical washer; 15 water inlet pipe; 16 upper flat cover; 17 octagonal gasket pad; 18 fixed pipe; 19 filter baffle; 20 reinforced pipe; 21 pedestal; 22 cone; valve interface; b bottom brine outlet; C desalted fluid outlet; d1 cooling water inlet; d2 cooling water outlet; f inlet; g slag outlet; h1–h3 thermocouple interface

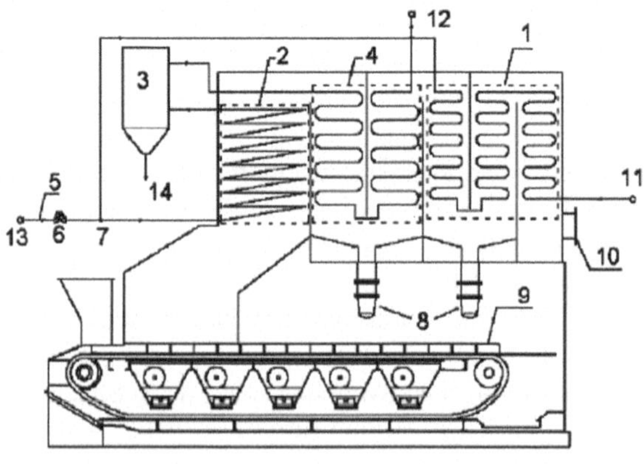

Fig. 7.10 High-salt high-concentration organic wastewater supercritical heating furnace [96]. 1 Economizer, 2 water wall, 3 desalination unit, 4 superheater, 5 oxidant line, 6 auxiliary valve, 7 tee, 8 cold ash bucket, 9 chain grate, 10 flue gas outlet, 11 economizer inlet end, 12 superheater outlet, 13 oxidant line inlet, 14 slag outlet

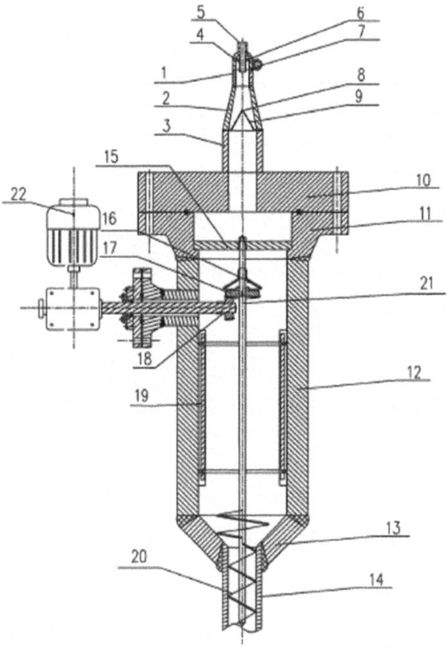

Fig. 7.11 Hydrostatic cyclone desalination device for waste organic matter supercritical treatment [97]. 1 Cylindrical section, 2 inverted vertebral body, 3 thick-walled tube, 4 spherical head, 5 overflow port, 6 deflector, 7 imported volute, 8 reflective screen top hole, 9 reflective screen, 10 flange cover, 11 flange, 12 buffer tank, 13 outlet vertebral body, 14 outlet pipe, 15 positioning bracket, 16 gear cover, 17 horizontal gear, 18 vertical gear, 19 scraper, 20 screw conveyor, 21 center shaft, 22 motor, 23 balance hole

can be directly heated without pretreatment, which is a major technical breakthrough of high-salt high-concentration wastewater heating furnace.

In order to solve the technical problems of efficient removal and continuous transport of solid inorganic salts under supercritical conditions, Wang Shuzhong et al. pioneered the development of a hydrocyclone desalination device under supercritical water conditions in 2012, as shown in Fig. 7.11. The high-efficiency separation of micron-scale solid inorganic salt is realized; the setting of strong inorganic salt transport mechanism realizes the continuous transportation of solid inorganic salt and overcomes the problem of continuous removal of high-viscosity solid inorganic salt in the system.

7.5.3 Particular Reactor Designs

As mentioned previously, tests in pilot-scale and commercial plants indicate that salt deposition seriously troubles SCWO equipments such as the preheater and the reactor. Organic wastewater suffers potential decomposition to produce inorganic salts with or without oxidant, due to the necessary high temperature (up to 400 °C) in the preheating stage 5 [98]. These salts gradually build up in the pipeline, and then the phenomenon of scaling and plugging eventually takes place. To cope with this problem, a particular reactor with a hydrothermal flame as the internal heat source has been proposed and built, which allows feedstock supplied at the low temperature even at the ambient temperature. The combustion products of a hydrothermal flame release enough energy to heat the low-temperature feedstock to the desired reaction temperature, so reducing the risk of salt deposition in the preheating exchanger. This type of reactor not only improves the reliability of the SCWO system, but also reduces its operation cost [70, 99–101]. But more significantly, salt deposition is mainly concentrated to occur in the reactor, which can be overcome by reactor configuration. Obviously, generated salts adhere to the internal surface of the reactor, resulting in heat transfer deterioration and pressure fluctuation, even shutdown of the whole SCWO system. Thus, salt control in the reactor is crucial for the scale-up of the SCWO process. Particularly innovative reactor design is even considered as a decisive factor in commercialization of SCWO systems [102, 103], because it may effectively reduce reactor corrosion and plugging induced by salt deposition. The competitive reactor structure of avoiding plugging includes reverse-flow reactor, transpiring wall reactor, cool wall reactor, centrifuge reactor, etc. Table 7.5 summarizes some typical SCWO reactors developed for salt deposition control.

(1) **Reverse-flow tank reactor**

The reverse-flow tank reactor evolves from the MODAR reactor with a vertical pressure vessel which forms two different thermal regions as shown in Fig. 7.12 [115, 116]. The upper part of the reactor is kept at a supercritical temperature (e.g., about 600 °C) and is the main region where organic matters are oxidized. The lower part of the vessel is kept at a subcritical temperature (e.g., 300 °C) [14]. The SCWO reaction region remains intense disturbance and agitation caused by a certain feed velocity through a top nozzle jet. Note that this type of reactor has a good phase separation effect. Salts precipitating in the supercritical zone are directed to the bottom of the vessel by gravity, inertia, and forced convection, and then re-dissolve in the subcritical zone. The formed brines are discharged from the bottom of the vessel. The desalted effluent is withdrawn from the top of the vessel (the reversal flow) in order to achieve the purpose of desalination in the reactor. However, salt deposition easily occurs on the internal surface due to the low velocity and strong disturbance at the vertical direction. Moreover, the reactor may take place corrosion because the internal surface is directly exposed to the corrosive reaction fluid [103].

(2) **Transpiring wall reactor**

The transpiring wall reactor (see Fig. 7.13) has attracted much attention in the recent decades due to its good performance in overcoming salt deposition and corrosion [63, 116, 117]. This reactor consists of a pressure-bearing wall and a porous transpiring wall, forming an annular mezzanine between these two walls [13, 21, 99, 117–122]. The porous wall is manufactured by sintered metal or ceramic, or several thin metal sheets or platelets that are bonded together, allowing heated transpiring water to pass through it. Transpiring water is pumped into the reactor to form a continuous thin protective film over the porous wall surface, where the water dilutes, re-dissolves, or sweeps away precipitated salt particles. Simultaneously, corrosive species are also kept away from the internal surface,

Table 7.5 Summary of SCWO reactors for salt deposition control

Reactor	Category	Control principle	Characteristics	Patents and literatures
Reversè-flow reactor	Tubular type	The reactor successively forms supercritical and subcritical regions along the reactor inlet direction. The two zones can intermittently counterchange to achieve salt removal according to the solubility feature of salts in SCW	The kind of long reactor is usually unsuitable for the high content of salt feedstock. Temperature alternation leads to thermal stress, so reducing the reactor life	Whiting [104]
	Tank type	Salts precipitating in the upper supercritical zone are directed to the reactor bottom by gravity, inertia, and forced convection, and then re-dissolve in the subcritical zone. The main clean effluent is discharged from the reactor top	The reactor has a good phase separation. Salt deposition easily forms on the internal surface at the vertical direction of the reactor	Patented by Killilea et al. [92], U.S., NO. 4822497
Transpiring wall reactor	Reverse-flow tank type	The porous transpiring wall is covered with a protective water film, keeping the wall away from salt particles. The brine is discharged from the reactor bottom to achieve salt removal, and clean effluent flows out from reactor's top outlet	The type of reactor exhibits good advantages in minimizing salt deposition and corrosion problems. Transpiring water easily causes temperature fluctuation and energy efficiency reduction	Xu et al. [34], Bermejo et al. [105], Abeln et al. [106]
	With a hydrothermal flame	A hydrothermal flame is utilized as the internal heat source instead of the high-temperature preheating of feedstock so as to avoid salt deposition in the preheater. The approach of overcoming salt deposition in the reactor is similar to that of the common transpiring wall reactor	Due to the extremely high reaction temperature, the required residence time can be less than one second, and a large amount of transpiring water is consumed. The special pressure-bearing wall needs to be designed properly to withstand such a high temperature	Wellig et al. [99], Cabeza et al. [100], Príkopský et al. [13], Bermejo et al. [107]
	With air as the	Air is used as the transpiring fluid instead of water to restrict inorganic salts in the	It can reduce the consumption of both energy and pure water and alleviate	Chen et al. [108, 109], Yang et al. [110]

(continued)

Table 7.5 (continued)

Reactor	Category	Control principle	Characteristics	Patents and literatures
	transpiring fluid	inner space of the porous tube and achieve dynamic gas seal	temperature fluctuation. It is easier to form an isotropic protective film compared with water	
Cool wall reactor	–	The feedstock solution cools the shell by heat diffusion to maintain the reactor's inner wall at a subcritical temperature, so salt deposition will not occur due to the dissolving salt action of the subcritical water film on the wall	The reactor can withstand high reaction temperatures and easily achieve automatic control and energy self-sufficiency. The salt deposition chamber effectively collects salts, but deposition still occurs in the case of the high-salt concentration feed	Cocero and Martınez [111], patented by Deaton et al. [112], U. S., NO. 5552039 Baur et al. [113]
Centrifuge reactor	–	Salt separation is achieved by the centrifugal force in the reactor, and then metal brushes are adopted to remove salts from the reactor wall	The distribution of reactants is not uniform, thus resulting in uncompleted oxidation reactions. A suitable velocity is required to accomplish salts separation	Patented by Halff and Reid [114], U.S., NO. 5688399

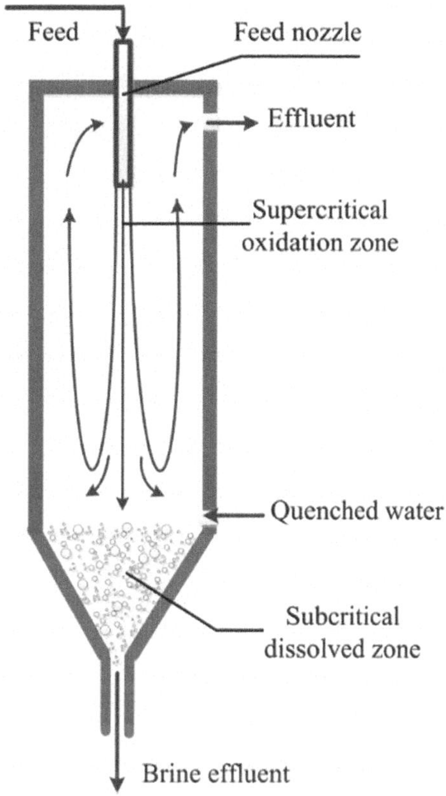

Fig. 7.12 Sketch of reverse-flow tank reactor (adapted from Ref. [91])

reducing the corrosion rate. Besides, transpiring water also balances the pressure between the pressure-bearing wall and the porous wall and supports a long-time continuous security operation of the SCWO plants. The temperature of transpiring water can be either subcritical or supercritical, but a higher temperature exists in the center of the reactor where oxidation reactions take place.

It is an interesting design that the porous wall functions as corrosion and salt deposition resistance while the pressure-bearing wall only needs to withstand the high pressure. The smart separation of salt control and strength requirement from the same material greatly improves the applicability of material (especially for the material with high corrosion resistance but poor strength properties) and reduces the cost of the reactor [5]. However, it cannot be ignored that transpiring water reduces the reaction temperature and dilutes the reaction fluid, which significantly influences waste destruction kinetics and its removal efficiency [107]. Moreover, these negative impacts weaken wastewater degradation capacity, resulting in a very low efficiency of energy recovery [123]. Salt deposition is also observed on the transpiring wall, although its surface is protected by the water film [120]. A loss of transpiring water is the biggest risk for any transpiring wall reactor. This is because obstruction of transpiring water may leave the porous wall and the reactor directly

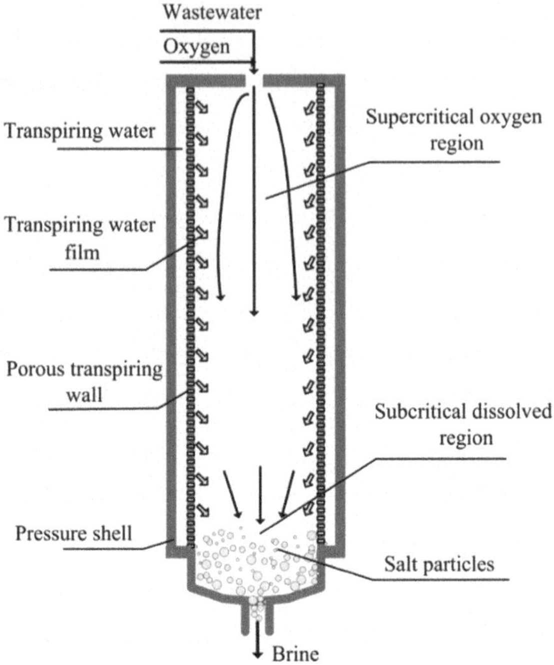

Fig. 7.13 Schematic diagram of transpiring wall reactor (adapted from Ref. [63])

exposed to severe corrosive attacking [98]. Some relevant designs are subsequently proposed, such as transpiring wall reactor with a hydrothermal flame as the internal heat source [13, 70, 99], transpiring wall reverse-flow tank reactor [34, 105, 106], and transpiring wall reactor with air as cooling medium [108–110]. Corresponding laboratory experiments or pilot-scale plants have been built, since the earliest example of transpiring wall reactor was patented by Daman [124].

In 2009, Wang et al. [125] developed a countercurrent tank reactor with a porous evaporating wall. The protective water film on the inside of the evaporation wall and the longitudinal and lateral temperature gradient zones in the reactor form uniform and stable around and at the bottom of the reactor. The subcritical water film separates the inorganic salt from the reactor pressure wall, effectively avoiding salt deposition in the reactor. Based on the characteristics of the internal flow in the reactor, on the one hand, the residence time of the reactants is prolonged, and on the other hand, the inorganic salt enters the catalyst tank after passing through the counterflowing strontium salt process. Xu et al. [126] designed and manufactured a novel reactor in 2010, which is utilized to treat sewage sludge by SCWO with a feed rate of about 100 L/h. The reactor is the core equipment of the first pilot-scale SCWO plant constructed in China. Its design concepts for salt removal are presented concisely, as shown in Fig. 7.14.

The novel reactor owns characteristics of transpiring wall reactor and MODAR reactor for the first time, which is simply nominated as “TWM reactor.” Similar to MODAR reactor, TWM reactor is also divided into supercritical zone above and

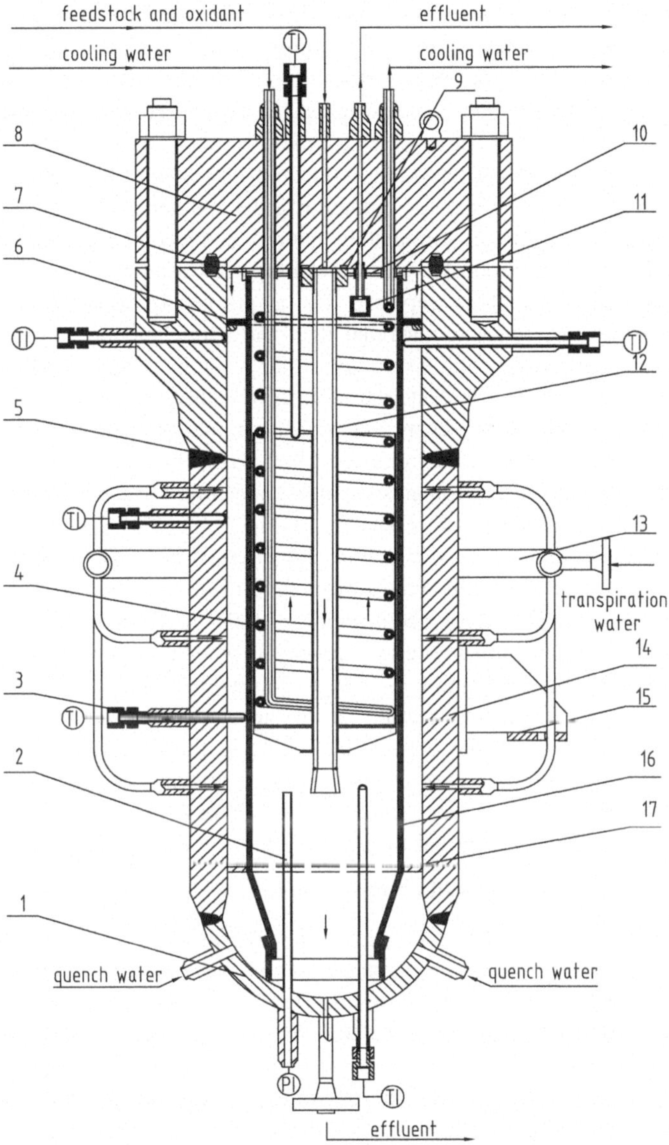

Fig. 7.14 Schematic diagram of TWM reactor for treating sewage sludge by SCWO [126]. 1 Spherical head; 2 pressure measurement tube; 3 temperature measurement tube; 4 cooling spiral coil; 5 catalyst box; 6 support ring; 7 eight-square cushion; 8 top cover; 9 fixed platform; 10 cooling cover; 11 filter; 12 center pipe; 13 annular water box; 14 pressure-bearing wall; 15 ear base; 16 transpiring wall; 17 locating ring; TI: temperature indication; PI: pressure indicate

subcritical zone below. Precipitated salts under supercritical condition fall down into the subcritical zone (formed by pumping into quench water) and are re-dissolved. Clean fluid after removed salts flows toward top outlet of reactor by the differences of density and flow resistance. The valves equipped on pipelines of top and bottom outlets of reactor, which can be seen in pilot-scale plant, are used to regulate flow rate of fluid at top outlet. Dirt fluid flows out TWM reactor from its bottom outlet, which contains large amount of salts and solids. The flow pattern of TWM reactor is helpful for eliminating natural convection effects which were mentioned by Wellig's group [13] for transpiring wall reactor.

MODAR reactor wall is replaced by a porous transpiring wall and a pressure-bearing wall is set to surround it to form fundamental configuration of TWM reactor. Clean and cool transpiration water is pumped into TWM reactor and forms a protective film on the internal surface of transpiring wall. TWM reactor configuration is expected to prevent salt deposition and decrease corrosion rate. In 2011, Wang Shuzhong et al. developed a fuel auxiliary thermal reactor for the coking and salt deposition problems of high-concentration and high-salt organic wastewater during preheating, as shown in Fig. 7.15. Through the innovative research and development of the structure, the jet collision of fuel and oxidant, the replacement of the cooling wall and the evaporation wall, and the end cover cooling are realized, so that the energy consumption of the reaction fluid preheating is reduced, and the over temperature and corrosion prevention and control are prevented. The reaction temperature is used to thoroughly treat difficult-to-degrade substances such as ammonia nitrogen. At the same time, a cooling water inlet is arranged at the bottom of the reactor. By adjusting the flow rate of the cooling water, the temperature of the salt discharge zone can be precisely controlled, and when the desalination pipe at the bottom of the reactor is clogged, the flow rate of the cooling water can be adjusted to increase the solubility of the bottom salt.

(3) **Transpiring wall reverse-flow tank reactor**

As illustrated in Table 7.5 and Fig. 7.16, the transpiring wall reverse-flow tank reactor also consists of a pressure-bearing wall and a porous transpiring wall, which possesses characteristics of both the transpiring wall reactor and the reverse-flow tank reactor. Similar to the transpiring wall reactor, this special reactor configuration is capable of forming a protective water film to minimize salt deposition and corrosion problems. The reactor can also form two distinct thermal regions and achieve salt separation like the reverse-flow tank reactor. Moreover, the water flowing in spiral coil takes away excess energy in the oxidation region, avoiding the danger of reactor overheating. Thus, this reactor resistance to corrosion and salt deposition is expected to be better than those of the above two reactors. Salts precipitating in the supercritical zone re-dissolve in the subcritical zone. Ultimately, these salts are discharged as a high-salt-containing effluent at the bottom of the reactor. Although this kind of reactor has advantages of both the reverse-flow tank reactor and the transpiring wall reactor, it certainly shares their partially common disadvantages. Therefore, plenty of experiments and a long-term test require to be conducted to optimize its configuration and operation parameters.

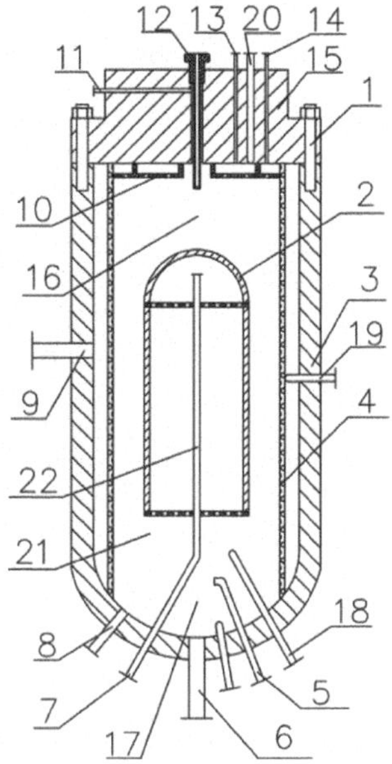

Fig. 7.15 Supercritical water treatment reactor using auxiliary fuel to recharge heat [127]. 1 Fastening bolt; 2 catalyst tank; 3 kettle body; 4 porous evaporation wall; 5 liquid salt discharge port; 6 solid salt discharge port; 7 product outlet; 8 cooling water inlet; 9 evaporation wall water inlet; 10 tank; 11 hot air inlet; 12 fuel core tube; 13 cold air inlet; 14 material inlet; 15 end cap; 16 high-temperature combustion zone; 17 subcritical salt dissolution zone; 18 desalination zone temperature measuring casing; 19 cavity temperature measuring sleeve; 20 reactor end cover temperature measuring sleeve; 21 salt separation area; 22 central tube

(4) Transpiring wall reactor with a hydrothermal flame as the internal heat source

The transpiring wall reactor with a hydrothermal flame as the internal heat source can form five regions, i.e., a, b, c, d, and e, as shown in Fig. 7.17. This reactor combines the characteristics of the hydrothermal flame releasing plenty of energy with the good wall protection concept of the transpiring wall reactor. The particular configuration may overcome salt deposition and plugging of the reactor. Simultaneously, the wastewater is heated to the desired reaction temperature by mixing with the combustion products of the hydrothermal flame. Thus, this type of reactor is able to destroy the wastewater supplied at the low temperature even at the ambient temperature, reducing the risk of salt deposition in the preheating exchanger. However, this kind of reactor is restricted by flame stability, over temperature of vessel walls in the oxidation reaction zone, as well as parts of other drawbacks of the transpiring wall reactor described before. For example, the flame stability is strongly influenced by both the mass flow and the temperature of transpiring water, wastewater, fuel, and oxygen. These four mixed streams may

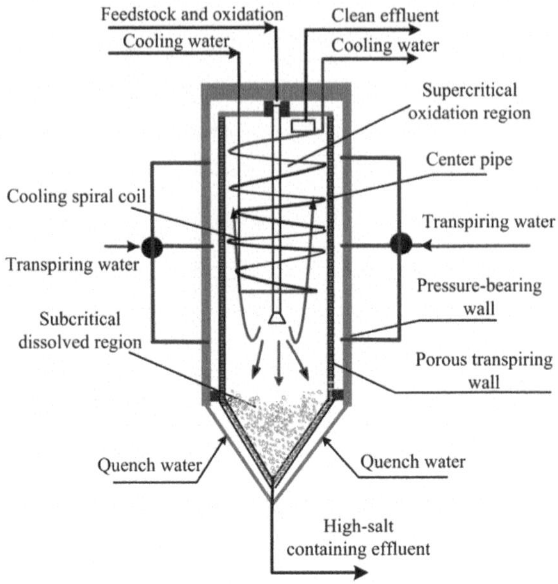

Fig. 7.16 Schematic diagram of transpiring wall reverse-flow tank reactor (adapted from Refs. [10, 34])

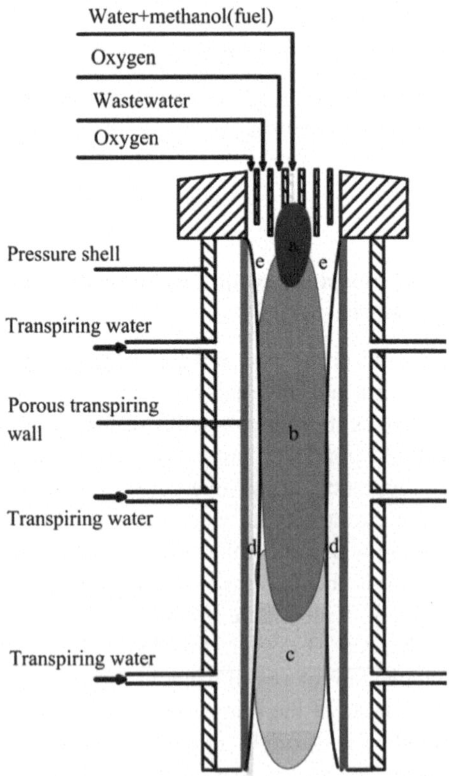

Fig. 7.17 Transpiring wall reactor with a hydrothermal flame as the internal heat source (adapted from Ref. [90]). a Hydrothermal flame; b oxidation reaction zone; c subcritical dissolving salt zone; d protective water film; e mixing zone of wastewater and oxygen

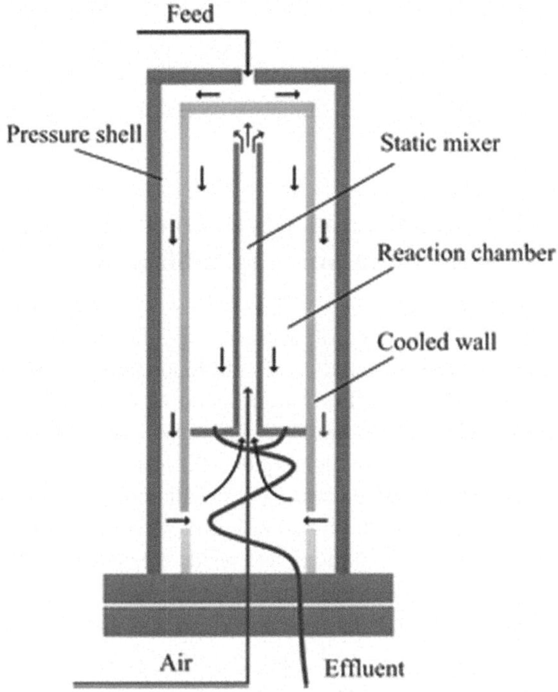

Fig. 7.18 Sketch of cool wall reactor (adapted from Ref. [111])

form complicated flow fields, which seem to be difficult to calculate the ignition temperature, concentration, and flow velocity of the mixed fluid. To maintain a stable burning hydrothermal flame, plenty of experiments and experiences are required to obtain suitable flow intensity and temperature.

(5) **Cool wall reactor**

The cool wall reactor (see Fig. 7.18) consists of a pressure shell and a cooled wall. Pure water that is introduced in the mezzanine (the annular space between the above two walls) cools the cooled wall to maintain the subcritical temperature of its internal surface. Thus, precipitated salt particles can re-dissolve in a subcritical film over the internal surface instead of depositing on the cooled wall, so avoiding reactor plugging induced by salt deposition. The relevant reactor is developed in the Chemical Engineering Department at the University of Valladolid and proved that the design is feasible and TOC removal of pollutant is over 99.9 wt%. This kind of reactor can withstand high reaction temperatures and is usually operated at autothermal conditions due to its great compact structure and good energy integration. Baur et al. proposed that this type of reactor could withstand a highest pressure up to 70 MPa and a possible temperature of 700–850 °C. This is because the density of the cold fluid is higher than that of the thermal fluid, and the external temperature of the reaction fluid is lower than that of the central region. The temperature of the cold fluid is determined according to the salt solubility curve and heat transfer of the reactor wall. However, a relatively low temperature of the cold wall may affect the reaction process, making reactions near the wall not complete. Moreover, corrosion and salt deposition may occur on the internal surface of the cooled wall, which is directly exposed to the reaction fluid.

7.6 Problems and Developmental Tendency of Salt Deposition and Crystallization

Salt deposition control achieves significant improvement in the innovation of research methods, studies of salt deposition and separation performance, the evolution of salt control techniques, and new reactor configuration designs. However, several challenges, such as transport properties, deposition mechanisms, and practical salt control techniques, still restrict the technology application. Operation parameters play important roles in salt deposition processes, and various ions in the mixture may exhibit mutual influence or common-effect. Thus, the prediction of salt deposition is quite difficult and needs extensive information on salts phase behavior in SCW. Moreover, the mechanism of salt deposition in the multi-component system have not been recognized deeply, and the entire evolution process of salts from microscopic nucleation to macroscopic deposition in SCW requires to be further revealed. Hence, it is necessary to gain more phase behavior data for analysis in depth. The combination of the experiment and numerical simulation is a desired method for the investigation of salt deposition.

In future, salt deposition control in a SCWO system requires global considerations and designs instead of previous single or local protection. Certainly, reactor configuration design has the priority to the whole SCWO system due to its crucial roles in salt control. Wise reactor concept combines with appropriate operation techniques before the reactor, in the reactor and/or after the reactor, forming multiple control and security units. Notably, the determination of salt control schemes should also consider the characteristics of wastewater and each technique, as well as the economic costs.

Reactor plugging caused by salt deposition is one of key technologies that need to be addressed in the commercialization of SCWO. This paper systematically summarizes the current status of salt deposition research at home and abroad. The research on salt deposition has made great progress in terms of sedimentation theory and source control of research methods and the development of engineering techniques to avoid reactor blockage caused by salt deposition. However, researchers have not been able to fully reveal the complex evolution of salt from micro-nucleation to macroscopic deposition in the SCWO multi-system. The mechanism of salt deposition is still unclear. There is no technology for solving salt deposition. Therefore, it is necessary to further explore the nature and deposition mechanism of salt and study the salt deposition kinetics and the phase behavior of multi-component systems. In order to maintain the continuous transport of salt in the reactor, it is necessary to further establish phase diagrams of different salts and mixtures, investigate the interaction mechanism between multi-component salts, avoid salt precipitation from the source, or ensure that the salt can be maintained after precipitation. With the solution of salt deposition problems and the development of reactor corrosion-resistant materials, SCWO will be able to achieve large-scale commercial applications.

References

1. M. C. Kutney, Thermodynamic and transport property modeling in super critical water. 2005, Massachusetts Institute of Technology.
2. R. W. Shaw, T. B. Brill, A. A. Clifford, C. A. Eckert, E. U. Franck, Supercritical water: A medium for chemistry. Chem. Eng. News, 1991, 69: 26–39.
3. P. A. Marrone, Supercritical water oxidation—Current status of full-scale commercial activity for waste destruction. Journal of Supercritical Fluids, 2013, 79: 283–288.
4. T. Tassaing, Y. Danten, M. Besnard, Infrared spectroscopic study of hydrogen-bonding in water at high temperature and pressure. Journal of Molecular Liquids, 2002, 101: 149–158.
5. V. Vadillo, J. Sanchez-Oneto, J. Ramon Portela, E. J. Martinez de la Ossa, Problems in Supercritical Water Oxidation Process and Proposed Solutions. Industrial and Engineering Chemistry Research, 2013, 52: 7617–7629.
6. J. Tester, H. Holgate, F. Armellini, P. Webley, W. Killilea, G. Hong, H. Barner, Emerging technologies in hazardous waste management III. In ACS Symposium Series. 1993. Am. Chem. Soc.: Washington DC.
7. G. Zhang, I. Hua, Supercritical water oxidation of nitrobenzene. Industrial and Engineering Chemistry Research, 2003, 42: 285–289.
8. D. Zou, Y. Chi, J. Dong, C. Fu, M. Ni, Supercritical water oxidation of MSW leachate: Factor analysis and behavior of heavy metals. Environ. Prog. Sustain. Energy, 2013.
9. P. E. Savage, Heterogeneous catalysis in supercritical water. Catalysis Today, 2000, 62: 167–173.
10. P. Kritzer, E. Dinjus, An assessment of supercritical water oxidation (SCWO): Existing problems, possible solutions and new reactor concepts. Chemical Engineering Journal, 2001, 83: 207–214.
11. P. Kritzer, Corrosion in high-temperature and supercritical water and aqueous solutions: a review. Journal of Supercritical Fluids, 2004, 29: 1–29.
12. P. A. Marrone, G. T. Hong, Corrosion control methods in supercritical water oxidation and gasification processes. Journal of Supercritical Fluids, 2009, 51: 83–103.
13. K. Príkopský, B. Wellig, P. R. Von Rohr, SCWO of salt containing artificial wastewater using a transpiring-wall reactor: experimental results. J. Supercrit. Fluids, 2007, 40: 246–257.
14. P. A. Marrone, M. Hodes, K. A. Smith, J. W. Tester, Salt precipitation and scale control in supercritical water oxidation—part B: commercial/full-scale applications. Journal of Supercritical Fluids, 2004, 29: 289–312.
15. M. Hodes, P. A. Marrone, G. T. Hong, K. A. Smith, J. W. Tester, Salt precipitation and scale control in supercritical water oxidation—Part A: fundamentals and research. Journal of Supercritical Fluids, 2004, 29: 265–288.
16. C. A. LaJeunesse, S. F. Rice, R. G. Hanush, J. D. Aiken, Salt deposition studies in a supercritical water oxidation reactor, in PBD: Oct 1993. 1993. p. Medium: ED; Size: 18 p.
17. Z. M. C. X. L. Shan, P. Q. Yuan, Precipitation of salt in near and supercritical water. J. East China Univ. Sci. Technol. (Natural Science Edition), 2007, 33: 606–609.

18. M. V. Fedotova, Effect of temperature and pressure on structural self-organization of aqueous sodium chloride solutions. Journal of Molecular Liquids, 2010, 153: 9–14.
19. M. Hodes, K. A. Smith, P. Griffith, A natural convection model for the rate of salt deposition from near-supercritical, aqueous solutions. Journal of Heat Transfer-Transactions of the Asme, 2003, 125: 1027–1037.
20. M. Schubert, J. W. Regler, F. Vogel, Continuous salt precipitation and separation from supercritical water. Part 1: Type 1 salts. Journal of Supercritical Fluids, 2010, 52: 99–112.
21. E. Fauvel, C. Joussot-Dubien, P. Guichardon, G. Charbit, F. Charbit, S. Sarrade, A double-wall reactor for hydrothermal oxidation with supercritical water flow across the inner porous tube. Journal of Supercritical Fluids, 2004, 28: 47–56.
22. F. J. Armellini, J. W. Teste, Experimental methods for studying salt nucleation and growth from supercritical water. Journal of Supercritical Fluids, 1991, 4: 254–264.
23. U. Hegde, M. Hicks. Salt Precipitation and Transport in Near-Critical and Supercritical Water. In 40th International Conference on Environmental Systems. 2010.
24. F. J. Armellini, Phase equilibria and precipitation phenomena of sodium chloride and sodium sulfate in sub- and supercritical water. 1993, Massachusetts Institute of Technology.
25. V. M. Valyashko, Phase equilibria of water-salt systems at high temperatures and pressures, in Aqueous Systems at Elevated Temperatures and Pressures, D. A. Palmer, R. Fernández-Prini, and A. H. Harvey, Editors. 2004, Academic Press: London. p. 597–641.
26. V. M. Valyashko, I. M. Abdulagatov, J. M. Levelt Sengers, Vapor-liquid-solid phase transitions in aqueous sodium sulfate and sodium carbonate from heat capacity measurements near the first critical end point. 2. Phase boundaries. J. Chem. Eng. Data, 2000, 45: 1139–1149.
27. V. M. Valyashko, Phase behavior in binary and ternary water-salt systems at high temperatures and pressures. Pure and Applied Chemistry, 1997, 69: 2271–2280.
28. M. S. Khan, S. N. Rogak, Solubility of Na_2SO_4, Na_2CO_3 and their mixture in supercritical water. Journal of Supercritical Fluids, 2004, 30: 359–373.
29. V. Valyashko, Phase equilibria in water-salt systems: some problems of solubility at elevated temperature and pressure. High Temperature High Pressure Electrochemistry in Aqueous Solutions, 1973, 4: 153–157.
30. M. S. Khan, Deposition of sodium carbonate and sodium sulfate in supercritical water oxidation systems and its mitigation. 2005.
31. M. Bermejo, P. Cabeza, J. Queiroz, C. Jiménez, M. Cocero, Analysis of the scale up of a transpiring wall reactor with a hydrothermal flame as a heat source for the supercritical water oxidation. J. Supercrit. Fluids, 2011, 56: 21–32.
32. V. I. Anikeev, A. Yermakova, Technique for complete oxidation of organic compounds in supercritical water. Russian Journal of Applied Chemistry, 2011, 84: 88–94.
33. E. Asselin, A. Alfantazi, S. Rogak, Thermodynamics of the corrosion of alloy 625 supercritical water oxidation reactor tubing in ammoniacal sulfate solution. Corrosion, 2008, 64: 301–314.
34. D. Xu, S. Wang, X. Tang, Y. Gong, Y. Guo, Y. Wang, J. Zhang, Design of the first pilot scale plant of China for supercritical water oxidation of sewage sludge. Chemical Engineering Research & Design, 2012, 90: 288–297.
35. K. Karakama, S. N. Rogak, A. Alfantazi, Characterization of the deposition and transport of magnetite particles in supercritical water. Journal of Supercritical Fluids, 2012, 71: 11–18.
36. A. A. Peterson, P. Vontobel, F. Vogel, J. W. Tester, Normal-phase dynamic imaging of supercritical-water salt precipitation using neutron radiography. Journal of Supercritical Fluids, 2009, 49: 71–78.
37. S. V. Makaev, T. M. Bitokhov, K. G. Kravchuk, M. A. Urusova, V. M. Valyashko, Salt deposition from hydrothermal solutions in a flow reactor. Russian Journal of Physical Chemistry B, 2011, 5: 1045–1055.
38. M. Schubert, J. W. Regler, F. Vogel, Continuous salt precipitation and separation from supercritical water. Part 2. Type 2 salts and mixtures of two salts. Journal of Supercritical Fluids, 2010, 52: 113–124.

39. M. Schubert, J. Aubert, J. B. Müller, F. Vogel, Continuous salt precipitation and separation from supercritical water. Part 3: Interesting effects in processing type 2 salt mixtures. Journal of Supercritical Fluids, 2012, 61: 44–54.
40. B. L. Haroldsen, B. E. Mills, D. Y. Ariizumi, B. G. Brown, Transpiring wall supercritical water oxidation reactor salt deposition studies. Nuclear Fuels, 1996.
41. S.-I. Kawasaki, T. Oe, S. Itoh, A. Suzuki, K. Sue, K. Arai, Flow characteristics of aqueous salt solutions for applications in supercritical water oxidation. Journal of Supercritical Fluids, 2007, 42: 241–254.
42. F. Demoisson, M. Ariane, A. Leybros, H. Muhr, F. Bernard, Design of a reactor operating in supercritical water conditions using CFD simulations. Examples of synthesized nanomaterials. Journal of Supercritical Fluids, 2011, 58: 371–377.
43. F. Masoodiyeh, M. R. Mozdianfard, J. Karimi-Sabet, Solubility estimation of inorganic salts in supercritical water. Journal of Chemical Thermodynamics, 2014, 78: 260–268.
44. N. Lummen, B. Kvamme, Properties of aging $FeCl_2$ clusters grown in supercritical water investigated by molecular dynamics simulations. Journal of Chemical Physics, 2010, 132.
45. N. Lümmen, B. Kvamme, Kinetics of NaCl nucleation in supercritical water investigated by molecular dynamics simulations. Physical Chemistry Chemical Physics PCCP, 2007, 9: 3251–3260.
46. M. Hovland, H. G. Rueslatten, H. K. Johnsen, B. Kvamme, T. Kuznetsova, Salt formation associated with sub-surface boiling and supercritical water. Mar. Petrol. Geol., 2006, 23: 855–869.
47. M. Hovland, T. Kuznetsova, H. Rueslatten, B. Kvamme, H. K. Johnsen, G. E. Fladmark, A. Hebach, Sub-surface precipitation of salts in supercritical seawater. Basin Research, 2006, 18: 221–230.
48. A. R. Hunt, Numerical simulation of hydrothermal salt separation process and analysis and cost estimating of shipboard liquid waste disposal. 2007, Massachusetts Institute of Technology.
49. V. Vadillo, M. Belen Garcia-Jarana, J. Sanchez-Oneto, J. R. Portela, E. J. Martinez de la Ossa, Simulation of Real Wastewater Supercritical Water Oxidation at High Concentration on a Pilot Plant Scale. Industrial and Engineering Chemistry Research, 2011, 50: 12512–12520.
50. I. Leusbrock, S. J. Metz, G. Rexwinkel, G. F. Versteeg, The solubility of magnesium chloride and calcium chloride in near-critical and supercritical water. Journal of Supercritical Fluids, 2010, 53: 17–24.
51. J. P. Chan, C. A. LaJeunesse, S. F. Rice, Experimental techniques to determine salt formation and deposition in supercritical water oxidation reactors. 1994, Sandia National Labs., Livermore, CA (United States).
52. A. Aimable, H. Muhr, C. Gentric, F. Bernard, F. Le Cras, D. Aymes, Continuous hydrothermal synthesis of inorganic nanopowders in supercritical water: Towards a better control of the process. Powder Technol., 2009, 190: 99–106.
53. F. J. Armellini, J. W. Tester, Solubility of sodium-chloride and sulfate in subcritical and supercritical water-vapor from 450–550°C and 100–250 bar. Fluid Phase Equilibria, 1993, 84: 123–142.
54. M. Urusova, V. Valyashko, Solubility of NaF and Li_2CO_3 salts of the 2nd type in supercritical water and solutions of aqueous electrolytes (NaCl, Na_2WO_4, and Li_2SO_4). Russ. J. Phys. Chem. B, 2014, 8: 919–923.
55. J. Rincón, R. Camarillo, A. Martín, Solubility of Aluminum Sulfate in Near-Critical and Supercritical Water. J. Chem. Eng. Data, 2012, 57: 2084–2094.
56. M. Hodes, P. Griffith, K. A. Smith, W. S. Hurst, W. J. Bowers, K. Sako, Salt solubility and deposition in high temperature and pressure aqueous solutions. AIChE Journal, 2004, 50: 2038–2049.
57. S. N. Rogak, P. Teshima, Deposition of sodium sulfate in a heated flow of supercritical water. AIChE Journal, 1999, 45: 240–247.

58. L. Zhou, S. Wang, H. Ma, Y. Gong, D. Xu, Oxidation of Cu (II)-EDTA in supercritical water—Experimental results and modeling. Chemical Engineering Research and Design, 2013, 91: 286–295.
59. A. Yoko, Y. Oshima, Recovery of silicon from silicon sludge using supercritical water. Journal of Supercritical Fluids, 2013, 75: 1–5.
60. J. Sabet, S. Jafarinejad, A. Golzary, Supercritical water oxidation for the recovery of dysprosium ion from aqueous solutions. 2014.
61. D. Zou, Y. Chi, J. Dong, C. Fu, F. Wang, M. Ni, Supercritical water oxidation of tannery sludge: Stabilization of chromium and destruction of organics. Chemosphere, 2013, 93: 1413–1418.
62. W. Zhu, Z. R. Xu, L. Li, C. He, The behavior of phosphorus in sub- and super-critical water gasification of sewage sludge. Chemical Engineering Journal, 2011, 171: 190–196.
63. E. Fauvel, C. Joussot-Dubien, V. Tanneur, S. Moussière, P. Guichardon, G. Charbit, F. Charbit, A porous reactor for supercritical water oxidation: experimental results on salty compounds and corrosive solvents oxidation. Industrial and Engineering Chemistry Research, 2005, 44: 8968–8971.
64. A. Kruse, D. Forchheim, M. Gloede, F. Ottinger, J. Zimmermann, Brines in supercritical biomass gasification: 1. Salt extraction by salts and the influence on glucose conversion. Journal of Supercritical Fluids, 2010, 53: 64–71.
65. W. T. Wofford III, J. W. Griffith, R. W. Humphries, J. W. Lawrence, Apparatus and method for applying an oxidant in a hydrothermal oxidation process, in U.S. Patent. 2002.
66. K. Karakama, Methods for the Characterization of Deposition and Transport of Magnetite Particles in Supercritical Water. 2011, University of British Columbia.
67. K. A. Smith, M. Hodes, P. Griffith, On the potential for homogeneous nucleation of salt from aqueous solution in a natural convection boundary layer. Journal of Heat Transfer, 2002, 124: 930–937.
68. J. C. W. P. Zhang, X. D. Zhang, X. W. Liu, Y. J. Xia, Research Progress in Wastewater Treatment by Supercritical Water Oxidation. J. Environmental Science, 2003, 29: 15–17.
69. R. N. McBrayer Jr., J. G. Swan, J. S. Barber, Method and apparatus for reacting oxidizable matter with particles. 1997, U.S. Patents.
70. B. Wellig, M. Weber, K. Lieball, K. Príkopský, P. Rudolf von Rohr, Hydrothermal methanol diffusion flame as internal heat source in a SCWO reactor. Journal of Supercritical Fluids, 2009, 49: 59–70.
71. J. A. Titmas, Method and apparatus for conducting chemical reactions at supercritical conditions, in U.S. Patent. 1986.
72. C. S. Laspidou, D. F. Lawler, E. F. Gloyna, B. E. Rittmann, Heater effects on cyclone performance for the separation of solids from high temperature and pressure effluents. Separation Science and Technology, 1999, 34: 3059–3076.
73. B. R. Foy, K. Waldthausen, M. A. Sedillo, S. J. Buelow, Hydrothermal Processing of Chlorinated Hydrocarbons in a Titanium Reactor. Environmental Science & Technology, 1996, 30: 2790–2799.
74. H. Schmidt, S. Baur, V. Casal. The SCWO-destruction of organic compounds in the presence of salt in leachates from dump sites in the SUWOX-facility. In Proc. of the GVC-Meeting: High Pressure Chemical Engineering, Forschungszentrum Karlsruhe. 1999.
75. M. W. Dassel, D. C. Matter, D. H. Rennie, R. N. McBrayer Jr., J. E. Deaton, F. W. Thompson, Reactor for supercritical water oxidation of waste. 1997, U.S. Patents.
76. J. S. Barber, Methods of controlling flow of fluids reacting at supercritical conditions. 1995, U.S. Patents, NO. 5427764.
77. D. A. Hazlebeck, Hydrothermal processing with phosphate additive. 2001, U.S. Patents, NO. 6238568.
78. M. Modell, Processing methods for the oxidation of organics in supercritical water. 1985, U. S. Patent, No. 4543190.
79. P. Whiting, A. H. Mehta, Supercritical water oxidation of organics using a mobile surface. 1996, U.S. Patent, No. 5543057.

80. M. Modell, Design of suspension flow reactors for SCWO. In Proceedings of Second International Conference on Solvothermal Reactions, Takamatsu, Japan. 1996.
81. M. Modell, E. F. Kuharich, M. R. Rooney, Supercritical water oxidation process and apparatus of organics with inorganics. 2001, U.S. Patent, No. 6264844.
82. P. C. Dell'Orco, L. Li, E. F. Gloyna, The Separation of Particulates from Supercritical Water Oxidation Processes. Separation Science, 1993, 28: 625–642.
83. G. W. Nauflett, R. E. Farncomb, M. L. Kumar, Supercritical water oxidation reactor with a corrosion-resistant lining. 1995, U.S. Patent, No. 5461648.
84. S. A. Gairns, J. Joustra, Apparatus for the self-cleaning of process tubes. 1999, U.S. Patent, No. 5890531.
85. M. Modell, E. F. Kuharich, M. R. Rooney, Supercritical water oxidation process of organics with inorganics. 1993, U.S. Patent, No. 5252224.
86. J. Elliott, D. Hazlebeck, D. Ordway, A. Roberts, M. Spritzer, J. Hurley, S. Rising, Update on hydrothermal oxidation developments on DARPA/ONR and air force projects at General Atomic [C]. In Proceedings of the International Conference on Incineration and Thermal Treatment Technologies, Portland, OR. 2000.
87. D. A. Hazlebeck, K. W. Downey, M. H. Spritzer, Downflow hydrothermal treatment. 2000, U.S. Patent, No. 6054057.
88. L. D. Bond, C. C. Mills, P. Whiting, S. L. Koutz, D. A. Hazlebeck, K. W. Downey, Method and apparatus to remove inorganic scale from a supercritical water oxidation reactor. 2000, U.S. Patent, No. 6056883.
89. L. D. Bond, C. C. Mills, P. Whiting, S. L. Koutz, D. A. Hazlebeck, K. W. Downey, Apparatus to remove inorganic scale from a supercritical water oxidation reactor. 1996, U.S. Patent, No. 5560822.
90. W. R. Killilea, G. T. Hong, K. C. Swallow, T. B. Thomason, Supercritical water oxidation: microgravity solids separation. SAE Transactions, 1988: 339–350.
91. G. T. Hong, W. R. Killilea, T. B. Thomason, Method for solids separation in a wet oxidation type process. 1989, U.S. Patent, No. 4822497.
92. W. R. Killilea, G. T. Hong, K. C. Swallow, T. B. Thomason, Supercritical water oxidation: microgravity solids separation. 1988, SAE Technical Paper.
93. A. H. Halff, A. F. Reid, Method for separation and removal of impurities from liquids. 1994, U.S. Patent, No. 5372725.
94. Y. X. S. Z. Wang, Y. M. Gong, H. Zhang, D. H. Xu, Y. Guo, Desalination and slag removal device for waste supercritical water treatment, in China Patent. NO. CN101570357. 2009.
95. Y. M. G. S. Z. Wang, X. Y. Tang, H. H. Ma, J. Zhang, Z. Lu, Pre-desalter for supercritical water treatment of waste organic matter, in China Patent. NO. CN101987750A. 2011.
96. L. Z. S. Z. Wang, G. Y. Meng, D. H. Xu, X. Y. Tang, Coal-fired supercritical DC heating furnace for organic wastewater, in China Patent. NO. CN102633313A. 2012.
97. L. Z. S. Z. Wang, H. H. Ma, X. Y. Tang, Y. M. Gong, D. H. Xu, Hydrostatic cyclone desalination device for waste organic matter supercritical treatment, in China Patent. NO. CN102642909A. 2012.
98. F. Zhang, S. Chen, C. Xu, G. Chen, J. Zhang, C. Ma, Experimental study on the effects of operating parameters on the performance of a transpiring-wall supercritical water oxidation reactor. Desalination, 2012, 294: 60–66.
99. B. Wellig, K. Lieball, P. Rudolf von Rohr, Operating characteristics of a transpiring-wall SCWO reactor with a hydrothermal flame as internal heat source. Journal of Supercritical Fluids, 2005, 34: 35–50.
100. P. Cabeza, J. Queiroz, S. Arca, C. Jiménez, A. Gutiérrez, M. Bermejo, M. Cocero, Sludge destruction by means of a hydrothermal flame. Optimization of ammonia destruction conditions. Chemical Engineering Journal, 2013, 232: 1–9.
101. J. P. S. Queiroz, M. D. Bermejo, F. Mato, M. J. Cocero, Supercritical water oxidation with hydrothermal flame as internal heat source: Efficient and clean energy production from waste. The Journal of Supercritical Fluids, 2015, 96: 103–113.

102. G. Brunner, Near and supercritical water. Part II: Oxidative processes. Journal of Supercritical Fluids, 2009, 47: 382–390.
103. D. Xu, S. Wang, C. Huang, X. Tang, Y. Guo, Transpiring wall reactor in supercritical water oxidation. Chemical Engineering Research and Design, 2014, 92: 2626–2639.
104. P. Whiting, Reversible flow supercritical reactor and method for operating same. Journal of Cleaner Production, 1996, 5: 159.
105. M. Bermejo, F. Fdez-Polanco, M. Cocero, Experimental study of the operational parameters of a transpiring wall reactor for supercritical water oxidation. Journal of Supercritical Fluids, 2006, 39: 70–79.
106. J. Abeln, M. Kluth, M. Pagel, Results and rough cost estimation for SCWO of painting effluents using a transpiring wall and a pipe reactor. J. Adv. Oxid. Technol., 2007, 10: 169–176.
107. M. D. Bermejo, P. Cabeza, J. P. S. Queiroz, C. Jimenez, M. J. Cocero, Analysis of the scale up of a transpiring wall reactor with a hydrothermal flame as a heat source for the supercritical water oxidation. Journal of Supercritical Fluids, 2011, 56: 21–32.
108. Z. Chen, G. Wang, Z. A. Mirza, S. Yang, Y. Xu, Study of transpiring fluid dynamics in supercritical water oxidation using a transparent reactor. Journal of Supercritical Fluids, 2014, 88: 117–125.
109. Z. Chen, G. Wang, F. Yin, H. Chen, Y. Xu, A new system design for supercritical water oxidation. Chemical Engineering Journal, 2015, 269: 343–351.
110. S. Yang, G. W. Wang, Y. J. Xu, New design of supercritical water oxidation reactor for sewage sludge treatment. Advanced Materials Research, 2013, 774: 212–215.
111. M. Cocero, J. Martınez, Cool wall reactor for supercritical water oxidation: modelling and operation results. Journal of Supercritical Fluids, 2004, 31: 41–55.
112. J. E. Deaton, J. M. Eller, R. N. McBrayer Jr., Turbulent flow cold-wall reactor, in U.S. Patent. 1996.
113. S. Baur, H. Schmidt, A. Krämer, J. Gerber, The destruction of industrial aqueous waste containing biocides in supercritical water—development of the SUWOX process for the technical application. Journal of Supercritical Fluids, 2005, 33: 149–157.
114. A. H. Halff, A. F. Reid, Method for separation and removal of impurities from liquids, in U. S. Patent. 1994.
115. G. T. Hong, W. R. Killilea, T. B. Thomason, Method for solids separation in a wet oxidation type process, in U.S. Patent. 1989.
116. H. Barner, C. Huang, T. Johnson, G. Jacobs, M. Martch, W. Killilea, Supercritical water oxidation: an emerging technology. Journal of Hazardous Materials, 1992, 31: 1–17.
117. M. D. Bermejo, M. J. Cocero, Supercritical water oxidation: A technical review. AIChE Journal, 2006, 52: 3933–3951.
118. M. D. Bermejo, F. Fdez-Polanco, M. J. Cocero, Effect of the transpiring wall on the behavior of a supercritical water oxidation reactor: modeling and experimental results. Industrial and Engineering Chemistry Research, 2006, 45: 3438–3446.
119. F. Zhang, S. Chen, C. Xu, G. Chen, C. Ma, Research progress of supercritical water oxidation based on transpiring wall reactor. Chemical Industry and Engineering Progress, 2011, 8: 003.
120. W. J. Gong, F. Li, D. L. Xi, Supercritical water oxidation of acrylic acid production wastewater in transpiring wall reactor. Environ. Eng. Sci., 2009, 26: 131–136.
121. G. T. Hong, V. A. Zilberstein, Iridium material for hydrothermal oxidation environments, in U.S. Patent. 1996.
122. M. Cocero, E. Alonso, M. Sanz, F. Fdz-Polanco, Supercritical water oxidation process under energetically self-sufficient operation. Journal of Supercritical Fluids, 2002, 24: 37–46.
123. F. Zhang, S. Chen, C. Xu, G. Chen, C. Ma, Experimental studies on supercritical water oxidation of glucose with a transpiring wall reactor. In Power and Energy Engineering Conference (APPEEC), 2011 Asia-Pacific. 2011. IEEE.
124. E. L. Daman, Process and apparatus for supercritical water oxidation, in U.S. Patent. 1998.

125. X. H. S. Z. Wang, H. Zhang, D. H. Xu, N. Wang, Porous evaporative wall device for supercritical water treatment of waste organic matter, in China Patent. NO. CN101555059. 2009.
126. D. H. Xu, S. Z. Wang, Y. M. Gong, Y. Guo, X. Y. Tang, H. H. Ma, A novel concept reactor design for preventing salt deposition in supercritical water. Chemical Engineering Research and Design, 2010, 88: 1515–1522.
127. S. Z. W. Y. Guo, D. H. Xu, Y. M. Gong, X. Y. Tang, J. Zhang, Supercritical water oxidation reactor using auxiliary fuel to recharge heat, in China Patent. NO. CN202131145U. 2012.

Chapter 8 Study on Key Technologies of Supercritical Water Gasification/ Oxidation

8.1 Innovative Research and Development of Novel Reactor

Supercritical water (SCW: $T > 374.15$ °C, $P > 22.1$ MPa) possesses specially physical and chemical properties such as high diffusivity, low viscosity, low dielectric constant, and small amount of hydrogen bonds. Organic matter and oxygen can be dissolved in SCW with any ratio under high enough pressure condition while the solubility of inorganic salt in SCW is extremely low. Therefore, these properties are beneficial to reduce resistance of mass and heat transfer and to separate salts from SCW. As an attractive media for regulating chemical reaction process and performing salt separation process, SCW has attracted more attentions since the early 1980s. The definition of supercritical water oxidation (SCWO) is that the organic matter is oxidized into harmlessly small molecular compounds such as carbon dioxide, nitrogen, water, and inorganic salt by excess oxidant in SCW. Supercritical water gasification (SCWG) means that organic matter undergoes hydrolysis, pyrolysis reaction and so on in SCW to form a hydrogen-rich mixing gas. SCW is not only a solvent for organic matter but also a reactant for hydrogen production in SCWG [1]. SCWO and SCWG have been identified as promising technologies for harmless treatment and resource utilization of organic wastes, respectively, by previous research.

Some inorganic salt additives are added into reactor to decrease its corrosion rate [2]. Partial alkali compounds are introduced from reactor inlet to neutralize acids formed in SCWO or employed as catalysts for hydrogen production in SCWG [3]. These salts together with those contained in feedstock may result in reactor plugging and increase corrosion rate of reactor because of their depositions. Therefore, it is required to prevent these salts from accumulating on the internal surface of reactor. Furthermore, there is hardly a long-life catalyst for organic waste containing plenty of salts in SCWO or SCWG till now. Kruse [4] reported it needed to remove salts precipitated before catalytic gasification in SCWG for decreasing

S. Wang et al., *Supercritical Water Processing Technologies for Environment, Energy and Nanomaterial Applications*,
https://doi.org/10.1007/978-981-13-9326-6_8

catalyst deactivation (poison and/or pollution) rate. It is also necessary to remove salts precipitated before catalytic oxidation in SCWO for the same reason.

Except for above different sources, salts may also come from reaction products. They will precipitate from SCW due to extremely low solubility and indicate complicated phase behaviors. Kawasaki's group [5] reported that NaCl tended to agglomerate while KCl and $CaCl_2$ tended to disperse and deposit on the internal surface of reactor. In terms of whether or not the solid-state salt precipitated is promised to accumulate on reactor's internal surface, the precipitated salt and the internal surface of reactor have two kinds of relations in reactor [6]. For the former case, salt removal can be accomplished by special instruments such as mechanical brushing, rotating scraper, reactor flushing, and so on. For the latter case, it can be achieved by particular reactor designs and operation means. They include reverse-flow tank reactor with brine pool, reversible flow tubular reactor, transpiring wall reactor, centrifuge reactor, cross-flow filtration, density separation, additive, high velocity flow, and homogeneous precipitation. Moreover, extreme pressure is provided to avoid salt precipitation in SCW. Marrone et al. [6] have objectively reviewed the properties of the above specific reactor configurations and operational approaches in commercial applications. Furthermore, there are some other special reactor configurations such as cool wall reactor reported by Cocero and Martínez [7] and two pipes reactor introduced by Baur et al. [8]. Whiting and Mehta [9] and Calzavara et al. [10] set a moving surface and a stirrer in their reactors, respectively. Príkopský [11] installed a protective metal sleeve replaced easily to prevent salt from depositing on the internal surface of their axial reactor in SCWO. There is no doubt that no one reactor design or operation mean has been proven to be clearly superior to the others in all aspects.

In this part, previous investigations about salt deposition problems and three kinds of main reactor configurations for preventing salt deposition are further reviewed objectively. A novel reactor is designed and manufactured, which is utilized to treat sewage sludge by SCWO with a feed rate of about 100 L/h. The reactor is the core equipment of the first pilot-scale SCWO plant constructed in China. Its design concepts for salt removal are presented concisely.

8.1.1 Three Kinds of Main Reactor Configurations for Preventing Salt Deposition

MODAR reactor [12] or its variation [13] belongs to a reverse-flow tank reactor with brine pool, which separates/removes salts by SCW property, as shown in Fig. 8.1. Precipitated salts under supercritical condition fall down into the subcritical zone at the bottom of reactor by gravity, inertia, and forced convection. MODAR reactor has overcome the three disadvantages of tubular reactor, which are too long structure size, not suitable for dealing with salt-containing feedstock and easy to be plugged. However, due to the low falling velocity of small particle salts

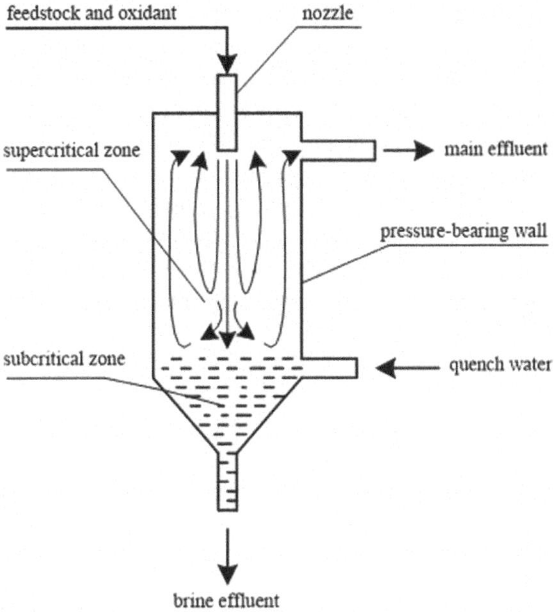

Fig. 8.1 Schematic diagram of reverse-flow tank reactor [6]

and the disturbance at the vertical direction, salt deposition is easy to take place on the internal surface of MODAR reactor in supercritical zone. Moreover, MODAR reactor may be eroded severely because its internal surface directly contacts corrosive reaction fluid.

Cool wall reactor [7] and transpiring wall reactor [14] can help to solve salt deposition problems existing in MODAR reactor. These two kinds of reactors are divided into pressure-bearing wall and non-pressure-bearing wall. For cool wall reactor, precipitated salts are re-dissolved in a subcritical water film on the internal surface of no-pressure-bearing wall. The film is formed by low-temperature feedstock cooling the no-pressure-bearing wall. However, severe corrosion still takes place because of the contraction between corrosive reaction fluid and internal surface of no-pressure-bearing wall. When scale on the internal and external surfaces of no-pressure-bearing wall is left uncontrolled, cool wall reactor will not continue to work normally since heat transfer efficiency reduces drastically.

Transpiring wall reactor can not only prevent salt deposition but also decrease reactor corrosion rate through a porous transpiring wall. Clean water flows across the porous wall to form a protective film to continuously dilute corrosive species, re-dissolve precipitated salt particles and/or sweep them away from the internal surface of reactor. The transpiring wall (no-pressure-bearing wall) can be made of porous ceramic [15], sintered metal, or many thin and porous metal layers/platelets bonded together [16]. The operational characteristics of transpiring wall reactor have been tested for a long time by some researchers. It is found that the concentrations of iron, nickel, and chromium ions in reactor effluent were very low, proving that corrosion rate of transpiring wall reactor was low. Fauvel et al. [17]

and Haroldsen et al. [18] also verified that transpiring wall reactor had a good corrosion resistance property. Thus, it can be claimed that transpiring wall reactor is the best reactor configuration for minimizing reactor corrosion rate at present. Nonetheless, plenty of energy is required to preheat the transpiration water entering into reactor [19]. The pressure difference of two sides of transpiring wall should be adjusted at a suitable range since it has a bad mechanical property.

On the basis of summarizing characteristics of previously main reactor configurations, together with our experimental results and design experiences of bench-scale SCWO facility for sewage sludge treatment reported elsewhere, a novel concept reactor is designed and manufactured for treating sewage sludge by SCWO, as shown in Fig. 8.2.

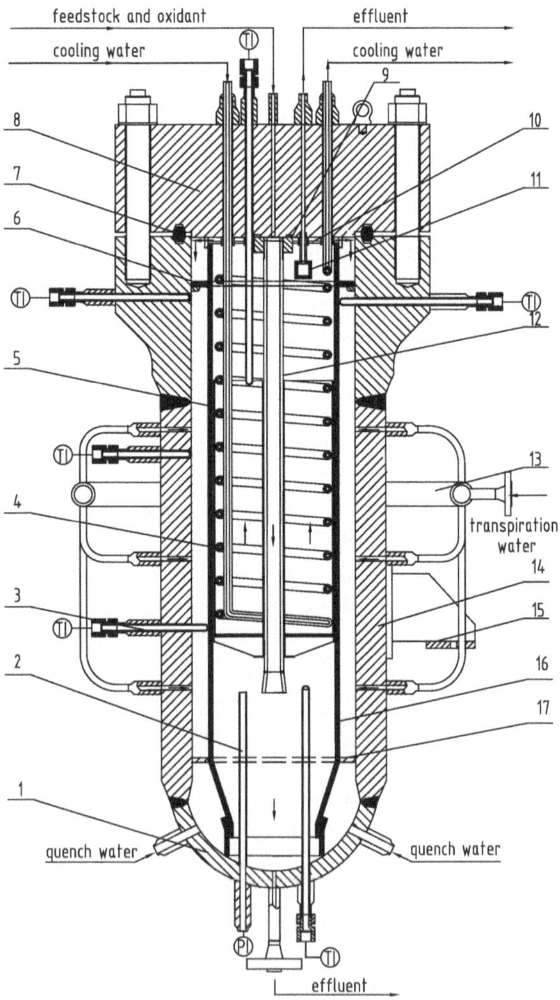

Fig. 8.2 Schematic diagram of TWM reactor for treating sewage sludge by SCWO. 1 Spherical head; 2 pressure measurement tube; 3 temperature measurement tube; 4 cooling spiral coil; 5 catalyst box; 6 support ring; 7 eight-square cushion; 8 top cover; 9 fixed platform; 10 cooling cover; 11 filter; 12 center pipe; 13 annular water box; 14 pressure-bearing wall; 15 ear base; 16 transpiring wall; 17 locating ring; TI: temperature indication; PI: pressure indicate

8.1.2 TMW Reactor Configuration

The novel reactor owns characteristics of transpiring wall reactor and MODAR reactor for the first time, which is simply nominated as "TWM reactor". Similar to MODAR reactor, TWM reactor is also divided into supercritical zone above and subcritical zone below. Precipitated salts under supercritical condition fall down into the subcritical zone (formed by pumping into quench water) and are re-dissolved. After removing salts, clean fluid flows toward top outlet of reactor by the differences of density and flow resistance. The valves equipped on pipelines of top and bottom outlets of reactor, which can be seen in pilot-scale plant, are used to regulate flow rate of fluid at top outlet. Dirt fluid flows out TWM reactor from its bottom outlet, which contains large amount of salts and solids. The flow pattern of TWM reactor is helpful for eliminating natural convection effects which were mentioned by Wellig's group [20] for transpiring wall reactor.

MODAR reactor wall which is replaced by a porous transpiring wall and a pressure-bearing wall is set to surround it to form fundamental configuration of TWM reactor. Clean and cool transpiration water is pumped into TWM reactor and forms a protective film on the internal surface of transpiring wall. TWM reactor configuration is expected to prevent salt deposition and decrease corrosion rate.

8.1.3 Main Parts in TWM Reactor

Center pipe of MODAR reactor is comparatively thin and long, while transpiring wall reactor has no center pipe. Transpiring wall reactor sets a complicated nozzle for mixing of oxidant and organic matter [21]. For TWM reactor, the mixing of oxygen and sewage sludge preheating is accomplished by a specially simplified mixer equipped on the pipeline of reactor inlet. Center pipe is inserted into the reactor deeply and its diameter is big enough to avoid plugging. Reactants in center pipe are further preheated by high-temperature fluid around or by reaction heat yielded through partial oxidation. Ideally, the mixture of oxygen and sewage sludge just reaches supercritical temperature at the outlet of center pipe. It is better to maintain subcritical temperature in center pipe because salts will not deposit on the internal surface of center pipe. The center pipe is set on a fixed platform by screw thread, which is easy to be replaced when plugged or eroded severely.

Generally, traditional transpiring wall reactor is a tubular reactor with a hydrothermal burner [22]. However, no hydrothermal burner is set in TWM reactor in order to decrease design temperature of reactor and operation cost of sewage sludge SCWO treatment. When transpiring wall configuration is combined with MODAR reactor, one problem is to design a suitable top cover for preventing salt deposition. In TWM reactor, top cover is cooled by low-temperature water in a particular chamber formed between top cover and cooling cover, as displayed in Fig. 8.3. The cooling cover is not porous so that the low-temperature water will not

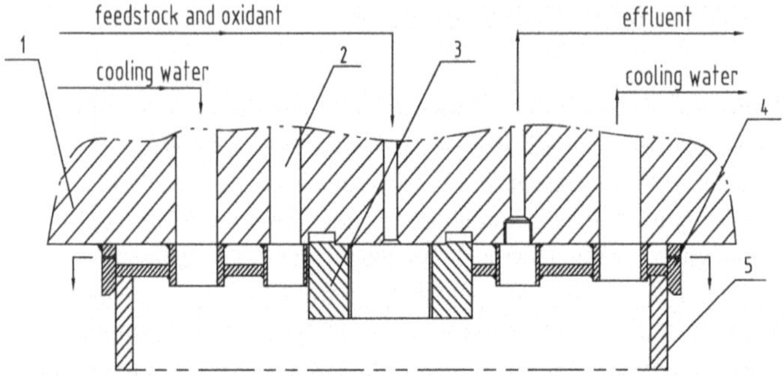

Fig. 8.3 Schematic diagram for cooling top cover of TWM reactor. 1 Top cover; 2 temperature measurement hole; 3 fixed platform; 4 cooling cover; 5 transpiring wall

fall into top reaction zone of reactor vertically. The low-temperature water is transported into the chamber from two pores at top cover to quench the cooling cover and from a subcritical water film on the lower surface of cooling cover. Then it enters into top annular space between transpiring wall and pressure-bearing wall to be further used as transpiration water. In Fig. 8.3, the two pores for transporting low temperature are not be seen because they are not set at the direction of top cover profile.

Wellig [21] has checked the percolation rates of transpiring wall consisting of sintered metal at two different voidages and pressure conditions. In TWM reactor, transpiring wall with filtering precision of 30 m and surface density of 20% is made of sintered wire netting. Approximately 100 floors of C276 metal nettings with pore size of 44 μm and thickness of 0.1 mm are sintered together at high-temperature condition to form the required transpiring wall. The inner diameter and length of cylindrical transpiring wall are 300 and 1180 mm, respectively. Moreover, the inner diameter of small end and length of conical transpiring wall are 200 and 245 mm, respectively. The cylindrical transpiring wall and the conical transpiring wall are connected through jointing by an especially subminiature electric welding machine to compose the whole transpiring wall. The whole transpiring wall is fixed in reactor by a support ring and a locating ring, and its expansion space is also provided. Transpiration water is adjusted at a subcritical temperature for re-dissolving precipitated salt, decreasing design temperature of pressure-bearing wall, and minimizing reactor's manufacture cost. Transpiration water with a low velocity is required to avoid itself shaking possibly. Thus, four inlets are arranged in one of three floors on the pressure-bearing wall to form a perfect film, and an annular water box is equipped for transpiration water distribution. It can be believed that the distribution is more homogeneous than that of reactor with a spindle transpiring wall and one transpiration water inlet. In TWM reactor, the transpiration water in the gap between transpiring wall and pressure-bearing wall doesn't contain oxidant (oxygen), which is different from the reactor designed by Mcguinness [23]. Moreover, it is also helpful for decreasing corrosion rate of transpiring wall.

A check valve will be equipped on the total pipeline of transpiration water inlet to ensure that reaction fluid will not conversely enter into the gap between pressure-bearing wall and transpiring wall.

In TWM reactor, an annular and porous catalyst box is fixed above the spout of center pipe to load catalyst. The catalyst box has no cover and its outer side is near to transpiring wall for avoiding salts deposition. Inert substance (like silicon dioxide) and catalyst are loaded at the bottom and top of the catalyst box, respectively. Reaction fluid enters into catalyst bed after salts have been removed at center pipe outlet and inert substance bed, which is beneficial to slow the deactivation rate of catalyst. Adding inert substance mainly aims to filtrate and capture partly precipitated salts as well as to provide large surface for salt deposition. TWM reactor also can be developed to utilize as a fluidized bed reactor in the future.

A cooling spiral coil is set in TWM reactor to control reaction temperature, which removes surplus heat in SCWO process by cooling water. The spiral coil is also equipped near to transpiring wall so that salt deposition on its outer surface can be prevented as much as possible.

8.1.4 Other Aspects About TWM Reactor

Components of feedstock should be confirmed in order to choose proper material or materials cooperation for reactor manufacture. Because one kind of material is impossible to be suitable for all varieties of feedstocks [24]. Hence, in TWM reactor, sewage sludge is chosen as a constant feedstock which contains moderate amount of salts. It is also found that nickel base alloy and titanium have benign corrosion resistance performances under supercritical and subcritical conditions, respectively [25]. Accordingly, the inner parts of TWM reactor such as transpiring wall, center pipe, sintered metal filter, catalyst box, fixed platform, and cooling cover are made of Hastelloy C276. The parts in subcritical zone are coated with titanium, and the other outer pressure-bearing parts consist of stainless steel 316.

In TWM reactor, removing salts will be conducted by SCW property before catalytic reaction. In order to ensure that reactor effluent of top outlet is clean enough and can be recovered, a small sintered metal filter with pore size of 30 m is fixed at the top outlet of reactor to further filtrate small salt particles. The fluid at the bottom of reactor contains plenty of salts and solids, and proper gradient on the internal surface of spherical head is provided for draining the dirt fluid. Moreover, proper inner diameter and velocity are used to prevent plugging of the inlet and outlet pipelines of TWM reactor.

In the following processes, reactor effluent of top outlet will be cooled and depressurized, then introduced into a gas–liquid separator for carbon dioxide and water separating. Ultimate liquid effluent will be recovered to utilize partially as transpiration water which is preheated up to a proper temperature by reactor effluent with high temperature. Quench water is pumped into reactor bottom to provide a protective water film, to form a subcritical zone re-dissolving precipitated salts, and to neutralize acids formed during reaction. Kawasaki et al. [5] presented that the bases such as $KHCO_3$, K_2CO_3, and KOH were fit to utilize as neutralizing agents in SCWO, but NaOH was unsuitable for many conditions due to its precipitation characteristics. Hence, in TWM reactor, KOH is employed as the neutralizing agent, which is added into the quench water at the bottom of reactor.

Of course, reactor temperature distribution should be measured and controlled exactly for reaction process of safety control and salt removal. In the reactor, some thermocouples are equipped at different positions to measure temperatures of reaction fluid, pressure-bearing wall, and transpiring wall. As indicated in Fig. 8.2, temperature measurement hole on pressure-bearing wall is drilled to the position approximate 1–2 mm distance to the internal surface of pressure-bearing wall. Then a smaller hole is drilled to contact transpiration water for accurate measurement of pressure-bearing wall temperature. The temperatures of supercritical zone are adjusted by interior reaction heat and/or exterior heat sources provided by cooling water in cooling spiral coil and/or by electric heater set on the pipeline of reactor inlet. The temperatures of pressure-bearing wall are controlled and regulated by temperature and/or flow rate of transpiration water. The temperature of subcritical zone is controlled at the range of 360–370 °C by quench water. In addition, the seal of top cover is accomplished by eight-square cushion. It can be assembled and dismantled conveniently to replace catalyst, inert substance, and inner parts of reactor. There is doubt that TWM reactor configuration also can be referenced by reactor design for sewage sludge SCWG. The shell of the newly constructed reactor without top cover is shown in Fig. 8.4.

TWM reactor is fit to deal with salt-containing feedstock but not the organic matter including plenty of heteroatoms. Because produced salts in reaction may plug inert substance bed and deactivate catalyst due to their deposition. Otherwise, inert substance bed and catalyst bed need to be rinsed or replaced frequently. Furthermore, the characteristics of heat and mass transfer about the reactor are relatively complicated.

8.2 Corrosion Prevention and Control Technology

8.2.1 On-Line Desalination during the Supercritical Water Oxidation Progress

The existence of aggressive anions in supercritical water oxidation environment is one of the key factors that cause the severe corrosion problem in the supercritical water oxidation system. Thus, the concentration of inorganic salts in the feed for SCWO treatment should be limited to reduce the corrosion of facilities materials. However, large amounts of inorganic salts are involved in the effluent after SCWO

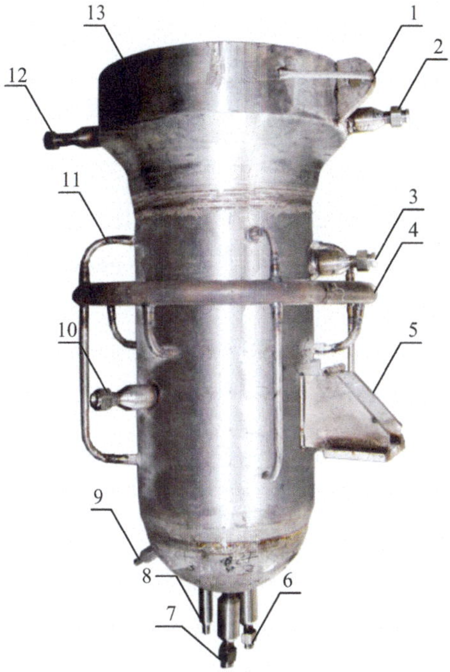

Fig. 8.4 Photograph of the shell of the newly constructed reactor without top cover. 1 Hoisting ring; 2, 3, 6, 10, 12 thermocouple joint; 4 annular water box; 5 ear base; 7 effluent outlet; 8 pressure gauge joint; 9 quench water inlet; 11 branch pipe; 13 pressure-bearing wall

reaction, which are newly formed during the reaction by heteroatoms in the organic matters, including Cl, S and P. Therefore, some methods should be used to remove the inorganic salts in the feed for SCWO treatment. Unfortunately, the conventional desalination methods (such as electrical dialysis, reverse osmosis, ion exchange, and electrical adsorption) are difficult to be applied in the SCWO system for an organic wastewater with a high salinity. In this section, an on-line desalination technique is introduced, which is able to effectively separate the inorganic salts from the feeds during the SCWO process.

Figure 8.5 shows the sequence flow diagram of the on-line desalination units. It is noteworthy that this diagram is just a part of a whole SCWO system, and the on-line desalination units can be added to any SCWO systems, which has no effects on the other parts of the system. The mixed wastewater is pressurized by a high-pressure metering pump and then sent to the heating furnace for preheating. A heat exchanging coil tube of the heating furnace in the system is arranged in two sections (including a high-temperature section and a low-temperature section). Fluid at an outlet of the low-temperature section (namely the middle outlet of the heating furnace) reaches a supercritical temperature. After entering the hydrocyclone separator, most solid salts with a particle size more than 10 μm in reaction fluid are separated by centrifuging. After salts removal, the fluid enters the high-temperature section of the heating furnace through the top outlet of the

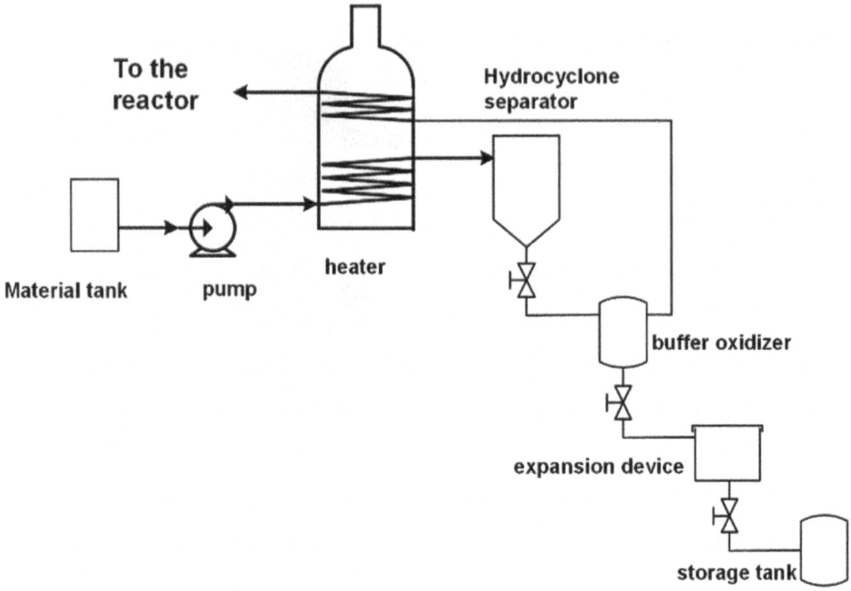

Fig. 8.5 Schematic diagram of the on-line desalination units

hydrocyclone, so as to ensure a heat exchanging coefficient of the heat exchange coil at the high-temperature section, and effectively prevent pipes and reactors following the hydrocyclone from plugging. At the same time, solid inorganic salts separated at the bottom of the hydrocyclone are spirally sent to the buffer oxidizer by a motor installed on the hydrocyclone. When the buffer oxidizer is full of the solid inorganic salts, a cut-off valve at the top inlet of the buffer oxidizer is closed, the motor installed on a top portion of the hydrocyclone is closed, a bottom cut-off valve of the buffer oxidizer is slowly opened, and a spirally transporting motor installed on the buffer oxidizer is started for transporting the solid inorganic salts in the buffer oxidizer to the expansion device. A fluid containing the solid inorganic salts expands in the expansion device. Steam generated enters the storage tank. Heat generated is recycled. The separated solid inorganic salts enter the salt storage, and are taken out for further landfill at a certain period. In addition, after being spirally sent into the buffer oxidizer, the fluid with a high salinity separated by the hydrocyclone reacts with oxygen previously injected through an oxygen pipe, which harmlessly removes organic pollutants. After the supersaturated fluid with a high salinity enters the buffer oxidizer, the solid inorganic salt will deposit at the bottom of the buffer oxidizer. Supercritical fluid at a top of the buffer oxidizer, which hardly contains solid inorganic salts, enters the top outlet of the hydrocyclone. Hereto, the on-line desalination process is finished.

8.2.2 Innovative Design of Anti-corrosion Reactors

A reactor is the core equipment in a SCWO system. The oxidizing agents (gas oxygen or compressed air) and organic feeds are mixed in the reactor, undergoing the supercritical water oxidation reaction. Large amounts of heat are released during the reaction, drastically raising the temperature of the reactor and promoting the extent of corrosion. Additionally, the heteroatoms in organic matters, such as Cl, S and P, are transferred into aggressive anions after the oxidation reaction. What's more, the solubility of the inorganic salt is significantly low in the supercritical water (typically less than 100 mg/L). Viscous salts in supercritical water oxidation of organic wastewater will precipitate and deposit on the reactor surface, forming a micro-environment between the salt deposit layer and the inner surface of the reactor. The existence of this salt deposit layer will obviously change the local corrosion environment of the reactor, increasing the corrosion rates of the facility materials.

Nowadays, it seems that the most effective approach to overcoming reactor corrosion and plugging is to design an appropriate reactor. Reactor configuration design is considered as the key problem of SCWO commercialization. Although tubular reactor is most widely used due to its simplicity and reliability properties, it is not fit to dispose high salt-containing feedstock or highly corrosive feedstock. Moreover, rapid exothermic reactions may result in uncontrolled hot-spots. It is very expensive to coat the high-temperature- and high-corrosion-resistant material on the inner surface of the tubular reactor. Corrosion and plugging problems are still big risks in the real operation, and even to make commercial SCWO plants inactive. The detail of the novel design of the anti-corrosion reactors is introduced in Sect. 8.1.

8.2.3 Feed Modification

When the feed stream is acidic at ambient temperature, Ni and Fe are selectively dissolved, and a de-alloyed oxide layer forms at the subcritical temperatures. Cr and Mo can form stable oxides in oxygen-free environment, however, hexavalent compounds of Cr and Mo form in oxidizing supercritical water environment, which are soluble and have no protectiveness on base alloy. Due to the fact that the structure of the de-alloyed oxide layer is highly defective, the growth rate of the de-alloyed oxide layer is extremely fast and cannot be accepted. Thus, the acidic feeds are recommended to be modified to avoid the severe corrosion of facilities materials.

Dilution is an effective method to reduce the risk of corrosion of an aggressive feed by decreasing the concentration of corrosive anions. However, the concentration of organic matter will also be diluted by this method, leading to a procedure economically unacceptable. Notably, 5% of organic maters concentration should be reached to maintain the self-heated balance of the SCWO progress. Thus, dilution of the feed could just be used in some specific occasions, such as treating some toxic materials that have the potential to do great harm and must be thoroughly decomposed with a neglection of cost.

Feed neutralization is another method to reduce the risk of corrosion from two aspects. First, ammonia would be formed by nitrogenous organic compounds during a supercritical water oxidation reaction performing at a temperature less than 600 °C. It is noteworthy that a hydrolysis of ammonia can undergo in a high-temperature and high-pressure environment, forming large amounts of hydrogen ions. And the hydrogen ion will combine with the aggressive anions in the effluence, such as chloride ion and sulfate ion, forming a strong acidic environment. Additionally, nickel ammine may form at the relatively low temperature, leading to a preferentially dissolution of nickel in facilities materials, which are mainly nickel-based alloys. Second, based on the E-pH analysis, the major alloying elements are able to form stable oxides in both the subcritical and supercritical temperature. Namely, Ni and Cr would form NiO and Cr_2O_3, respectively. Generally, the pH of feed should be adjusted to neutral, or even alkaline to prevent the severe corrosion caused by acid.

Caustic soda is recommended to be used as the neutralizer, which is relatively cheap and easy to get. Specially, if the feed is sludge, the neutralizer is better to be injected into the SCWO system after hydrolysis of the sludge feed, namely after the preheating process. This is because that the sludge is an inert feed which will consume large amounts of the neutralizer and improve the operation cost. However, a high-pressure pump is needed if the neutralizer is injected after the preheating process, which will raise the investment cost. Nevertheless, a neutralizer is important for a SCWO system due to its obvious effect on corrosion reduction.

8.2.4 New Operation Scheme with Reduced Operating Parameters

A high temperature and a high oxygen concentration are needed in order to achieve a satisfied removal efficiency of organic matters in a SCWO system. However, the corrosion of structural materials is also apparently promoted by the increasing temperature and oxygen concentration. Therefore, a new operation scheme should be used to achieve a high organic matters removal efficiency, but not at the expense of the facilities materials corrosion.

The core concept of the new operation scheme is to reduce the operation temperature and oxygen concentration in a reactor. As a result, the chemical oxygen demand (COD) removal efficiency cannot reach the standard value, and the concentration of ammonia nitrogen in the effluence was also unsatisfied. To work around this problem, some secondary treating methods were used for ammonia nitrogen and COD thoroughly removal, including an ammonia stilling process, a phosphorous removal process, and a biochemistry process, which are all conventional water treatment techniques and can be operated easily with a relatively low operation cost.

8.2.5 Novel Mixer

The reactor is the core facility in a SCWO plant because the SCWO reaction occurs in the reactor. In a reactor, the heated feed and oxygen first meet in a mixer, where the local oxygen concentration is much higher than that in other areas in the reactor. There are two technical drawbacks to this process. First, drastic oxidation reaction occurs once the sludge and oxygen are mixed, producing a dangerous increase in temperature in the reactor. Second, the local high oxygen concentration and local high temperature in the mixer and the forepart of the reactor cause severe corrosion of structural materials, leading to unpredictable corrosion effects, such as pitting corrosion and stress corrosion cracking. Thermal control of the SCWO process is needed to avoid the formation of hot-spots. Several researchers have developed new types of reactors to avoid local overheating, including multi-oxidant injection and cool-water injection systems.

A new mixer is developed to optimize the oxygen injection process in the reactor. As seen in Fig. 8.6, this mixer consists of six oxidant injection holes in the core tube and helical mixing elements are used to enhance mixing. The size number of the holes is determined by computational fluid dynamics (CFD) simulation to ensure that the input rate and flow distribution of oxygen are appropriate. Both the core tube and the helical mixing elements are made of nickel-based alloy 625, and the outer tube, which is also made of alloy 625, is thickened in order to improve the corrosion resistance. Through this method, severe corrosion occurrence and hot-spot formation are avoided.

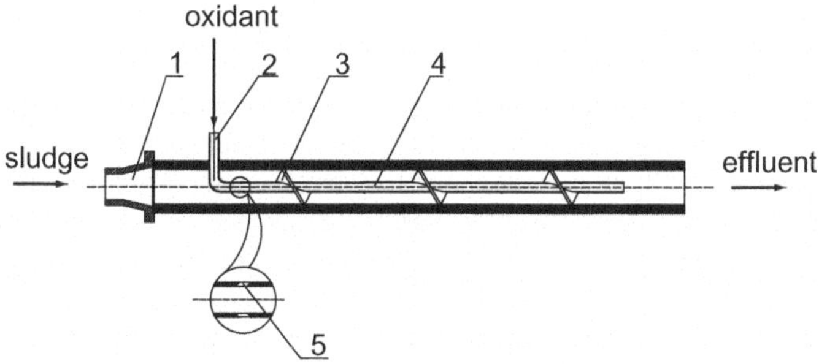

Fig. 8.6 Schematic diagram of the novel mixer. 1 Adapter flange; 2 core tube; 3 helical mixing element; 4 and 5 injection holes

8.3 Salt Deposition Prevention and Control Integrated Technology

Salt precipitation is a direct result of the poor salt solubility in SCW. These salts are either originated from feedstock or released from the products of organics oxidation, which may precipitate and deposit inside SCWO equipment. In many traditional fields, some methods have been used to avoid or decrease salt deposition, including distillation, ions exchange, electro dialysis, reverse osmosis, etc. However, these methods are not suitable for the scale-up SCWO system due to the limitation of the low removal efficiency for complex feedstock components at harsh operation conditions. Measures of solving salt deposition problems in SCWO plants are using some special operation techniques.

The reliable way to overcome salt deposition is subsection control along the process of wastewater treatment. The primary consideration is to minimize the salt content entering into the reactor. The objective can be achieved through restricting the content of salts in the feedstock and further setting a salt separator to remove salts before the reactor. Besides, cold feed injection restrains organics decomposition to salts and avoids salt deposition in the preheater, so it can prevent feed lines from salt deposition and plugging. The second stage of salt control is inside the reactor, by inhibiting salt precipitation or making precipitated salts away from the inner surface of the reactor. The present techniques mainly include additive, extreme pressure, low turbulence, homogeneous precipitation, etc. Nonetheless, salt deposition on the reactor's internal wall seems to be not inevitable for a long-term running of a SCWO system in the deteriorating case, so the techniques of mechanical brushing, rotating scraper, and reactor flushing can be employed to deal with the salt layer attaching to the internal wall. Eventually, in order to overcome salt deposition and plugging occurring in cooling and depressurization downstream lines, salts carried by the reactor effluent need further managements. Minimizing the salt content of the reactor effluent is one of the fundamental and effective approaches to overcoming salt deposition appearing in the cooling heat exchanger. The industrial techniques of density separation and filtration are used to remove salt particles from the reactor downstream fluid. Although many methods have been developed for the control of salt deposition, each method has its own characteristics, and no "right" approach exhibits better than all the others or properly works in all cases. For instance, filtration is easy to be plugged and corroded, and difficult to select appropriate filtration materials for the actual heteroatoms-containing wastewater. Increasing pressure to minimize salt deposition may also lead to the increase of solubilities of both corrosive substances and protective oxides, which accelerates the corrosion rate of the reactor. The appropriate means for a given SCWO system mainly depends upon specific combination of operation conditions and properties of the feedstock. It is noted that more than one of these methods can be employed in a SCWO process, and they are also generally used in conjunction with a specific reactor form.

As mentioned above, there are a variety of operation techniques that have been developed for overcoming salt deposition, but few of them can solve the problem thoroughly. A valuable idea, source control, is proposed here to overcome or decrease salt deposition in the whole SCWO system. The primary purpose of the source control is to minimize the amount of salts in the feedstock and in the reaction fluid. Generally, the wastewater treated by SCWO tubular reactors requires to satisfy certain requirements of salt concentrations (e.g. $NaCl < 200$ ppm, $Na_2SO_4 < 1$ ppm and $Na_2CO_3 < 1$ ppm, etc.). Nevertheless, the salt content of the wastewater is difficult to reduce and control on-line due to its complex component features and harsh operation conditions. A desired result is that these salts can be removed before entering into the reactor. Thus, salt separation seems to be a good means of achieving salt removal, completely solving salt deposition problems both in the reactor and in downstream pipelines. Salt separation before the reactor has been described in Sect. 7.2.1. By using the on-line desalination during the supercritical water oxidation progress, the salt can be removed and will not be struck in the reactor.

8.4 Supercritical Water Oxidation System Safety and Automatic Control Technology

In this Section, some safety and automatic control technology are introduced in detail, including new types of equipment design and a novel mode of operation.

8.4.1 Thermal Hydrolysis of Sludge

Thermal hydrolysis pretreatment (THP) is widely used for sludge pretreatment, which has been proven to be an efficient method to improve sludge disintegration. However, our previous studies have shown that thermal hydrolysis in SCWO plants reduces the viscosity of sludge, and it has little effect on the COD removal efficiency of organic matters.

To some extent, the use of thermal hydrolysis pretreatment is a safety control technology. First, the power of the diaphragm pump must be increased to transport viscous sludge, which will significantly increase the risk of overload of the pump's motor. Second, a large heat exchange area is needed if high-viscosity sludge is fed into the heat exchanger, which substantially increases the risk of plugging.

Figure 8.7 shows the schematic diagram of a SCWO system. In this system, the feed is heated to 85 °C in the preparation tank (3), and the thermal hydrolysis of the sludge occurs in the sludge buffer tank (8). The reaction time and reaction temperature of the sludge in the tank (8) are 180 min and 170 °C, respectively. Thus, the viscosity is as low as 33 mPa s and has no threaten on the overload of the pump and the plugging of the heat exchanger.

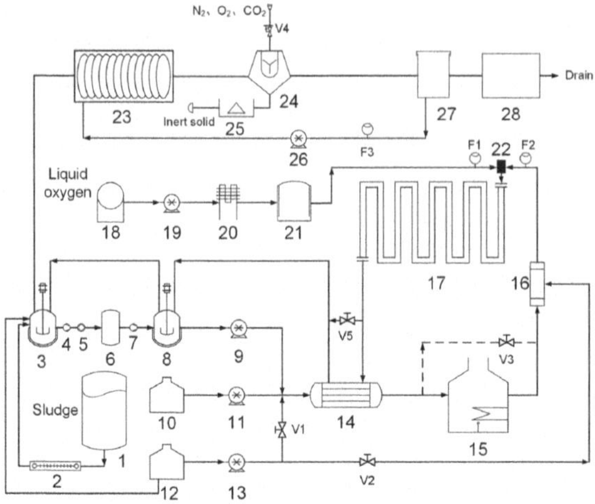

Fig. 8.7 Schematic diagram of the commercial plant for mixed sludge SCWO treatment. 1 Sludge tank; 2 screw conveyer; 3 preparation tank; 4 screw pump; 5 homogeneous emulsification pump; 6 sludge tank for sampling; 7 screw pump; 8 sludge buffer tank; 9 diaphragm pump; 10 methanol tank; 11 plunger pump; 12 water tank; 13 plunger pump; 14 heat exchanger; 15 heating furnace; 16 multiple nozzle desuperheater; 17 tube reactor; 18 liquid oxygen tank; 19 liquid oxygen pump; 20 vaporizer; 21 oxygen buffer; 22 mixer; 23 pressure relief device; 24 three phase separator; 25 centrifugal dehydrator; 26 plunger pump; 27 effluent buffer tank; 28 secondary treatment unit; F1–F3 flowmeter; V1–V5 electric valve

8.4.2 New Start-up Strategies with Aided by Methanol

The start-up process is one of the challenging steps in the commercialization of SCWO technology, since the feeds that containing a large amount of salt tend to stick to the inner walls of the pipes in heating modules.

In traditional SCWO plants, during start-up process, water is heated in a heating module to reach supercritical temperature (generally 400 °C) and then the feed is switched to organic waste. Once the process has started, the effluent is used to preheat the cold sludge in the heat exchanger, and the heat released by the oxidation reaction is enough to preheat the cold feed, which makes the process self-sufficient and permits the heating module to be switched off. Generally, the organic waste to be treated often contains some soluble salt that has low solubility in supercritical water. In the pipes of the heating module, the pipe walls are heated directly and are the first to reach supercritical temperature, forming a high-temperature micro-environment. Many studies have shown that inorganic salts tend to crystallize in the overheated micro-environment and stick to the inner walls of pipes, leading to clogging and interruption of the continuous process.

In order to avoid the risk of plugging and overheat in the heater, a new strategy is proposed, and this strategy can effectively decrease the equipment investment. During start-up, as shown in Fig. 8.7, not the sludge but the methanol solution in the methanol tank (10) is fed into the system and is heated by the heating furnace (15). When the final preheating temperature achieves to approximately 400 °C, the first step of start-up is finished. Then, the heating furnace is shut down, and oxidant is injected into the tube reactor (17) to react with methanol. A large amount of heat is released during the oxidation reaction, which maintains the energy balance of the whole system. After a few minutes, the electric valve (V3) is opened, and the heating furnace (15) is bypassed. Then, the feed is switched to sludge in the sludge buffer tank (8), which has already been heated to 170 °C. Through this method, no sludge passes the heating furnace, so effectively avoiding plugging in the heating facility. Additionally, only pure water, whose corrosivity is obviously lower than that of sludge at supercritical temperature and pressure, passes through the heating furnace. As a result, the risk of corrosion is also reduced.

8.4.3 Novel Mixer Design

The reactor is the core facility in a SCWO plant because the SCWO reaction occurs in the reactor. The safety control of the reactor is very important because the temperature, pressure, and concentration of oxygen are very high in the reactor. In a reactor, the heated feed and oxygen first meet in a mixer, where the local oxygen concentration is much higher than that in other areas in the reactor. There are two potential risks of this process. First, drastic oxidation reaction occurs once the feed and oxygen are mixed, producing a dangerous increase in temperature in the reactor. Second, the local high oxygen concentration and local high temperature in the mixer and the forepart of the reactor cause severe corrosion of structural materials, leading to unpredictable corrosion effects, such as pitting corrosion and stress corrosion cracking, or even burst of the reactor. Thermal control of the SCWO process is needed to avoid the formation of hot-spots. The design of the mixer has been introduced in Sect. 8.2.5. By using of this new mixer, the risk of localized corrosion and overheating are avoided.

8.4.4 Depressurization System with Capillary Devices and Valve Group

Depressurization of the effluent process is important in high-temperature and high-pressure plants because overpressure in the system may cause high-pressure pipes to crack or even burst and decompression of the system will cause vaporization of the high-temperature feed. Thus, a reliable pressure control system is needed in the commercial plant. Back pressure regulator valves are widely used for depressurization. However, feeds that contain suspended solid particles severely erode the internal parts of valves.

In this Section, a system with capillary devices and a series of valve group, which was designed by our group, is used for depressurization. The details of this system are shown in our patent. As shown in Fig.8.7, the back pressure of the SCWO reactor is regulated by a capillary device. The capillary pressure is precisely controlled by the fluid flow into the capillary. The fluid that enters into the capillary consists of the effluent from the outlet of the reactor (17) and the effluent from the outlet of the effluent buffer tank (27), which is re-pressurized to supercritical pressure by a plunger pump (26). During operation, if the back pressure of the reactor increases, the plunger pump (26) is adjusted, and the flow of effluent is decreased until the back pressure of the reactor returns to the set value. In contrast, if the back pressure of the reactor decreases, more effluent is pumped into the capillary to enhance the depressurization ability of the capillary.

8.4.5 Thermal Control of the SCWO Reactor

The temperature of the SCWO reactor is one of the key parameters that affect the COD removal efficiency of organic matter. Thus, various methods were designed to realize precise thermal control of the SCWO reactor in this commercial plant. There are several temperature measurement points in the reactor. If the temperature of the reactor increases, as shown in Fig. 8.7, the electric valve (V5) is slightly opened to allow some of the high-temperature effluent to bypass the heat exchanger (14), decreasing the temperature of the outlet of the heat exchanger (14). However, the opening of the electric valve (V5) should be limited to prevent a large amount of high-temperature effluent from directly passing through the coil heat exchanger in the sludge buffer tank (8) and causing unpredictable damage. If the temperature of the reactor increases sharply in a short time, the electric valve (V2) will be opened, and cooling water will be injected into the system via the multiple nozzle desuperheater (16). If the temperature of the reactor decreases, however, the electric valve (V5) will be slightly closed to allow more hot effluent to pass through the heat exchanger (14). Furthermore, if the heat released by the SCWO reaction is not sufficient, methanol in the tank (10) will be pumped into the reactor (17) and mixed with the feed in order to enhance the SCWO reaction.

8.5 Economic Improvement Technology of Supercritical Water Oxidation System

8.5.1 Catalytic Supercritical Water Oxidation

Catalyst is able to shorten reaction time, reduce reaction temperature, optimize reaction pathway, and improve conversion efficiency of organic matters, so it has a remarkable influence on improving the running cost of SCWO. Most of catalysts employed presently come from those used in wet air oxidation. Compared with homogeneous catalysts, heterogeneous catalysts are easier to be recovered and more effective. Thus, a catalyst box set in our TWM reactor can load suitable catalyst like bulk MnO_2.

8.5.2 Utilization of Liquid Oxygen

Liquid oxygen, gaseous oxygen, hydrogen peroxide, and air can be applied as oxidants in SCWO. Unfortunately, hydrogen peroxide is not commercially applicable because it is very expensive. Nitrogen in air does not react with feedstock, and costs a lot to separate them from gaseous products. Liquid oxygen is chosen as the oxidant of this plant due to its relatively cheaper price compared with the same amount of gaseous oxygen. Oxygen is regarded as one of the major costs, whereas high oxygen excess coefficient is not necessary because it is not realistic for a prospective commercial application. In this plant, oxygen excess coefficient is anticipated to control at about 1.1 to ensure complete decomposition of organic matters, to avoid recovering excessive oxygen as well as to decrease the running cost. The system uses liquid oxygen cold energy to realize the centralized liquefying and recycling of carbon dioxide, and greatly improves the system economy. In addition, the system is predesalted by the system of liquid oxygen cooling, reducing the subsequent desalting load and corrosivity of the system, reducing the grade of the equipment material, reducing the investment of the equipment, and recovering and selling valuable inorganic salts to improve the economy of the system.

8.5.3 Gas Recovery and Oxygen Reuse

The SCWO system innovatively adopts the process of combining gas phase product carbon dioxide recovery with oxygen recovery. By developing regular exhaust and compressor variable flow operation technology, high-rate oxidation and low-rate oxygen consumption are realized in the reaction process, that is, the reactor operates with high oxidation coefficient, and the gas phase after reaction is produced. The excess oxidants in the material are separated and recycled under variable flow conditions.

8.5.4 Waste Heat Recovery

Energy recovery is often achieved by preheating feedstock through reactor effluent of high temperature via heat exchangers [26]. Surplus energy is able to be utilized to produce steam for electricity generation or to form hot water for industrial application [27]. However, according to the computation conducted by the ProSimPlus (c) software, Marias et al. [28] confirmed that released energy in SCWO should be recovered in the form of heat but not electricity. For a larger scale SCWO system, energy recovery may potentially employ a supercritical steam turbine to generate electricity through the direct expansion of products [29]. However, it is not possible to design and construct a supercritical steam turbine in a SCWO plant presently due to the critical operation conditions and the very difficult separation of solids under SCWO conditions [28]. It is supposed that SCWO would be developed as a multifunctional technology for generating heat and electricity on the basis of removing organic pollutants in the future. There is no doubt that sufficient energy recovery and heat income must be adequately considered in order to decrease the running cost, otherwise an external or internal heat source is required. In this apparatus, heat is recovered enough by preheating feedstock and transpiring wall water to realize a low running cost. In addition, major byproducts such as hot water and CO_2 can be collected for sale or reuse.

References

1. M.B.G. Jarana, J. Sánchez-Oneto, J.R. Portela, E.N. Sanz, E.J.M.D. Ossa, Supercritical water gasification of industrial organic wastes. J Supercrit Fluid, 2008, 46: 329–334.
2. L.B. Kriksunov, D.D. Macdonald, Corrosion in supercritical water oxidation systems: A phenomenological analysis. Chemphyschem A European Journal of Chemical Physics & Physical Chemistry, 1995, 142: 4069–4073.
3. T. Yoshida, Y. Oshima, Partial oxidative and catalytic biomass gasification in supercritical water: A promising flow reactor system. Ind Eng Chem Res, 2004, 43: 4097–4104.
4. A. Kruse, Hydrothermal biomass gasification. J Supercrit Fluid, 2009, 47: 391–399.
5. S.I. Kawasaki, T. Oe, S. Itoh, A. Suzuki, K. Sue, K. Arai, Flow characteristics of aqueous salt solutions for applications in supercritical water oxidation. J Supercrit Fluid, 2007, 42: 241–254.
6. P.A. Marrone, M. Hodes, K.A. Smith, J.W. Tester, Salt precipitation and scale control in supercritical water oxidation—Part B: Commercial/Full-scale applications. J Supercrit Fluid, 2004, 35: 289–312.
7. M.J. Cocero, J.L. Martínez, Cool wall reactor for supercritical water oxidation: Modelling and operation results. J Supercrit Fluid, 2004, 31: 41–55.
8. S. Baur, H. Schmidt, A. Krämer, J. Gerber, The destruction of industrial aqueous waste containing biocides in supercritical water—development of the SUWOX process for the technical application. J Supercrit Fluid, 2005, 33: 149–157.
9. P. Whiting, A.H. Mehta, Supercritical water oxidation of organics using a mobile surface. US 5543057; 1996.
10. Y. Calzavara, C. Joussot-Dubien, H.A. Turc, E. Fauvel, S. Sarrade, A new reactor concept for hydrothermal oxidation. J Supercrit Fluid, 2004, 31: 195–206.
11. K.P. Príkopský, Characterization of continuous diffusion flames in supercritical water, 2007.
12. L.S. Cohen, J. Dan, G. Lee, D.W. Ordway, Hydrothermal oxidation of Navy excess hazardous materials. Waste Manage, 1998, 18: 539–546.
13. C.Y. Huang, Apparatus and method for supercritical water oxidation. US 5100560; 1992.

14. J. Abeln, M. Kluth, G. Petrich, H. Schmieder, K. Forschungszentrum, Waste treatment by SCWO using a pipe and a transpiring wall reactor, 2005.
15. H.C. Lee, J.H. In, S.Y. Lee, J.H. Kim, C.H. Lee, An anti-corrosive reactor for the decomposition of halogenated hydrocarbons with supercritical water oxidation. J Supercrit Fluid, 2005, 36: 59–69.
16. H.H. Mueggenburg, D.C. Rousar, M.F. Young, Supercritical water oxidation reactor with wall conduits for boundary flow control. US 5387398; 1995.
17. E. Fauvel, C. Joussotdubien, A. V. Tanneur, S. Moussière, P. Guichardon, A. G. Charbit, et al., A porous reactor for supercritical water oxidation: Experimental results on salty compounds and corrosive solvents oxidation. Ind Eng Chem Res, 2005, 44: 8968–8971.
18. B.L. Haroldsen, B.E. Mills, D.Y. Ariizumi, B.G. Brown, Transpiring wall supercritical water oxidation reactor salt deposition studies. Nuclear Fuels, 1996.
19. D. Su, Z. Zheng, Y. Wang, D. Tang, Supercritical water oxidation technology. Industrial Water Treatment, 2003 (in Chinese).
20. B. Wellig, K. Lieball, P.R.V. Rohr, Operating characteristics of a transpiring-wall SCWO reactor with a hydrothermal flame as internal heat source. J Supercrit Fluid, 2005, 34: 35–50.
21. B. Wellig, Diss, Transpiring wall reactor for supercritical water oxidation, 2003.
22. R.M. Serikawa, T. Usui, T. Nishimura, H. Sato, S. Hamada, H. Sekino, Hydrothermal flames in supercritical water oxidation: Investigation in a pilot scale continuous reactor. Fuel, 2002, 81: 1147–1159.
23. T.G. Mcguinness, Supercritical oxidation reactor. US 5558783; 1996.
24. P. Kritzer, N. Boukis, E. Dinjus, Factors controlling corrosion in high-temperature aqueous solutions: A contribution to the dissociation and solubility data influencing corrosion processes. Journal of Supercritical Fluids, 1999, 15: 205–227.
25. P. Kritzer, E. Dinjus, An assessment of supercritical water oxidation (SCWO): Existing problems, possible solutions and new reactor concepts. Chem Eng J, 2001, 83: 207–214.
26. M.D. Bermejo, D. Rincon, A. Martin, M.J. Cocero, Experimental performance and modeling of a new cooled-wall reactor for the supercritical water oxidation. Ind Eng Chem Res, 2009, 48: 6262–6272.
27. M.J. Cocero, E. Alonso, M.T. Sanz, F. Fdz-Polanco, Supercritical water oxidation process under energetically self-sufficient operation. J Supercrit Fluid, 2002, 24: 37–46.
28. F. Marias, F. Mancini, F. Cansell, J. Mercadier, Energy recovery in supercritical water oxydation process. Environ Eng Sci, 2008, 25: 123–130.
29. M.D. Bermejo, M.J. Cocero, Destruction of an industrial wastewater by supercritical water oxidation in a transpiring wall reactor. J Hazard Mater, 2006, 137: 965.

Chapter 9
Near-Zero-Emission System of Pollutants Based on Supercritical Water Oxidation

The potential of SCWO lies in the complete destruction of persistent, organic waste materials. The main obstacles in the technical application of SCWO include the construction materials degradation triggered by high-temperature and high-pressure conditions and some highly aggressive species (such as oxygen, acids, alkali, and some salts), the precipitation of salts and insoluble materials, especially piping plugging issues resulting from them, and high running cost. In the following, the above three key problems encountered in designing systematic processes and their potential solutions are reviewed roughly, more information is available in the above sections of this book. The emphasis is put on introducing the abundant and significant work from Energy and Environment Institute in Xi'an JiaoTong University, in order to overcome the mentioned key technological and/or economic obstacles occurring during the industrial designs and commercial implement of SCWO, from the first pilot-scale SCWO plant of China to environment-friendly economical SCWO comprehensive processes for harmless disposal of various organic wastes.

9.1 Key Problems and Process Development

9.1.1 Presence and Solutions of Main Problems

9.1.1.1 Corrosion Issues

Harsh operation conditions (high concentration of oxidant, extreme pH value, high temperature, and pressure) together with reaction intermediate/ultimate products (high concentrations of ionic species, free radicals, acids, and inorganic salts) result in severe corrosion problems in SCWO. Corrosion issue mainly occurs on reactor's inner wall [1]; it also appears in heat exchanger and cooler on inlet and outlet pipelines of reactor. Corrosion in SCWO circumstance depends on parameters

S. Wang et al., *Supercritical Water Processing Technologies for Environment, Energy and Nanomaterial Applications*,
https://doi.org/10.1007/978-981-13-9326-6_9

Table 9.1 Corrosion resistance performances of metals against different media at subcritical and supercritical temperatures [4])

Materials	$T < T_c$ [a]; high density		$T > T_c$ [a]; low density	
	Good resistance	Poor resistance	Good resistance	Poor resistance
Nickel-based alloys	H_3PO_4, HF, Alkaline solutions	HCl, HBr, H_2SO_4, HNO_3	All acids	$[H_3PO_4] > 0.1$ mol/kg, NaOH
Titanium	All acids	F^-	HCl	H_2SO_4, H_3PO_4

[a]The critical temperature of water

concerning both the aqueous solutions (chemical dissolution process, electrochemical dissolution process, and influence of anions) and the materials (alloy composition, surface condition, material purity, and heat treatment) [2, 3]. For a specific reactor material, corrosion is influenced by the dissociations of acids, salts and bases, the solubilities of corrosion products and gases, and the stabilities of protective oxide layers; here these characteristics are affected by density and ionic product of solution [4]. In order to decrease corrosion rate, it is better to ensure that the solution density everywhere in reactor is below 200 kg/m^3 [1]. Corrosion under supercritical water conditions is categorized into general corrosion, dealloying, pitting, stress corrosion cracking, intergranular corrosion, crevice corrosion, under-deposit corrosion, hydriding, galvanic corrosion, non-coupled corrosion [5], or parts of them [2, 4].

Corrosion is a key obstacle to limit the commercial application of SCWO, which not only shortens reactor life but also induces a bad treatment effect of feedstock due to the formations of corrosion products. Materials serving for SCWO include stainless steels, nickel-based alloys, titanium, tantalum, noble metals and ceramics, etc. [4]. Many researchers have studied the corrosion resistance properties of these materials under supercritical and/or subcritical conditions for the aim of minimizing corrosion rate [6–11]. It can be concluded from Table 9.1 that each material shows an individual corrosion resistance property under the same reaction conditions. No one kind of material can withstand corrosion at all conditions, but some have perfect corrosion resistance properties under specific conditions [4]. Thus, appropriate reaction conditions such as heteroatom types in feedstock, reaction temperature and pressure should be fixed in order to minimize corrosion rate for a chosen reactor construction material. Generally speaking, nickel-based alloys show a good corrosion resistance performance among all the acids listed in Table 9.1 under supercritical conditions. Titanium is fit to be employed under subcritical conditions, e.g., preheater and cooler installed before and after reactor.

9.1.1.2 Precipitation and Piping Plugging

Plugging is another main obstacle to hinder SCWO commercialization, which is induced by sticky salt agglomeration and deposition on the internal surface of reactor. Salt deposition also accelerates reactor corrosion rate and catalyst inactivation rate and reduces heat transfer coefficient of reactor wall. When plugging takes place, SCWO plant has to be shut down, washed, and restarted, which will directly decrease the reliability of the plant and increase its running cost. Inorganic salts, whether soluble or not, may come from feedstock or reaction byproducts [4], and their viscosities decide the tendency of depositing on reactor wall. In general, different salts own different deposition characteristics [12]. Armellini and Tester [13] have explored experimental methods to research salt nucleation and growth in supercritical water. Salt deposition principles and the corresponding solutions have been systemically presented in the previous literatures [14, 15]. In a commercialized SCWO plant, reactor plugging can be avoided by particular reactor designs, special instruments, and operation means [5]. Possible solutions are to adopt reverse-flow tank reactor, reverse-flow tubular reactor, transpiring wall reactor, centrifuge reactor, mechanical brushing, rotating scraper, reactor flushing, cross-flow filtration, density separation, additive, high-flow velocity, homogeneous precipitation, and extreme pressure [15]. In fact, precipitated salts are relatively difficult to be removed out of SCWO plant during operation process, thus most SCWO plants do not have salt removal function. Salts in SCWO are classified into type 1 and type 2 according to their solubilities [16, 17]; type 1 salt has a high solubility in the range of water's critical temperature, while type 2 salt has a low solubility in this region. Type 2 salt can be separated by properties of supercritical water [18, 19], and the suitable supercritical conditions should be controlled at approximately 400 °C and 25 MPa according to the solubilities of salts mentioned in the previous studies [14, 20–22]. Hydrocyclone or centrifugal reactor helps to remove soluble salts under the above supercritical conditions but severe wearing problem is inevitable. At present, no an approach or a reactor configuration can really solve reactor plugging problem effectively and economically. Of course, low salt concentration in feedstock is helpful for avoiding reactor plugging, whereas this may need to dilute feedstock. According to the investigations of Marrone et al. [5], aluminum concentration in energetic hydrolysates should be limited to 300–400 ppm to avoid plugging.

In addition, when the preheating temperature of feedstock is in the range of 200–450 °C [23, 24], some undesired intermediate products like tar and char will be generated, which may also plug preheating pipeline. To overcome this problem, the following options are taken into consideration, which include enlarging the inner diameter of the preheating pipeline, adding a small amount of oxidant into the preheating pipeline to inhibit the formations of tar and char, increasing the heating rate of feedstock, and preheating feedstock up to a lower temperature even not preheating [25]. However, the improvement of the heating rate is limited by setting space of heating part, heating methods and withstanding temperature of preheater wall. The above fourth approach means that a large amount of heat will not be recovered into SCWO system, which will undoubtedly reduce energy efficiency and increase the running cost. Meanwhile, supplemental fuel should be injected into reactor to maintain a high reaction temperature [26–30]. But this alternative is less favorable for the energetic balance of the process [27].

9.1.1.3 High Running Cost

High reaction temperature and pressure result in a relatively expensive running cost, which also further affects the development of SCWO. The effective means may be to load proper catalyst to moderate the harsh reaction conditions, to minimize oxidant consumption, to recover heat of reactor effluent as well as to increase byproduct income. First, catalyst can improve reaction conditions, but it is also easy to be inactive due to the critical reaction conditions. Savage et al. [31, 32] reported the employed homogeneous and heterogeneous catalysts are mainly heteropoly-acids, alkali carbonates, carbons, transition metal oxides, and bulk MnO_2. Besides, they also especially pointed out that the bulk MnO_2 has high activity, hydrothermal stability, ability maintenance and resistance to metal leaching under supercritical reaction conditions. Second, oxidant consumption has an important influence on the running cost [1]. Excessive oxidant should be recovered from SCWO process to improve the technology's sustainable potentialities [33]. Thus, particular recovery units are installed in some researchers' SCWO plants to collect surplus oxidant [34], correspondingly the equipment investment increases. If oxidant excess coefficient is controlled in the range of 1.0–1.3 in order to neglect oxidant recovery unit, organic matters might not be destructed thoroughly. So the oxidant excess coefficient should be minimized as much as possible and meanwhile make sure that organic matters are decomposed completely. Third, designs and combinations of heat exchangers need to be optimized for recovering the heat in reactor effluent because energy recycle utilization is beneficial to further reduce the running cost. Fourth, byproduct income is also of importance for economic improvement. For instance, when one ton of dry sewage sludge (DS) is treated by SCWO in Harlingen Waterworks System, 120 dollars are available via selling CO_2 and hot water to a nearby clothing factory [35].

Heat self-sufficiency is a very significant condition to guarantee a low running cost, which means that no supplemental energy is provided in the operation process. Using the Aspen Plus (C) software and considering a turbine as a supercritical water detent, Lavric et al. [36] simulated the SCWO process and found that it can achieve energetic self-sufficiency at above 600 °C, 23 MPa and slightly excessive oxygen. Cocero et al. [37] found that the minimum heat value required for feedstock is 930 kJ/kg (equal to 2 wt% n-hexane) to achieve a heat self-sufficient operation. In this case, reactor effluent can preheat the feedstock from room temperature to 400 ° C and generate electric power equivalent to that consumed by high-pressure pump and air compressor in the SCWO plant [37]. Therefore, heat in reactor effluent should be recycled into the reaction system as much as possible for accomplishing heat self-sufficiency. Of course, auxiliary energy such as electricity, gas fuel, or liquid fuel ought to be supplied to maintain reaction temperature for low concentration of organic matters feedstock or to start SCWO plant. Here, heat supplying methods can be divided into internal heat source form and external heat source form. The internal heat source form means liquid fuel including methanol [28], alcohol or isopropyl alcohol [2, 26], etc. is pumped into reactor through a separate pipeline to combust with oxygen to compensate for the heat requirement. However, a special reactor configuration design requires that the pressure-bearing wall can withstand a very high-flame temperature [30, 38]. The external heat source form without a combustion phenomenon in reactor means heat is supplied by an outer electric or gas-fired heater [35]. Obviously, the supplemental heat should be minimized as much as possible in order to obtain a low running cost.

9.1.2 Developments on Process

9.1.2.1 An Overview of Foreign Development

SCWO, which has been the subject of many patents since firstly proposed by Modell in MIT in the middle of 1980s [39], is an effective and advanced oxidation technology to destruct organic matters by taking advantage of the unique properties of supercritical water under the typical operation conditions of 450–600 °C, 24–28 MPa [40]. Up to now, SCWO plants have been commercialized by several famous companies [35, 41, 42], such as General Atomics, Chematur AB, HydroProcessing, Supercritical Fluids International, and SuperWater Solution. In 2001, two commercial-scale SCWO plants were installed in Harlingen (Texas, USA) to treat sewage sludge with the solid-containing rates of 7–8 wt% [33, 35]. Another SCWO plant with a treatment capacity of 7 m^3/h has also been built by Chematur Engineering AB (Sweden) to deal with sewage sludge [43]. Besides, there are some other pilots or commercial-scale SCWO plants for the harmless treatment of sewage sludge or organic pollutants. From 2009 to 2011, SuperWater Solution installed and successfully tested a 5 dry t/d SCWO system for the Iron Bridge Regional Water Reclamation Facility in Orlando, FL, USA [41]. Nevertheless, the three key problems concerning corrosion, plugging, and high running cost still exist [40], even make some commercial-scale SCWO plants inactive [5, 27]. In 2007, the SCFI Group acquired patented supercritical water oxidation technology (Aqua Critox®) from Chematur Engineering AB of Sweden. Further work directed by the SCFI group has emphasized the reduction of costs.

9.1.2.2 SCWO Technology in Xi'an Jiaotong University

Since the early 2000s, Energy and Environment Institute in Xi'an Jiaotong University, based on the supercritical water technologies, has committed to the related research on harmless disposal and energy-oriented utilization of high concentration organic wastewaters and municipal/industrial sludges and has accumulated abundant experience in water quality evaluation and products characterization.

The research areas are involved in organic molecule reaction mechanisms and kinetics, catalytic characteristics and mechanisms, mechanisms and theories of alloy corrosion, salt crystallization kinetics, fuel and auxiliary heating mechanisms, etc. More than 200 research papers have been published in international academic journals and been cited and supported widely.

The stepwise engineering amplifications have been performed from initial experimental facility to test equipment (4.5 L/h), to small pilot plant (0.5 t/d). To further verify the industrialization feasibility of supercritical water oxidation technology in China, in 2009, we successfully designed and constructed the first pilot-scale plant of China, actually of which the processing capacity (3 ton/day) has reached the small industrial scale. A series of key technologies urgently needed to be solved during the industrialization implement, such as reactor corrosion prevention and control, piping plugging control, continuous desalting, reactor and system security control, energy recycling, and comprehensive utilization, have been improved and/or resolved. The overall performance of these pilot plants developed in our institute reaches the international first-class level and fills the China blank in the field of supercritical water treatment technology.

Depending on the accumulated research, design, and construction experiences, in the view of specific characteristics of various municipal/industrial sludges and organic wastewaters such as printing and dyeing wastewater, phenol-containing wastewater, and pharmaceutical/pesticide-production wastewater, combining with their individual industrial production processes and customer-specific requirements, we have carried out the process integration innovation of large-scale supercritical water treatment system to maximize the efficiency, economy, and safety and have finished a series of mature, reliable SCWO process packages.

9.2 First Pilot-Scale Plant of China for SCWO

The first SCWO pilot-scale plant of China with the treatment capacity of 125 kg/h has been constructed for harmless treatment of sewage sludge, as seen in Fig. 9.1. Reverse-flow tank reactor configuration and a porous transpiring wall are combined to design a novel TWM reactor, which aims to overcome reactor corrosion and plugging problems. A salt-removing plant is connected with the reactor bottom outlet, making soluble and insoluble salts removed under high temperature and pressure conditions in operation process. To some extent, the pilot-scale plant can achieve organic matters destruction, ammonia nitrogen removal, salt separation, surplus energy recovery, automatic control, and safety protection. The minimized running cost of the plant likely depends on the milder reaction conditions, lower oxygen consumption, more sufficient energy recovery, and more byproduct income. Compared with sewage sludge incineration, SCWO has a lower running cost and preferably harmless treatment results.

Fig. 9.1 First pilot-scale plant for sewage sludge SCWO treatment in China

9.2.1 Apparatus and Procedure

The schematic diagram of the first pilot-scale plant is illustrated as Fig. 9.2. Liquid oxygen is pressured and its flow rate is regulated by the low-temperature pump (2), subsequently gasified in the vaporizer (3), and then gaseous oxygen is further preheated by the heat exchanger (4). About 80 wt% of oxygen directly enters the mixer (5), and the rest of them mixes and prereacts with feedstock from the outlet of the heat exchanger (8) to prevent the formations of tar and char. Homogenized sewage sludge is compressed by a high-pressure meter pump (7), preheated in the heat exchangers (8), heated by an electric heater (9) at plant starting and heat deficiency stages, further mixed with oxygen in the mixer (5), and finally entered into the reactor (10) for oxidation reaction. Precipitated salts at a supercritical temperature fall down into the subcritical zone of the reactor bottom to re-dissolve. A large amount of reaction fluid, from which inorganic salts have been removed, reversely enters a catalyst box and then flows out from the top outlet of the reactor.

Reactor top effluent is divided into two ways fluids. The first one flows into the salt-removing plant (11) and then enters the heat exchangers (8) to preheat feedstock. The second one is transferred into the heat exchanger (15) to preheat transpiring wall water. The two ways fluids are converged and cooled through the heat exchanger (16) and then depressurized by the back pressure valve (19). Gaseous products and liquid products are separated in the low-pressure gas–liquid separator (20). The separated liquid enters into the vaporizer (3) to preheat liquid oxygen, and then part of them is drained from the water box (21) without pollution. The rest served as transpiring wall water is transported into the heat exchanger (15), and its flow rate is adjusted by the high-pressure meter pump (22) with a transducer. At plant starting or heat deficiency stage, transpiring wall water is preheated by the electric heater (14) and then enters the reactor (10) to cool its pressure-bearing walls

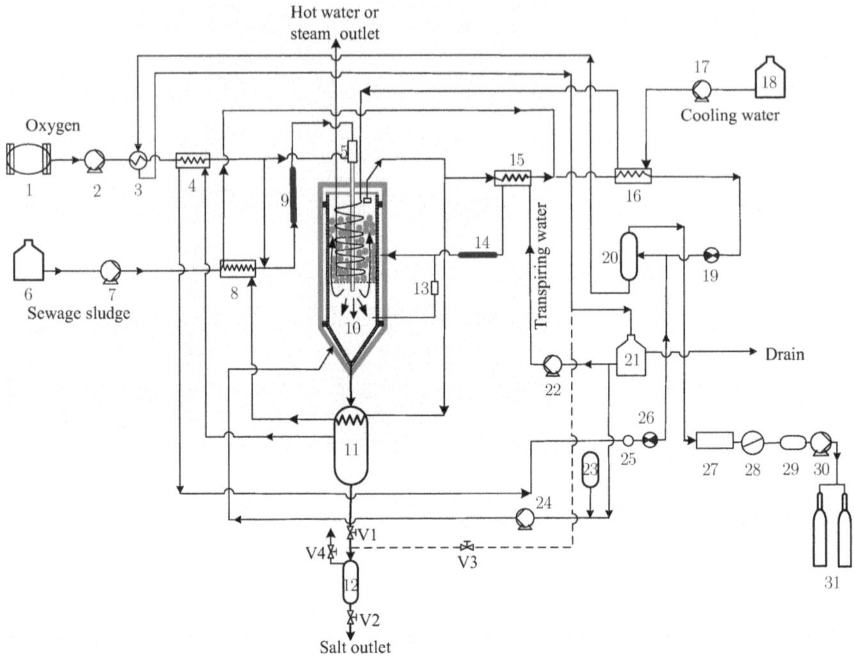

Fig. 9.2 Schematic diagram of the first pilot-scale plant for sewage sludge SCWO in China. (1) Liquid oxygen tank; (2) low-temperature pump; (3) vaporizer; (4, 8, 15, 16) heat exchanger; (5) mixer; (6) feedstock tank; (7, 22, 24) high-pressure meter pump; (9, 14) electric heater; (10) reactor; (11) salt-removing plant; (12) salt-collecting tank; (13) pressure balance plant; (17) high-pressure pump; (18, 21) water box; (19, 26) back pressure valve; (20) low-pressure gas–liquid separator; (23) caustic box; (25) filter; (27) drier; (28) flow rate meter; (29) buffer; (30) pressure pump; (31) gas cylinder; (V1-V4) electric valve

as well as form a protective water film which re-dissolves or washes off the precipitated salts and corrosive species under supercritical conditions.

Fluid with high concentration of salts entering the salt-removing plant (11) is preheated up to a supercritical temperature by high-temperature reactor effluent. The precipitated salts and insoluble salts fall down into the bottom of the salt-removing plant (11) by gravity and inertia and then enter the salt-collecting tank (12). These salts are periodically removed out of the plant and the methods will be detailed introduced in the following section. Clean fluid at the top of the salt-removing plant (11) flows into the heat exchanger (4) to be cooled down and then is filtered by the filter (25) and depressurized by the back-pressure valve (26). Gaseous and liquid products are similarly separated by the low-pressure gas–liquid separator (20). The flow rate meter (28) is installed to measure gas flow rate, and gas products can be collected by the gas cylinder (31) after compressed by the pressure pump (30). Liquid from the water box (21) combined with alkali compound (NaOH) solution from the caustic box (23) is compressed by the high-pressure meter pump (24) and then flows into the reactor bottom to form a protective water film and a subcritical zone to wash off corrosive species and re-dissolve precipitated salts. Additionally, cooling water from the water box (18) is pumped into the heat exchanger (16) and the reactor (10) by the high-pressure pump (17) to produce hot water or steam when heat in the reactor effluent is surplus.

9.2.2 Properties of the Pilot-Scale Plant

9.2.2.1 Avoidance of Corrosion and Plugging

In order to prevent reactor from corroding and plugging, many varieties of reactor configurations have been developed, which approximately include transpiring wall reactor, SUWOX reactor, downflow type reactor, cooled wall reactor, reverse flow tubular reactor, reverse-flow tank reactor with a brine pool (MODAR reactor), fluidized bed reactor, centrifugal reactor, double-wall stirred reactor, deep shaft reactor, and TWM reactor (Transpiring-wall MODAR reactor) [2, 4, 44]. Tubular reactor, which is proved to be unfit to dispose salt-containing wastewater, is often very long to confirm proper residence time and velocity for large flow feedstock. For instance, Pérez et al.'s reactor with an inner diameter of 6.22 mm and a treatment capacity of 2 L/min is 150 m long [45]. Transpiring wall reactor as an alternative to conventional reactors has attracted much attention due to its performance for minimizing corrosion and plugging problems [2, 46–48]. A pore size more than 24 μm is commonly recommended to avoid pore plugging for the porous transpiring wall [48]. However, transpiring wall reactor dilutes hot reaction fluid and decreases reactor effluent temperature, which makes heat recovery less efficient [2], thus transpiring wall temperature is usually supposed to control at 400 °C [49]. Nowadays, one potential development tendency of SCWO is to design a miniature commercial plant to deal with highly toxic organic pollutants such as chemical weapons [5]. Another one is to construct a large-scale plant with a low running cost to treat common pollutants like sewage sludge, etc. A novel TWM reactor is applied to our pilot-scale plant to avoid salt depositing and to decrease corrosion rate [44]. The reactor belongs to a reverse-flow tank reactor with a brine pool and, meanwhile, is combined with a porous transpiring wall made of sintered stainless steel 316 L net. It also consists of pressure-bearing wall and non-load-bearing wall and is expected to avoid reactor corrosion and plugging as well as to slower catalyst inactivation rate. The detailed information concerning the TWM reactor has been stated in our previous studies published in the Journal "Chemical Engineering Research and Design" [44].

The pressure balance plant (13) (seeing Fig. 9.2) with a taper sealing surface is installed to prevent the pressure difference between the two sides of the transpiring wall from exceeding its acceptable pressure limit. As soon as the pressure difference is near to the limit value because of pores plugging, the taper sealing surface would open to protect the transpiring wall. Additionally, the specially designed salt-removing plant, composed of a cylinder, a center pipe, a reflecting plate, and a heat exchanger coil, is connected with the reactor bottom outlet. First, salts separated by the reactor are transferred into the salt-removing plant for the secondary separation and finally stored in the salt-collecting tank in the form of concentrated brines. After a period of time of salts removal, close the electric valve (V1) and open the electric valve (V2) to depressurize up to environment pressure, then open the electric valve (V3) and utilize reactor effluent at the water box (21) inlet to wash and push salts out of the salt-collecting tank. After that, close the electric valve (V2) and open the electric valve (V4) for exhaust until reactor effluent fills full of the salt-collecting tank, finally close the electric valve (V4) and open the electric valve (V1). Thus the plant can remove soluble and insoluble salts in batches during the continuously running process under high temperature and pressure conditions.

In the plant, a pressure difference sensor is installed between the inlet and outlet of the heat exchanger (8) to determine plugging conditions. The heat exchanger (8) has to be washed when the pressure difference is higher than 3.0 MPa. Furthermore, the inner diameters of the heat exchanger (8) are enlarged gradually, and, meanwhile, 20 wt% oxygen is introduced from the outlet of the heat exchanger (8) to avoid the formations of tar and char. These methods are considered to help to prevent these feedstock preheaters from plugging due to organic matters decomposing into undesired intermediate products in the preheating process.

9.2.2.2 Minimization of Running Cost

Catalyst is able to shorten reaction time, reduce reaction temperature, optimize reaction pathway, and improve conversion efficiency of organic matters. So it has a remarkable influence on improving the running cost of SCWO. Most of catalysts employed presently come from those used in wet air oxidation. Compared with homogeneous catalysts, heterogeneous catalysts are easier to be recovered and more effective. Thus, a catalyst box set in our TWM reactor can load suitable catalyst like bulk MnO_2.

Liquid oxygen, gaseous oxygen, hydrogen peroxide, and air can be applied as oxidants in SCWO. Unfortunately, hydrogen peroxide is not commercially applicable because it is very expensive. Nitrogen in air does not react with feedstock and costs a lot to separate them from gaseous products. Liquid oxygen is chosen as the oxidant of this plant due to its relatively cheaper price compared with the same amount of gaseous oxygen. Oxygen is regarded as one of the major costs, whereas high oxygen excess coefficient is not necessary because it is not realistic for a prospective commercial application [1]. In this plant, oxygen excess coefficient is anticipated to control at about 1.1 to ensure complete decomposition of organic matters, to avoid recovering excessive oxygen as well as to decrease the running cost.

Energy recovery is often achieved by preheating feedstock through reactor effluent of high temperature via heat exchangers [42, 50]. Surplus energy is able to be utilized to produce steam for electricity generation or to form hot water for industrial application [2, 37, 42, 51]. However, according to the computation conducted by the ProSimPlus (c) software, Marias et al. [52] confirmed that released energy in SCWO should be recovered in the form of heat but not electricity. For a larger scale SCWO system, energy recovery may potentially employ a supercritical steam turbine to generate electricity through the direct expansion of products [53]. However, it is not possible to design and construct a supercritical steam turbine in a SCWO plant presently due to the critical operation conditions and the very difficult separation of solids under SCWO conditions [52]. It is supposed that SCWO would be developed as a multifunctional technology for generating heat and electricity on the basis of removing organic pollutions in future. There is no doubt that sufficient energy recovery and heat income must be adequately considered in order to decrease the running cost, otherwise an external or internal heat source is required. In this apparatus, heat is recovered enough by preheating feedstock and transpiring wall water to realize a low running cost. In addition, major byproducts such as hot water and CO_2 can be collected for sale or reuse.

9.2.2.3 Advancement of Control Methods

Taking into account several factors such as safe control, convenient operation, and commercial application, the plant can be almost automatically operated and controlled in a separate control room at a far distance. On the contrary, the equipment investment evidently rises owing to the installation of electric valves and instruments with remote display and/or control performances. The pressure difference between two sides of transpiring wall is balanced by continuous fluid through small pores to protect the non-load-bearing pressure wall, while reactor pressure is adjusted by back pressure valves. Reactor inner fluid temperature distributions are monitored and regulated to ensure reaction temperature conditions. Pressure-bearing wall temperature is measured and controlled to avoid exceeding its maximum work temperature. Reactor top fluid temperature is determined and maintained by changing the heating power of the electric heater (9) and the flow rate of cooling water entering reactor cooling coil. Flow rates of two outlets of reactor are distributed according to the opening degrees of the electric valves set at the reactor outlets. Moreover, reactor pressure-bearing wall can be prevented from over-heating, whose temperature is controlled in a normal range via regulating the opening degrees of the electric valves equipped on the reactor top and middle transpiring wall water pipelines, the heating power of the electric heater (14) and the flow rate of the high-pressure metering pump (22). Therefore, reaction pressure, reaction temperature and pressure-bearing temperature, and so on are automatically controlled to ensure the plant's safe running.

Opening degree of electric valve and major parameters like flow rates, pressures, temperatures on different sites in the equipment are indicated and/or adjusted through an automatic control interface based on four control cabinets. The plant is almost able to be started and shut down by clicking buttons on the interface indicated on a computer screen. Important equipment parameters involving reactor top fluid temperature, reactor bottom fluid temperature, mixer wall temperature, reactor top cover temperature, reactor wall temperature, salt-removing plant inner fluid temperature, reactor pressure, etc. are illustrated on the computer screen during the whole operation process. This helps to monitor and determine whether the plant works under safe conditions. As soon as any value exceeds its originally set maximum value, the control interface will alarm with a sound and a red indication, and fault shutdown procedure will be performed automatically if necessary. Thus, the plant with advanced automatic control properties is convenient and safe enough under high temperature and pressure conditions.

9.2.3 *Primary Experiment Results*

For the initial experiment tests of this pilot-scale plant, sewage sludge, as characterized in Table 9.2, was disposed at 400 °C, 25 MPa with the residence time of 6 min and the oxygen excess coefficient of 2.0.

Table 9.2 indicates the satisfied treatment results of sewage sludge SCWO in our apparatus at 400 °C, 25 MPa with the residence time of 6 min and the oxygen excess coefficient of 2.0. It can be observed that the removal rates of COD, TOC, and ammonia-nitrogen could reach up to approximately 99.6, 98.5 and 96.5%, respectively. The high total salinity separation rate (83.6%) was also obtained with the help of the salt separation function mentioned in the previous section. As demonstrated in Fig. 9.3, the chroma of liquid product is nearly similar to distilled water, and there are no suspended matters and solids existing in the collected liquid. Thus, the solid separation rate was 100%, which probably is attributed to the deposits of the newly formed and previous contained solids in the system, especially in the salt-collecting plant. The above liquid's indexes absolutely meet with "Sewage Comprehensive Discharge Standard" (GB8978-1996) in China. Gaseous products mainly included CO_2, N_2, a small amount of water vapor and excessive oxygen.

Table 9.2 Sewage sludge properties and its SCWO treatment results at 400 °C, 25 MPa with the residence time of 6 min and the excess-oxygen coefficient of 2.0 in the pilot-scale plant

Analysis items	COD (mg/L)	TOC (ppm)	Ammonia-nitrogen (mg/L)	Solid concentration (g/L)	Total salinity (mg/L)	pH value
Sewage sludge	5750	2081.5	237.5	4.05	397.7	7.45
Liquid product	24.9	30.2	8.3	0	65.4	7.53
Removal/separation rate	99.6%	98.5%	96.5%	100%	83.6%	–

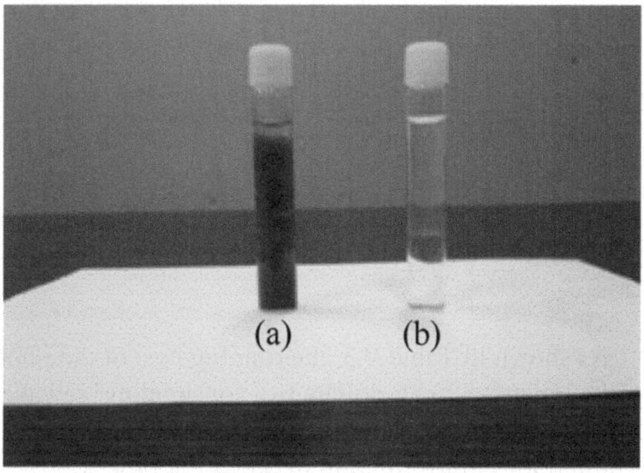

Fig. 9.3 Photo comparison between untreated fresh sewage sludge and its liquid product in the pilot-scale plant **a** fresh sewage sludge; **b** liquid product (400 °C, 25 MPa, 6 min residence time and the oxygen excess coefficient of 2.0)

9.2.4 Economical Comparison with Incineration

Incineration can also achieve quantitative reduction and harmless treatment of sewage sludge. Fluidized bed incinerator is usually adopted, which characterizes complex administration methods, high equipment investment, and expensive running cost. In an incineration process, partial heavy metal together with soot will diffuse into air, and some pollutants such as dioxins and NOx also will be produced. Dealing with these pollutants will remarkably increase equipment investment and running cost. Moreover, auxiliary fuel needs to be added to keep combustion going on when the water content of sewage sludge is higher than 60 wt%. The running cost of incineration is approximately 245.12–306.40 dollars/t_{DS}, which is about 2–4 times as much as those of other treatment methods such as landfill, drying, and anaerobic digestion [54]. Besides, SCWO does not require an energy-consuming drying procedure and can directly dispose sewage sludge with more than 90 wt% water content. It can accomplish heat self-sufficiency, and the lower reaction temperature ensures that no secondary pollutants such as dioxins and NO_X are formed.

The running cost of sewage sludge SCWO is influenced by adopted equipment, treatment capacity, concentrations and types of organic matters, operation conditions, and so on. An economical comparison between SCWO and incineration inEco Waste Technologies Company (EWT) reveals that the running cost of SCWO is about 50% of the latter, and the payback period of SCWO equipment investment is shorter [55]. MODAR Company presented that SCWO running cost is lower than that of incineration and is approximately 0.75 dollars/gal for a treatment capacity of 5000–100,000 gal/d. A SCWO plant (with a treatment capacity of 10 m^3/d) constructed by Organo (Japan) to dispose 1 wt% activated sludge can reduce the treatment cost of 20–30%. The net running cost of the SCWO plant for sewage sludge treatment in Harlingen Water System is approximately 60 dollars/t_{DS}. A cost evaluation performed by Aki and Abraham [56] through a computer indicates that SCWO has more technical and economic advantages than incineration. As it comes to a larger scale treatment capacity, SCWO will have a much lower running cost [57].

Table 9.3 Economical comparison of the SCWO pilot-scale plant and incineration plant in Shidongkou Wastewater Treatment Factory (Shanghai, China)

Items	SCWO pilot-scale plant	Incineration plant
Treatment capacity	3 t/d	213 t/d
Water content	92%	70%
Equipment investment	0.58 million dollars	12.26 million dollars
Running cost	76.56 dollars/t_{DS}	81.66 dollars/t_{DS}

As shown in Table 9.3, the running cost of the pilot-scale plant is estimated to be approximately 76.56 dollars/t_{DS} for treating sewage sludge with 92 wt% water content. The plant can achieve heat self-sufficiency, and only electric power is required for pumps running in operation process. The yield of hot water of 80 °C is approximately 1.2 5 t/h, and the corresponding income is about 1.92 dollars/h. If CO_2 income is considered, the total running cost will be less than 76.56 dollars/t_{DS}. As indicated in Table 8.3, the equipment investment of the SCWO plant is about 0.58 million dollars. It can be believed that the high-level automatic control system leads to the high equipment investment. It is also seen that Shidongkou Wastewater Treatment Factory (Shanghai, China) has spent 12.26 million dollars to build an incineration plant with a treatment capacity of 213 t/d to deal with sewage sludge with 70 wt% water content, and the plant's running cost is 81.66 dollars/t_{DS}. Therefore, it can be concluded that the equipment investment of the SCWO plant is higher than that of the incineration plant under the same treatment capacity condition. As the increase of treatment capacity and the improvement of byproduct income, the running cost of the pilot-scale plant will further decline, which will be evidently less than that of incineration. It is thought that byproduct income must be ensured in order to obtain a comparably low SCWO running cost. Several options can be adopted to further improve the economy of the pilot-scale plant, which includes optimizing the main equipment compositions to simplify the whole system, enlarging the treatment capacity to operate the plant under an optimal economical condition, conducting batch production to decrease equipment investment, and disinterring byproduct utilization value to increase income.

9.3 Development and Design of Environment-Friendly Economical SCWO Comprehensive Processes

Supercritical water oxidation technology can efficiently and thoroughly degrade the high-concentration and high-salinity industrial sludges. The experimental results of one kind of petrochemical sludge are shown in Table 9.4. It can be seen that when the reaction temperature and oxidation coefficient are 600 °C and 3, respectively, with a reaction pressure of 25 MPa and a 10 min residence time, the COD removal rate of the petrochemical sludge reaches to 99.82%, implying the degradation and

Table 9.4 Degradation of organic matter under several typical conditions

Item number	Temperature (°C)	Pressure (MPa)	Oxidation coefficient	Residence time (min)	Initial COD ($mg \cdot L^{-1}$)	Residual COD ($mg\ L^{-1}$)	Conversion of COD (%)
1#	600	25	1.1	10	47,140	140	99.70
2#	600	25	1.5	10	47,140	123	99.74
3#	600	25	3	10	47,140	86	99.82

removal of most organic matters in the sludge. However, for a variety of highly persistent pollutants, in order to achieve similar desirable results of the COD removal rate near or more than 99.9%, harsh reaction conditions are generally required, such as strict reaction temperature and pressure of 600–700 °C and 27–29 MPa, 3–10 times oxidant coefficient, and long reaction time (100–600 s). Unfortunately, these harsh reaction conditions enhance the requirements for expensive construction material, the excess-oxygen supplement and large reactor size, which significantly increase the investment of the whole plants and extremely reduce the economy.

At the same time, the experimental results also show that after preliminary SCWO treatment under relatively moderate reaction conditions of 520 °C and an oxidant coefficient of 1.1, biodegradability of the petrochemical sludge is significantly improved, as shown in Table 9.4. BOD_5/COD ratio of the petrochemical sludge increased from ~0 to 0.33–0.6, comparing with the original sludge. In fact, the above-mentioned phenomena are present widely for almost all potential organic wastes. Considering the higher safety and economy of SCWO plants serving at moderate conditions and relatively lower investment and running cost of regular biochemical treatment facility, one comprehensive SCWO process coupling moderate SCWO reaction with conventional biochemical technology is provided, which is expected to effectively improve the safety and economy of whole system while degrading the organic matters completely (Table 9.5).

However, supercritical water oxidation reactions happen at high temperature and pressure, meaning that the heating and pressuring of feedstock and the opposite process for the reaction effluents are necessary and unavoidable. In a conventional SCWO process, heat exchangers are usually used to cool the reaction effluents by untreated feedstock as cooling medium, meanwhile, the feedstock is preheated to the expected temperature and then entering the reactor and reacting with the oxidants. But the critical aqueous solutions' zones at temperature of 320–410 °C corresponding exactly to the heat exchanger, which owns high density, large dielectric

Table 9.5 Biodegradability indexes before and after SCWO treatment at moderate condition of 520 °C

Item	COD (mg/L)	BOD_5 (mg/L)	BOD_5/COD
Feedstocks	47,140	~0	0
Reaction effluents	500–600	200–300	0.33–0.6

constant, and high solubility of inorganic salts, are aggressive to the piping and equipment materials generally inducing serious corrosion issues by improper controls, posing a potential threat to the reliable operation of SCWO system. In addition, various organic wastes generally contain considerable amounts of heteroatom such as chlorine, sulfur, and phosphorus, which are transformed into corresponding inorganic acids after SCWO. Please note that the oxidant supplied into the system is often excessive to ensure the complete degradation of organic matters, causing a high level of residual oxidant in the reaction products. Therefore, reaction effluents are more corrosively than the initial feedstock. For the double-pipe heat exchanger used in conventional SCWO process, the cooling medium in the inner tube is feedstock while the reaction products as thermal fluid exists in the outer tube, causing internal side of the shell tube and both internal and external surfaces of the inner tubes facing serious corrosion problems and requiring expensive corrosion-resistant alloys, which greatly increases the investment of the equipment. Further, when the reaction effluents in the outer tube contain a large amount of inorganic salts or insoluble solids, plugging problems of the shell tube is extremely likely to occur due to its relatively complex multi-turn flow channel. In order to overcome or improve the potential higher investment, severe corrosion and blockage risk of the heat exchanging units, the indirect heat transfer unit is also proposed and implemented in recently developed SCWO comprehensive process here.

9.3.1 General Configuration

An indirect heat transfer supercritical water oxidation system is shown in Fig. 9.3, which mainly contains feedstock pretreatment system, oxygen supply system, SCWO reaction system, and follow-up processing system. The feedstock is filtered to remove over-sized substances through the filter (1), grinded to 50 μm by a low-pressure pump (2), and stored in feedstock tank (3) with added caustic to satisfy the feeding requirements by low-pressure caustic pump (22). Then, the feedstock is compressed by a high-pressure feedstock pump (4), preheated in the preheater (5) by high-temperature intermediate medium, and the method will be detailly introduced in the following section, then entered into reactor (6) when reaches the expected temperature. Liquid oxygen is pressured and its flow rate is regulated by the low-temperature oxygen pump (17), subsequently gasified in the vaporizer (18), and then gaseous oxygen is stored in the oxygen buffer tank (19), ultimately enters the reactor (6) after being depressurized by the oxygen pressure regulator (20) and reacting with the feedstocks. Preheated sludge reacts homogeneously with oxygen in the reactor (6) and decomposes most of the organic matters into CO_2, N_2 and H_2O. Meanwhile, the SCWO reaction releases a certain amount of heat and reaction effluent with high-temperature and high-pressure flows into the inner tube of heat exchanger (6) to heat the intermediate medium in the outer tube, and the effluents are also cooled through the intermediate medium at the same time. Then, the cooled effluents are reduced to an appropriate pressure by the

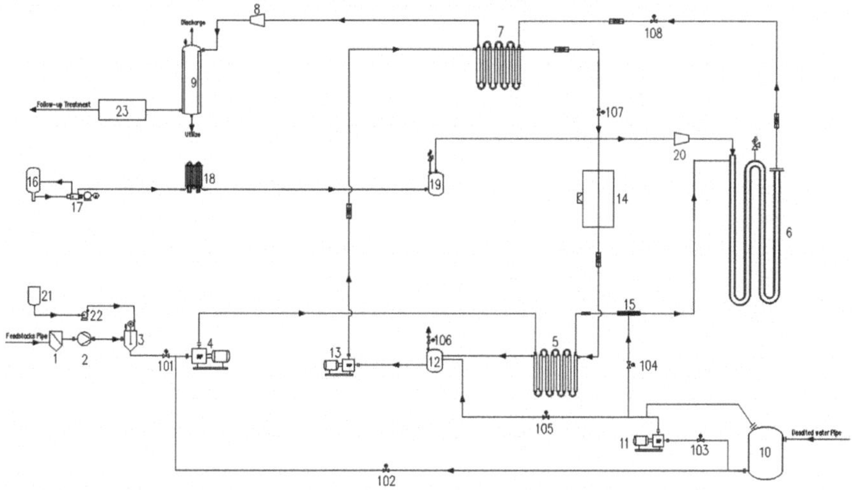

Fig. 9.4 Schematic diagram of the indirect heat transfer SCWO comprehensive system. (1) Filter (2) low-pressure pump (3) feedstock tank (4) high-pressure feedstock pump (5) preheater (6) reactor (7) heat exchanger (8) depressurization device (9) three-phase separator (10) water storage tank (11) high-pressure water pump (12) water buffer tank (13) booster pump (14) heater (15) spray desuperheater (16) liquid oxygen tank (17) low-temperature oxygen pump (18) vaporizer (19) oxygen buffer tank (20) oxygen pressure regulator (21) caustic tank (22) low-pressure caustic pump (101—108) valve (23) Biochemical treatment unit

depressurization device (8). Gaseous products, liquid products, and solid products are similarly separated in the low-pressure three-phase separator (9). Finally, the separated gas discharges directly to the atmosphere, the separated solid is used for brick-making or ceramsite, and the separated liquid enters into the follow-up biochemical treatment unit (23) to treat deeply (Fig. 9.4).

9.3.2 *Key Submodules*

9.3.2.1 Heat Transfer Between Supercritical Fluids

The schematic diagram of the indirect heat transfer unit by intermediate medium is illustrated as Fig. 9.5. The system contains feedstock loop, intermediate medium loop, and water supplement branch. The feedstock loop is comprised of high-pressure feedstock pump (4), preheater (5), spray desuperheater (15), reactor (6), regenerator (7), and back pressure valve (8), while the intermediate medium loop consists of water buffer tank (12), high-pressure circulating water pump (13), preheater (5), regenerator (7) and water coolers (H-1), and water supplement branch is divided into two ways by high-pressure water pump (11), the first one flows into the spray desuperheater (15), and the other is transferred into the water buffer tank (12).

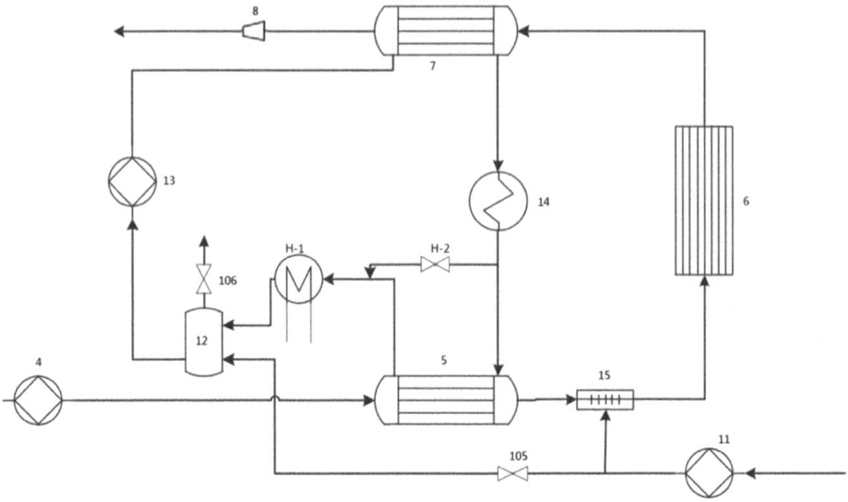

Fig. 9.5 Schematic diagram of the indirect heat transfer unit by intermediate medium. (H-1) water coolers (H-2) regulating valve (4) high-pressure feedstock pump (5) preheater (6) reactor (7) regenerator (8) back pressure valve (11) high-pressure water pump (12) water buffer tank (13) high-pressure circulating water pump (14) heater (15) spray desuperheater (105) regulating valve (106) back pressure valve

To ensure the operability and economy of the SCWO process, using supercritical water as intermediate medium is preferred. In the SCWO system, the feedstock before and after the reaction is the supercritical pressure fluid, and there is a large specific heat region near the critical point. Thus, in order to match enthalpy-temperature variation characteristics of the supercritical fluid, the pressure of the intermediate medium should be higher than that of the fluid after the reaction, which could ensure the heat exchange efficiency between medium and feedstock.

Therefore, when the indirect heat transfer system is started, clean water is poured into the SCWO reaction system by high-pressure feedstock pump (4) and gradually adjusts the back pressure valve (8) until the pressure after the reactor (6) increases to a target valve I which means the reaction loop has boosted successfully. And then boost the intermediate medium circuit, set a target pressure valve II by back pressure valve (106), inject water into buffer tank (12) through high-pressure water pump (11), then start high-pressure circulating water pump (13) for the circulation of medium and gradually adjust the back pressure valve (106) until the pressure inside the buffer tank increases to the target valve II. After the system has boosted, heat intermediate medium with a heater (14), adjust the power of the heater to ensure the constant heating rate of the reactor when the temperature of reactor input reaches to target temperature III, and then change feedstock instead of clean water.

The prepared feedstock enters the preheater (5) through high-pressure feedstock pump (4) for being preheated to a required temperature and then enters the reactor (6) reacting with oxidant which a certain amount of heat will release during the reaction process, then the high-temperature and high-pressure reaction effluent flows into the regenerator (7) for transferring the heat to the intermediate medium and is cooled simultaneously, depresses to low pressure by back pressure valve (8) and enters to a follow-up process. The intermediate medium from the water buffer tank (12) under the action of high-pressure circulating water pump (13), firstly absorbs the heat of the reaction effluent through the regenerator (7), then flows through the heater (14) if necessary to further increase the temperature, and eventually enters the preheater (5) for preheating the feedstock. When the system is running normally, the reactor operating temperature is adjusted by the power of the heater to ensure that it is within the normal range, what's more, the reactor is equipped with two-stage over-temperature protection measures which are used to stop heater and spray desuperheating. Moreover, a pressure interlocking control method is used to maintain normal pressures of the reaction process and the intermediate medium circuit, which could ensure the high efficiency of heat transfer effect by intermediate medium.

At this point, corrosive fluid only passes through the inner tube of the preheater and the regenerator, and both outer tubes of the heat exchanger are clean water. Therefore, only inner tubes of the preheater and regenerator require to use high-end corrosion-resistant alloys, and the outer tubes can use cheap carbon steel or low-alloy steel, which will greatly reduce the investment cost of heat exchanger for the SCWO process. In addition, the clean water in the outer tube also dramatically avoids the blockage risk compared with dirty fluid containing insoluble solids.

9.3.2.2 Depressurization of Fluids with High Content of Solids

SCWO plants are high-temperature and high-pressure systems, meaning that reaction effluent is still required to be cooled and depressed. As mentioned, cooling the high-temperature effluent can be achieved by the indirect heat transfer system. However, the conventional method normally used to maintain and decrease the pressure of SCWO system by adapting a pressure regulating valve. It means that the pressure drop of the whole system is achieved on one point, which will lead to extremely high velocities of feedstock and noise issues here, make the valve occur failure easily, and eventually affect the operation reliability of the SCWO system. In addition, if the feedstock contains solid particles, the existence of inert inorganic salt will result in severe abrasion or clogging around the regulating valve during the depressurization process of the SCWO system. Therefore, it is not advisable to achieve a complete depressurization in a single step by a regulating valve during the SCWO process, especially for the particles-containing feedstocks. Capillary pressure reduction is achieved by the frictional drag generating from the fluid flowing through the small diameter and high-length capillary, realizing the step-down of the system step by step, and effectively avoiding the problems about high velocities in tubes and corrosion damage of valves compared with depressurization by a single

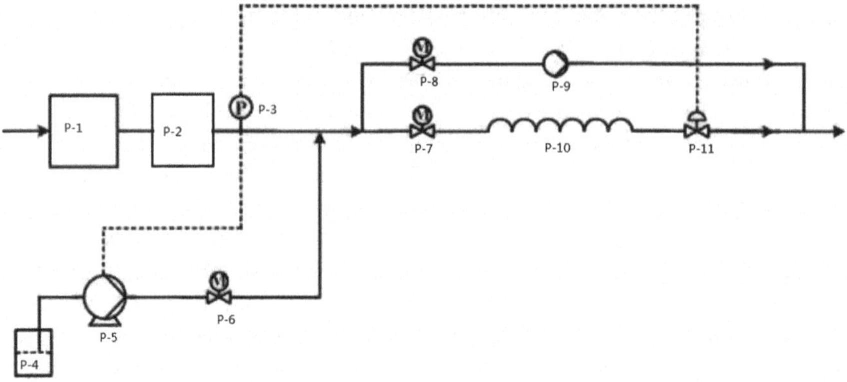

Fig. 9.6 Schematic diagram of the capillary depressurization system. (P-1) reactor (P-2) heat exchanger (P-3) pressure gauge (P-4) choke water buffer tank (P-5) high-pressure choke water pump (P-6,7,8) globe valve (P-9) back pressure valve (P-10) capillary pressure reduction (P-11) regulating valve

valve. What's more, adjusting the addition of choke water flowing into the capillary could accurately control the pressure of system, which ensures the reliability of the system operation.

In the present comprehensive SCWO system, a capillary system is proposed to control the pressure illustrated in Fig. 9.6, which comprises a capillary branch, a choke water branch and a back pressure valve branch. The capillary device is optimized by FLUENT simulation, ultimately determining that the size of the capillary is DN 18*6.5, which is capable of depressing the pressure of 23 MPa when the choke water is 3 t/d. Moreover, for optimized configuration, empirical formula of pressure drop between the system and choke water flow is optimally confirmed $\Delta P = 0.85641 \times \mathrm{F} - 20.363$ MPa, providing a theoretical basis for the regulation of the system pressure.

Obviously, the pressure balance and regulation of the system both achieved by the capillary device. The SCWO system is required to be boosted by clean water firstly before feedstock introduced into system, which is realized by turning on the globe valve (P-8) and regulating the opening degree of the back pressure valve (P-9) on the back pressure branch, ensuring the rear pressure of the reactor (P-1) gradually increases to a certain value I and then operating steadily for 30 min. Then, switch the back pressure valve branch and the capillary branch by closing the valve (P-8), opening the valve (P-7), and switching on the high-pressure choke water pump (P-5), afterward, gradually increase the pressure of the reactor until reaching the expected pressure II by adjusting the output flow of the choke water pump (P-5) from 30% of the charge, so as to boost pressure of the system successfully.

After the SCWO system operating normally for a while, high-pressure effluent from the reactor (P-1) flows into the capillary after cooling down by heat exchanger (P-2). Once the rear pressure of the reactor (P-1) changing, adjust the flow of choke water to regulate the pressure fluctuations by high-pressure choke water pump (P-5), moreover, adjust open degree of the regulating valve (P-11) ulterior if necessary. When the SCWO system need to be depressed during shutting down stage, first of all, open the regulating valve (P-11) fully and gradually decrease choke water flow of the high-pressure pump (P-5) until the water load reduced to 30%, then release interlock among the rear pressure of the reactor with the regulating valve (P-11) and the pump (P-5), and turn off the pump (P-5) and globe valve 5. Afterward, switch from capillary branch to back pressure branch, turn on the globe valve (P-8) and turn off the globe valve 6, establish the interlock between the rear pressure of the reactor (P-1) and the back pressure valve (P-9), gradually decrease opening degree of the back pressure valve (P-9) to make sure that the back pressure of the reactor gradually decreases to normal pressure.

9.3.2.3 Continuous High-Pressure Discharge Filter with Spiral Seal

At present, continuous filters are widely applied in chemical production due to various advantages such as simple structure, high filter rating, automatically washing the filter screen, and PLC intelligent control discharge. On the other hand, with merits of sensitive response, safe application, and low price, electromagnetic valves are generally used as the switch of discharging pipe about the conventional continuous filters. However, the electromagnetic valve has a poor reliability installed in the vertical discharging pipe, causing the service life of the electromagnetic valve greatly shortened; therefore, the discharging pipe in the conventional continuous filters is all horizontally installed. Liquid needed to be filtered passes through the filter screen and the clear liquid after filtered flows into the follow-up process, remaining the filter residue gradually accumulated on the internal wall of the filter screen. When the filter residue is collected to a certain amount, which implies that the filter reaches the cleaning differential pressure or the preset cleaning time, then the PLC issues the instruction to rotate the electric scraper, clean the residue on the internal wall of the filter screen. Due to the gravity accumulation, the residue is accumulated to the inlet of the discharging electromagnetic valve, meanwhile, PLC sends an instruction to open the valve, and the residue is discharged taking advantage of the differential pressure between the filter and the outlet of the electromagnetic valve. However, particles-containing feedstocks generally consist amounts of solid phase even after SCWO treatment, causing filter residue collected quickly in the internal wall of the filter in a short time and then accumulated in the discharge pipe and the inlet of the solenoid valve after washed by the electric scratch brush, forming a denser filter layer even a block. Therefore, the differential pressure between inside and outside of the filter is quite difficult to discharge smoothly even opening the solenoid valve, which will affect the normal operation of the filter to a certain extent. Furthermore, this intermittent discharge mode which relies on pressure or time to automatically control the discharge has a biggest drawback. The chamber of the filter is communicated with the atmosphere by the discharge pipe and the discharge solenoid valve, and the discharge valve constantly open and close, and the pressure of the filter chamber fluctuates frequently during the discharging process, which will inevitably lead to a pressure fluctuations of the whole SCWO system and affect the stability and security of the system (Fig. 9.7).

Therefore, a continuous discharge filter of spiral seal is designed and applied in present process to overcome these difficulties for achieving the high pressure continuous separation of three phase present in the SCWO effluents. The filter consists of a liquid inlet (S-2), a cylinder filter screen (S-3), and a liquid outlet (S-5) which provided on the external of the filter screen (S-3). When liquid passes through the filter screen, solids in the liquid are trapped in the internal wall of the filter screen, and the purified liquid after filtered is discharged via the liquid outlet. Moreover, the filter comprises a rotating drive rod where a plurality of electric scratch brushes (S-4) is installed, and a two-stage spiral blade unit is provided on the bottom of the drive rod, what's more, the drive rod is driven by the motor (S-1) provided on the central of the cylindrical chamber.

Meanwhile, a vertical discharge pipe (S-8) is provided below the discharge opening (S-9) of the continuous filter, making the residue become more easier to be discharged than horizontal discharge pipe due to the gravity and pressure, which effectively solves the problem of the blocking in the continuous filter during the discharge while filtering high solids liquid. Furthermore, the two-stage spiral blades comprise the mixing blades (S-6) and the spiral blades (S-7). The blades (S-6) for

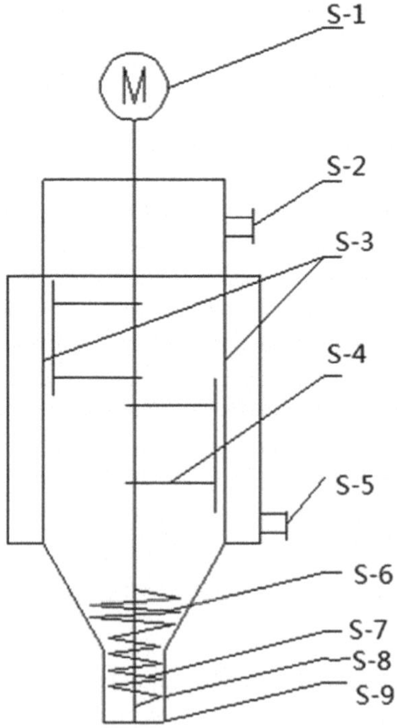

Fig. 9.7 Schematic diagram of continuous high-pressure discharge filter of spiral seal. (S-1) motor (S-2) liquid inlet (S-3) filter screen (S-4) electric scratch brushes (S-5) liquid outlet (S-6) mixing blades (S-7) spiral blades (S-8) vertical discharge pipe (S-9) discharge opening

mixing filter residue could prevent the residue from squeezing into blocks and overcome the difficultly of moving down while the blades (S-7) for sealing and pushing filter residue are installed to efficiently stabilize internal pressure of the filter chamber and push the residue out by sealing the filter residue during the discharge process. As a result, the continuous discharge filter with spiral seal could be widely applied to the separation of high-pressure three-phase fluids, benefiting from its capability of discharging smoothly, stabilizing pressure, and ensuring security for the SCWO system.

9.3.3 Conclusions

The comprehensive SCWO system with the treatment capacity near 100 ton per day, coupling with conventional follow-up processing system, has been designed optimally for harmless treatment of one petrochemical sludge, which can be almost directly used to degrade various organic wastes with a specific design optimizations. The heat transfer between supercritical fluids takes the measure of indirect heat transfer system by clear intermediate medium to preheat the feedstocks and cool down the reaction effluents, which effectively decreases the severe corrosion, dramatically avoids the blockage risk, and greatly reduces the investment cost of heat exchanger. Meanwhile, the depressurization of the SCWO system is achieved by capillary system, realizing the accurate control of system pressure and overcoming the problems about high velocities and corrosion damage compared with pressure reducing valves. Furthermore, SCWO could efficiently degrade organic matters in organic wastes into CO_2, N_2, and H_2O; the reaction products of gaseous, liquid, and solid phase are separated in the high-pressure three-phase separator with continuous discharge filter of spiral seal at the same time. The continuous discharge filter of spiral seal installed here is to solve the difficultly of filter residue blocking and efficiently stabilize internal pressure of the filter during the discharge process. In addition, the SCWO system is connected with conventional follow-up treatment such as ammonia stilling and biochemistry technologies to completely remove the ammonia and organic matters. It has proved that the reaction effluents could reach first level of Chinese national discharge standard and could totally be recycled. On the other hand, with the follow-up aid of biochemical treatment unit, the reaction conditions of SCWO can be relatively mild in these circumstances, which obviously decrease the cost of system equipment investment. What's more, the plant can achieve heat self-sufficiency, and only electric power is required for pumps running in operation process. The excess heat generated in the process of reaction is used to produce steam providing for industry, and the CO_2 separated from the gaseous products is sold after collected. As a result, with the improvement of the reaction system and the comprehensive utilization of follow-up treatment units and byproduct, it can be expected that the equipment investment and the running cost of the SCWO system will further decline comparing with conventional technology for sludge treatment. In a word, this comprehensive process provides a novel, environment-friendly economical implementation method of SCWO technology for harmless disposal of various organic wastes.

References

1. P. Kritzer, E. Dinjus, An assessment of supercritical water oxidation (SCWO): Existing problems, possible solutions and new reactor concepts. Chem Eng J, 2001, 83: 207–214.
2. M. D. Bermejo, M. J. Cocero, Destruction of an industrial wastewater by supercritical water oxidation in a transpiring wall reactor. J Hazard Mater, 2006, 137: 965–971.
3. P. Kritzer, Corrosion in high-temperature and supercritical water and aqueous solutions: a review. J Supercrit Fluid, 2004, 29: 1–29.
4. G. Brunner, Near and supercritical water. Part II: Oxidative processes. J Supercrit Fluid, 2009, 47: 382–390.
5. A. M. Philip, S. D. C. And, D. W. Dalton, SCWO System Designs for Waste Treatment: Application to Chemical Weapons Destruction. Ind Eng Chem Res, 2015, 44: 9030–9039.
6. E. Asselin, A. Alfantazi, S. Rogak, Corrosion of nickel–chromium alloys, stainless steel and niobium at supercritical water oxidation conditions. Corros Sci, 2010, 52: 118–124.
7. S. H. Son, J. H. Lee, C. H. Lee, Corrosion phenomena of alloys by subcritical and supercritical water oxidation of 2-chlorophenol. J Supercrit Fluid, 2008, 44: 370–378.
8. G. S. Was, P. Ampornrat, G. Gupta, S. Teysseyre, E. A. West, T. R. Allen, et al., Corrosion and stress corrosion cracking in supercritical water. J Nucl Mater, 2007, 371: 176–201.
9. I. Betova, M. Bojinov, P. Kinnunen, S. Penttilä, T. Saario, Surface film electrochemistry of austenitic stainless steel and its main constituents in supercritical water. J Supercrit Fluid, 2007, 43: 333–340.
10. Q. Zhang, R. Tang, K. Yin, X. Luo, L. Zhang, Corrosion behavior of Hastelloy C-276 in supercritical water. Corros Sci, 2009, 51: 2092–2097.
11. M. Sun, X. Wu, Z. Zhang, E. H. Han, Oxidation of 316 stainless steel in supercritical water. Corros Sci, 2009, 51: 1069–1072.
12. S. I. Kawasaki, T. Oe, S. Itoh, A. Suzuki, K. Sue, K. Arai, Flow characteristics of aqueous salt solutions for applications in supercritical water oxidation. J Supercrit Fluid, 2007, 42: 241–254.
13. F. J. Armellini, J. W. Teste, Experimental methods for studying salt nucleation and growth from supercritical water. J Supercrit Fluid, 1991, 4: 254–264.
14. M. Hodes, P. A. Marrone, G. T. Hong, K. A. Smith, J.W. Tester, Salt precipitation and scale control in supercritical water oxidation—Part A: fundamentals and research. J Supercrit Fluid, 2004, 29: 265–288.
15. P. A. Marrone, M. Hodes, K. A. Smith, J. W. Tester, Salt precipitation and scale control in supercritical water oxidation—part B: commercial/full-scale applications. J Supercrit Fluid, 2004, 29: 289–312.
16. Marshall, W. L, Water and its solutions at high temperatures and pressures., 1975.
17. M. Schubert, J. W. Regler, F. Vogel, Continuous salt precipitation and separation from supercritical water. Part 1: Type 1 salts. J Supercrit Fluid, 2010, 52: 99–112.
18. M. Schubert, J. W. Regler, F. Vogel, Continuous salt precipitation and separation from supercritical water. Part 2. Type 2 salts and mixtures of two salts. J Supercrit Fluid, 2010, 52: 113–124.
19. F. Vogel, S. Stucki, T. B. Truong, M. H. Waldner, Process for Generating Methane and/or Methane Hydrate From Biomass., 2013.
20. S. N. Rogak, P. Teshima, Deposition of sodium sulfate in a heated flow of supercritical water. Aiche J, 2010, 45: 240–247.

21. K. Stendahl, S. Jäfverström, Phosphate recovery from sewage sludge in combination with supercritical water oxidation. Water Science & Technology A Journal of the International Association on Water Pollution Research, 2003, 48: 185.
22. K. Príkopský, B. Wellig, P. R. V. Rohr, SCWO of salt containing artificial wastewater using a transpiring-wall reactor: Experimental results. J Supercrit Fluid, 2007, 40: 246–257.
23. Y. Matsumura, T. Minowa, B. Potic, S. R. A. Kersten, W. Prins, W. P. M. V. Swaaij, et al., Biomass gasification in near- and super-critical water: Status and prospects. Biomass Bioenerg, 2005, 29: 269–292.
24. S. Baur, H. Schmidt, A. Krämer, J. Gerber, The destruction of industrial aqueous waste containing biocides in supercritical water—development of the SUWOX process for the technical application. J Supercrit Fluid, 2005, 33: 149–157.
25. J. R. Portela, J. López, E. Nebot, D. L. O. E. Martínez, Elimination of cutting oil wastes by promoted hydrothermal oxidation. J Hazard Mater, 2001, 88: 95–106.
26. M. D. Bermejo, F. Cantero, M. J. Cocero, Supercritical water oxidation of feeds with high ammonia concentrations: Pilot plant experimental results and modeling. Chem Eng J, 2008, 137: 542–549.
27. M. D. Bermejo, M. J. Cocero, Supercritical water oxidation: A technical review. Aiche J, 2010, 52: 3933–3951.
28. A. Sobhy, R. I. L. Guthrie, I. S. Butler, J. A. Kozinski, Naphthalene combustion in supercritical water flames. P Combust Inst, 2009, 32: 3231–3238.
29. C. Narayanan, C. Frouzakis, K. Boulouchos, K. Príkopský, B. Wellig, P. R. V. Rohr, Numerical modelling of a supercritical water oxidation reactor containing a hydrothermal flame. J Supercrit Fluid, 2008, 46: 149–155.
30. R. M. Serikawa, T. Usui, T. Nishimura, H. Sato, S. Hamada, H. Sekino, Hydrothermal flames in supercritical water oxidation: investigation in a pilot scale continuous reactor. Fuel, 2002, 81: 1147–1159.
31. P. E. Savage, J. B. Dunn, J. Yu, Recent advances in catalytic oxidation in supercritical water. Combust. Sci. Technol, 2006, 178: 443–465.
32. P. E. Savage, A perspective on catalysis in sub- and supercritical water. J Supercrit Fluid, 2009, 47: 407–414.
33. M. Svanström, M. Fröling, M. Olofsson, M. Lundin, Environmental assessment of supercritical water oxidation and other sewage sludge handling options. Waste Manag Res, 2004, 41: 321–338.
34. T. Cai, D. J. Yang, Development of Wastewater Treatment Technology in Papermaking Industry. Paper Chemicals, 2006.
35. J. W. Griffith, D. H. Raymond, The first commercial supercritical water oxidation sludge processing plant. Waste Manag, 2002, 22: 453–459.
36. E. D. Lavric, H. Weyten, J. D. Ruyck, V. Pleşu, V. Lavric, Delocalized organic pollutant destruction through a self-sustaining supercritical water oxidation process. Eng Conv Manag, 2005, 46: 1345–1364.
37. M. J. Cocero, E. Alonso, M. T. Sanz, F. Fdz-Polanco, Supercritical water oxidation process under energetically self-sufficient operation. J Supercrit Fluid, 2002, 24: 37–46.
38. B. Wellig, K. Lieball, P. R. V. Rohr, Operating characteristics of a transpiring-wall SCWO reactor with a hydrothermal flame as internal heat source. J Supercrit Fluid, 2005, 34: 35–50.
39. M. Modell, Processing methods for the oxidation of organics in supercritical water., 1982.
40. B. Veriansyah, B. Prabowo, J. D. Kim, Supercritical Water Oxidation for Toxic Organic Wastewater Disposal. Environ Eng Sci, 2006, 19: 513–522.
41. D. S. Sloan, R. A. Pelletier, M. Modell, Sludge Management in The City Of Orlando. It's Supercritical!. Proceedings of the Water Environment Federation, 2008, 2008: 7762–7774.
42. M. D. Bermejo, D. Rincon, A. Martin, M. J. Cocero, Experimental Performance and Modeling of a New Cooled-Wall Reactor for the Supercritical Water Oxidation. Ind Eng Chem Res, 2009, 48: 6262–6272.
43. A. Gidner, L. Stenmark, Supercritical water oxidation of sewage sludge-state of the art. Blood, 2001, 108.

44. D. H. Xu, S. Z. Wang, Y. M. Gong, Y. Guo, X. Y. Tang, H. H. Ma, A novel concept reactor design for preventing salt deposition in supercritical water. Chem Eng Res Des, 2010, 88: 1515–1522.
45. I. V. Pérez, S. Rogak, R. Branion, Supercritical water oxidation of phenol and 2,4-dinitrophenol. J Supercrit Fluid, 2004, 30: 71–87.
46. W. J. Gong, F. Li, D. L. Xi, Oxidation of industrial dyeing wastewater by supercritical water oxidation in transpiring-wall reactor. Water Environment Research A Research Publication of the Water Environment Federation, 2008, 80: 186.
47. Y. M. Gong, S. Z. Wang, X. Y. Tang, D. H. Xu, H. H. Ma, Supercritical water oxidation of acrylic acid production wastewater. Environ Technol, 2014, 35: 907–916.
48. M. D. Bermejo, A. Fernando Fdezpolanco, M. J. Cocero, Effect of the Transpiring Wall on the Behavior of a Supercritical Water Oxidation Reactor: Modeling and Experimental Results. Ind Eng Chem Res, 2006, 45: 3438–3446.
49. J. Abeln, M. Kluth, G. Petrich, H. Schmieder, K. Forschungszentrum, Waste Treatment by SCWO Using a Pipe and a Transpiring Wall Reactor..
50. V. Lavric, V. Pleşu, J. D. Ruyck, Chemical reactors energy integration through virtual heat exchangers—benefits and drawbacks. Appl Therm Eng, 2005, 25: 1033–1044.
51. M. J. Cocero Alonso, E. Alonso Sánchez, F. Fernandez Polanco, Supercritical Water Oxidation of Wastewater and Sludges—Design Considerations. Eng Life Sci, 2015, 2: 195–200.
52. F. Marias, F. Mancini, F. Cansell, J. Mercadier, Energy recovery in supercritical water oxydation process. Environ Eng Sci, 2008, 25: 123–130.
53. M. D. Bermejo, M. J. Cocero, Destruction of an industrial wastewater by supercritical water oxidation in a transpiring wall reactor. J Hazard Mater, 2006, 137: 965.
54. D. Xu, S. Wang, X. Tang, Y. Gong, Y. Guo, Y. Wang, et al., Design of the first pilot scale plant of China for supercritical water oxidation of sewage sludge. Chem Eng Res Des, 2012, 90: 288–297.
55. L. Zhang, X. Lu, Treatment of Organic Pollutants by Supercritical Water Oxidation Process. Environ Pro Chem Ind, 2002.
56. S. N. V. K. Aki, M. A. Abraham, An economic evaluation of catalytic supercritical water oxidation: Comparison with alternative waste treatment technologies. Environ Pro, 2010, 17: 246–255.
57. C. A. Eckert, J. G. V. Alsten, T. Stoicos, Supercritical fluid processing. Environ Sci Technol, 1986, 20: 319–325.

The manufacturer's authorised representative in the EU is Springer Nature Customer Service Centre GmbH, Europaplatz 3, 69115 Heidelberg, Germany. If you have any concerns regarding our products, please contact ProductSafety@springernature.com

Printed and bound by CPI Group (UK) Ltd, Croydon, CR0 4YY
24/07/2026
02175975-0007